Practical Use of Mathcad®

Springer
London
Berlin
Heidelberg
New York
Barcelona
Hong Kong
Milan
Paris
Santa Clara
Singapore
Tokyo

Hans Benker

Practical Use of Mathcad®

Solving Mathematical Problems with a Computer Algebra System

Translated by Anthony Rudd

Springer

Prof. Dr.Hans Benker
Martin-Luther-Universität
Fachbereich Mathematik und Informatik
06120 Halle (Saale)
Germany

ISBN 1-85233-166-6 Springer-Verlag London Berlin Heidelberg

British Library Cataloguing in Publication Data
Benker, Hans
 Practical use of MATHCAD : solving mathematical problems
 With a computer algebra system
 1. Mathcad (Computer file) 2. Mathematics – Computer programs
 I. Title
 510.2'85'5369
ISBN 1852331666

Library of Congress Cataloging-in-Publication Data
Benker, Hans, 1942-
  [Mathematik mit MATHCAD. English]
  Practical use of MATHCAD : solving mathematical problems with a
computer algebra system / Hans Benker : Translated by Anthony Rudd.
     p.   cm.
  Revised, extended edition of the original German title "Mathematik
mit MATHCAD".
  Includes index.
  ISBN 1-85233-166-6 (alk. paper)
  1. Mathematics—Data processing.   2. MathCAD.   I. Title.
QA76.95.B4613   1999   99-33922
510'.0285'5369—dc21   CIP

Printed in Great Britain

Originally published in German under the title Mathematik mit Mathcad 2nd edition by Hans Benker
The use of registered names, trademarks etc. in this publication does not imply, even in the absence of a specific statement, that such names are exempt from the relevant laws and regulations and therefore free for general use.

Typesetting: Camera ready by author
Printed and bound at the Athenæum Press Ltd., Gateshead, Tyne & Wear
34/3830-543210 Printed on acid-free paper SPIN 10688737

For my wife Doris
and my daughter Uta

Preface

This *book*, which is a *revision* and *extension* of the original edition published in 1996 (see [2]) with the German title *Mathematik mit MATHCAD (Mathematics Using MATHCAD)*, discusses the use of the *program system* MATHCAD® to *solve mathematical problems* with *computers*. The book is based on the *current MATHCAD Version 8 Professional* for WINDOWS 95/98 (see [5]).

Whereas MATHCAD and MATLAB (see [4]) were originally conceived as purely *systems* for *numerical mathematical calculations*, the more recent versions of both products have licensed a *minimum variant* of the *symbolic processor* of the MAPLE *computer algebra system* for *exact (symbolic) calculations*.

Thus, MATHCAD has been developed to be an equal partner to the established *computer algebra systems* AXIOM, DERIVE, MACSYMA, MAPLE, MATHEMATICA, MuPAD and REDUCE. However, because these *systems* contain *numerical methods* as well, they are no longer just pure *computer algebra systems*.

Consequently, MATHCAD can also be designated as being a *computer algebra system* (or just: *system*).

MATHCAD possesses some *advantages*:

- Better *numerical capabilities* more than compensate for the somewhat limited capabilities provided for exact (symbolic) calculations.

- The *calculations* are performed in the MATHCAD worksheet using the usual *mathematical symbols (standard notation)*.

- Thanks to the *superior layout capabilities* in the worksheet, MATHCAD can be used to create treatises directly.

- All *calculations* can be performed using *units of measurement*.

- The more than 50 *electronic books* that exist for various *mathematical, technical, scientific* and *economics disciplines* contain all the *standard formulae, equations* and *mathematical methods* relevant for the subject. Detailed *text* and *graphics* explain these entries, which can be integrated without difficulty into your own calculations.

These advantages have been a significant factor in making MATHCAD one of the *preferred systems* for *engineers* and *scientists*.

MATHCAD exists for *various computer platforms*, such as for IBM-compatible personal computers, workstations under UNIX and APPLE computers.

We have used for this book the *Version 8 Professional* for IBM-*compatible personal computers* with a Pentium processor (generally referred to as PCs) running under WINDOWS 95/98. Because the *form* of the *user interface* and the *menu* and *command structure* does not significantly differ for other *types* of *computer*, the discussions made in this book are generally applicable.

Because the user interface for Version 8 described in the book has changed only slightly from Version 7, most of the problems discussed in the book can also be calculated using Version 7. The same applies for Versions 5 and 6. The author's previous books [2,3,4] contain additional details. However, all earlier versions have reduced capabilities and have fewer functions than those supplied with Version 8.

In the author's opinion, the MATHCAD *system* has unjustifiably only in the last few years received the attention it deserves. This makes itself apparent, for example, in the limited number of books that have been published on MATHCAD. This book aims to help close this gap. Its audience addresses both *students* and *employees* at *polytechnics* and *universities*, as well as *engineers* and *scientists* working in industry and institutes.

This book provides solutions to *basic mathematical problems* for *engineering* and *natural science*. It can serve both as *reference* and *tutorial* for lecturers and students as well as a *manual* for *experts*.

This book is based on lectures that the author held at Halle University for students of mathematics, computer science and engineering.

In order to keep the book compact, the *mathematical theory* of the discussed problems is handled only to the extent needed to produce a solution using MATHCAD. Consequently, all proofs have been omitted. However, this does not mean that *mathematical rigour* is ignored, as is the case in many books concerned with computer algebra systems.

This *book focuses* on *transforming* the *basic mathematical problems* that are to be solved into the *language* of MATHCAD and on the *interpretation* of the *returned results*. To save the reader from tedious *searching* to find the solution to a problem, value is placed on a *concise description*.

If the readers have *mathematical difficulties*, they should consult the appropriate *mathematical books* (see Bibliography).

This *book* is divided into *two parts*:

I. Chapters 1 - 11 :

   Introduction: Preface, installation, structure, user interface and basic properties of MATHCAD

II. Chapters 12 - 27 :

   Main part: Solution of mathematical problems from engineering and natural science using MATHCAD

The copious examples that illustrate all the discussed *mathematical problems* should be used as exercises. These examples show the reader the *capabilities* and *limitations* involved with the *use* of MATHCAD.

Because computers on which *computer algebra systems* have been installed will replace hand-held calculators, the *use* of such systems to solve *mathematical problems* using *computers* will *gain* in *importance* in future.

Because these *systems* can be used without requiring extensive programming and computer knowledge, and can solve most of the standard mathematical problems, their role will increase for the user. Before you use or write *software* in a conventional programming language such as BASIC, C, FORTRAN or PASCAL, because it is much simpler to solve the problem using an available *computer algebra system*, you will try this approach first. Only when this approach fails will you need to return to using *numerical software*.

This book shows how MATHCAD as a *computer algebra system* can be used to easily solve many *mathematical problems*. The capabilities of an *exact solution* are tried first before an attempt is made to use the *numerical solution*. This approach is warranted because the MATHCAD *system* represents a successful *mix* of *computer algebra methods* and *numerical methods* under a common WINDOWS user interface. This *user interface* permits interactive work and provides a *clear display* of the *calculations*.

The author hopes that this introduction to MATHCAD stimulates the readers not only to make use of MATHCAD to obtain a quick solution to their problems, but also to consider the mathematical background.

As with all *computer algebra systems*, MATHCAD cannot replace mathematics. However, it can relieve the user from making long-winded calculations and avoid computational errors, and so provide the user with time to consider the creative use of mathematics.

At this point I would like to *thank* all those people who supported me in the *realization* of this *book project*:

- Dr. Merkle, Ms. Grünewald-Heller and Ms. Mowat from Springer Verlag Heidelberg and London for accepting the book suggestion in the publisher's program and for their technical assistance.

- My wife Doris who showed much understanding for my work in the evenings and weekends.

- My daughter Uta who critically read the manuscript many times and gave many useful suggestions.

- The European branch of *MathSoft International* in Bagshot (England) for the free use of the new Version 8 of MATHCAD and the new *electronic books*.

Finally, a few *comments* on the *form* of this *book*:

- In addition to *headings*, MATHCAD *functions*, *commands* and *menus/submenus* are shown *bold*. The same also applies for *vectors* and *matrices*.

- The *names* of *programs*, *files* and *directories*, and the *names* of *computer algebra systems* are shown in capitals.

Examples and *figures* are *numbered* starting at 1 in each chapter; the *chapter number* is *prefixed.* Thus, *Fig. 2.3* and *Example 3.12* represent *figure 3* from *Chapter 2* and *example 12* from *Chapter 3*, respectively.

- The *symbol*

  is used to indicate the end of an *example.*

- The *symbol*

  prefixes important text passages in the *book* which are also ended with the *symbol*

- The *symbol*

  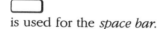

  is used for the *space bar.*

- *Important terms* and *designations* are written *italicized.* This also applies for *explanatory text* for the *calculations* performed by MATHCAD that are taken directly from the MATHCAD worksheet.

- An ⇒ *arrow* that *separates* the individual *menus* of a MATHCAD *menu sequence* also represents a *mouse click.*

Merseburg, July 1999   Hans Benker

Contents

1 Introduction

MATHCAD at the *beginning* of its development was a purely *numerical system*, i.e., it consisted of a collection of *numerical methods* (*approximation methods*) with a standardized *user interface* for the *numerical* (*approximate*) *solution of mathematical problems* using computers. The *more recent versions* (starting at version 3) added a licensed *minimum variant* of the *symbolic processor* of the *computer algebra system* MAPLE for *exact* (*symbolic*) *calculations* (*formula manipulation*).

All existing universal *computer algebra systems* make use of the following facilities for the *solution* of mathematical problems using a computer:

*   *Exact* (*symbolic*) calculation (*formula manipulation*) based on *computer algebra*
*   Provision of *numerical standard methods* (*approximation methods*) for an *numerical* (*approximate*) *solution*.

Consequently we can also include MATHCAD in the class of *computer algebra systems* , which we will *designate* just as *systems*.

♦

Whereas the *methods* of the *computer algebra* predominate in the AXIOM, DERIVE, MACSYMA, MAPLE, MATHEMATICA, MuPAD and REDUCE, the *numerical methods* are more evident in the MATHCAD and MATLAB *systems* and the *computer algebra* taken from *another system* (MAPLE).

♦

The requirement to include *methods* for the *numerical calculation* in the *systems* is based on the fact that the *exact solution* of a *mathematical problem* demands a *finite algorithm*, which cannot be found for many problems. A *finite solution algorithm* is understood to be a *method* that provides the *exact solution* to a problem in a *finite number of steps*.

The *difference* between *exact* and *numerical calculations* is described and illustrated with examples in Section 1.2 and Chapter 6.

♦

[image marker]

This *book* is divided into *two parts*:

I.   Chapters 1 - 11:

Introduction: Introduction, installation, structure, user interface and basic characteristics of MATHCAD.

II. Chapters 12 - 27:

Main part: Use of MATHCAD to solve mathematical problems from engineering and science.

♦

The scope of this book obviously does not permit a comprehensive coverage of all mathematical domains. Solutions are provided using MATHCAD for *mathematical basic problems* and important special problems that occur in *engineering* and *science*.

The detailed discussion of special areas (e.g., statistics) must be reserved for other dissertations on MATHCAD that should also include a discussion of the available *electronic books*.

♦

To avoid exceeding the scope of the book, the *mathematical theory* of the considered problems is handled only to the extent needed to provide a solution using MATHCAD. Consequently, all proofs are omitted. However, this does not mean that *mathematical rigour* is ignored, as is often the case in books that discuss computer algebra systems.

This *book emphasizes* the *conversion* of the *mathematical problems* to be solved into the *language* of MATHCAD and the *interpretation* of the *returned results*. Value is placed on a *brief representation* and so permits the reader to achieve a quick use.

Readers should consult the appropriate *specialized books* if they have *questions* of a *mathematical nature* (see Bibliography).

♦

All the *calculations* in the book performed with MATHCAD have been taken directly from the MATHCAD *worksheet*; any *explanatory text* supplied by MATHCAD is shown *italicized*. This makes it easy for the reader to follow the calculations, because the expressions to be computed are entered in the worksheet in the same manner.

♦

1.1 Structure and Operation of Computer Algebra Systems

As with all *computer algebra systems*, MATHCAD has the *following structure:*

*   *Front end*

The front end is used for the *interactive* work with the *system*.

* *Kernel*

  The *kernel* is *loaded* into the computer main storage when the system starts. The *kernel* contains the programs used to *solve* basic mathematical problems.

* *Packages*

  They contain advanced applications and only need to be loaded if they are required.

  MATHCAD designates these as *worksheets, documents* or *electronic books.*

The *structure* of the *computer algebra systems* has a significant effect on saving memory in main storage and permits the users to make continuous extensions in the form of their own *packages* that they can write to solve problems.

♦

Computer algebra and the use of *numerical methods characterize* the operation of MATHCAD and other computer algebra systems.

The *exact (symbolic) processing* of *mathematical expressions* with *computers* is called *computer algebra* or *formula manipulation*. Although both terms are used as synonyms, the formula manipulation term for the following reasons is more appropriate:

- Because it is not just used to investigate the solution of algebraic problems, the term computer algebra could easily be misunderstood.

- The term algebra represents the associated methods for the *symbolic manipulation* of *mathematical expressions*, i.e., the algebra supplies principally the tool:

    * *Manipulation* of *expressions,* such as by
        - simplification
        - partial fraction decomposition
        - multiplication
        - factorization
        - substitution

    * *Construction* of *finite solution algorithms,* e.g., for the
        - solution of equations and differential equations
        - differentiation
        - integration

Computer algebra can only be used to compute those mathematical problems that can be processed in a finite number of steps (*finite algorithm*) to produce the exact solution; all *calculations/transformations* in *computer algebra* are performed exactly (*symbolically*). In contrast to computer algebra, the *numerical methods* (*approximation methods*) compute using *rounded decimal numbers* (*floating-point numbers*) and produce only *approximate values* for the solution. The *following errors* can occur here :

* *Rounding errors* resulting from the limited *computing accuracy* of the computer, because it can process only *finite decimal numbers*.

* *Termination errors* can occur, because the methods must be *terminated* after a *finite number of steps*, although in most cases the solution has not yet been attained. The familiar *regula falsi* used to determine the zeros for functions is an example.

* Totally *incorrect results* can be supplied, because *numerical methods* are *not always convergent*.

   ♦

These properties for *numerical methods* just discussed show that the *exact calculations should* always *be favoured* provided they are successful.

However, exact computational rules (finite algorithms) cannot be specified for a number of mathematical problems, such as in the following Example 1.1 we discuss for integral calculus.

♦

Example 1.1:

It is not possible to *compute exactly* every *definite integral* using the known *theorem* from the differential and integral calculus (see Chapter 20):

$$\int_a^b f(x)\,dx = F(b) - F(a)$$

You can only *compute* those *integrals* for which the *primitive* $F(x)$ of the *function* $f(x)$, i.e.

$$F'(x) = f(x)$$

can be specified in *analytical form* after a *finite number of steps*, e.g. using known *integration rules,* such as *integration by parts, substitution* or *partial fraction decomposition.*

   ♦

As with all *computer algebra systems*, MATHCAD *works interactively*. This is a major advantage over *numerical programs* created using conventional *programming languages* (BASIC, C, FORTRAN, PASCAL, etc.).

Interactive operation permits a continuous *dialogue* between the user and the *computer* using the computer monitor in which the following *cycle* continuously repeats itself:

I. *Input* by the *user* of the *expression* to be *calculated* in the system's *work-sheet*.

II. *Calculation* of the *expression* by the system.

III. *Output* of the *calculated results* in the system's *worksheet.*

IV. The *calculated result* is *available* for further computations.

♦

We now consider in more detail the two *basic tasks* for the *interactive operation* using *computer algebra systems*, i.e., the *input* of the *expression* to be computed and the *initiation* of the *calculation*:

- The *input* of a *mathematical expression* in the *current worksheet* can be made in the following two *ways*:

  I. Using the *keyboard*, some *systems* (DERIVE, MATHCAD, MATHEMA-TICA and MAPLE) provide additional *mathematical symbols* in their user interface.

  II. By *copying* from other worksheets (using the clipboard).

- The *calculation* for a *mathematical expression* can be *initiated* in the following two ways:

  I. *Input of commands or functions* into the *worksheet*, where the *expression to be computed* is entered in the *argument*. The arguments are enclosed within parentheses.

  II. *Selection* with the *mouse* of a *menu sequence* from the menu bar of the user interface after entering the *expression* to be *computed*. We will write the required *menu sequences* in this book in the form

  menu_1 ⇒ menu_2 ⇒ ... ⇒ menu_n

  where each ⇒ *arrow* represents a *mouse click*, and a *mouse click* also *closes* the complete menu sequence.

Because the individual *computer algebra systems* have been written by different software companies, you cannot expect any standard form for the designation of the menus/commands/functions used to solve a problem. This notwithstanding, a designation standard would be desirable for all systems, because this would significantly simplify the work with different systems.

♦

Both I and II *calculation possibilities* can be used in MATHCAD for a number of problems, i.e., you can use either *functions* or *menu sequences* to compute problems. We will indicate this in the appropriate sections.

♦

An intuitive WINDOWS user interface, similar to other systems, simplifies the work with MATHCAD. This interface is used to perform the *dialogue* with the *system* (see Chapter 3).♦

☞

The MATHCAD *user interface* can be used to arrange in the *work window* the processed calculations in the form of worksheets as typically used for manual calculations (see Chapters 3 and 4).

Because MATHCAD represents the mathematical expressions to be computed in *standard mathematical notation* (see Chapter 4), MATHCAD is *superior* to all other systems in the *arranging* of the *worksheet*.

The computer algebra systems use *terms* such as *document, notebook, scratchpad and worksheet* for these worksheets. MATHCAD uses the two synonymous terms, *worksheet* and *document*.

◆

1.2 Application Areas for Computer Algebra Systems

Although *computer algebra* has been greatly *influenced* by *algebra* and *solves* many *problems* in this area, *computer algebra* can also *solve problems* from *mathematical analysis* and its applications. This includes, in particular, problems that can be *solved* using *algebraic methods*, i.e., through *symbolic manipulation* of mathematical expressions. We discuss this in the following Example 1.2 for the differential calculus.

◆

Example 1.2:

The *differentiation* of *functions* (see Section 19.1) provides a *typical example* for the *use* of *computer algebra*:

The *knowledge* of the *rules* for the *differentiation* of *elementary functions*

x^n, $\sin x$, e^x, etc.

and the known *differentiation rules*

*   *product rule*

*   *quotient rule*

*   *chain rule*

permits the *differentiation* of differentiable functions that are composed of elementary functions. You can interpret this as being an *algebraic handling* of the *differentiation*.

In a similar manner as for the differentiation, methods for the *calculation* of certain classes of *integrals* and *differential equations* can be formulated and performed as part of the *computer algebra*.

◆

Because of the characteristics just described, *problems* from the *following areas* can be *solved exactly* with *computer algebra systems* and thus also with MATHCAD:

- *Basic arithmetic operations* (see Chapter12)
    - * Because the calculation is performed exactly, the *addition* of *fractions,* e.g.,

        $$\frac{1}{3}+\frac{1}{7}$$

        returns the result as a fraction

        $$\frac{10}{21}$$

    - * If *real numbers* can only be represented exactly as symbols, no further change is made when such real numbers, such as

        $$\sqrt{2} \text{ and } \pi$$

        are input.

    - * MATHCAD can *approximate rational* and *real numbers* using *finite decimal numbers* with *defined accuracy:*
        The *input* of the *numerical equal sign* = after the entered number returns the *approximation* as shown in the following examples:

        $$\frac{10}{21} = 0.476190476190476 \qquad \sqrt{2} = 1.414213562373095$$

        $$\pi = 3.141592653589793 \qquad e = 2.718281828459045$$

- *Manipulation* of *expressions* (see Chapter 13)

    This belongs to the *basic functions* of *computer algebra systems*. These include:
    - * *simplification*
    - * *partial fraction decomposition*
    - * *expansion*
    - * *multiplication*
    - * *factorization*
    - * *reduction to a common denominator*
    - * *substitution*
    - * *transformation of trigonometric expressions*
- *Linear algebra* (see Chapters 15 and 16)

In addition to *addition, multiplication, transposition* and *inversion* of *matrices, eigenvalues, eigenvectors* and *determinants* can be computed and systems of *linear equations* can be solved.

- *Functions*

  All *computer algebra systems*:

  * recognize the complete palette of *elementary functions* and a number of *special functions* (see Section 17.2)
  * permit the *definition* of *user functions* (see Section 17.2.3)
  * can *differentiate* functions and *integrate* as part of the provided integration rules (see Chapters 19 and 20)
  * can solve some classes of *differential equations* (see Chapter 23).

- *Graphical representations* (see Chapter 18)

  Because both *2D* and *3D-graphs* can be created, you can *represent graphically functions* having one or two variables. *Animated graphs* are also possible.

- *Programming*

  Because the AXIOM, MACSYMA, MAPLE, MATHEMATICA, MuPAD and REDUCE *systems* have their own *programming languages*, programs can be created to solve complex problems. Compared with the use of conventional programming languages, this programming has the advantage that all available commands and functions of the systems can also be used.

  Although the *programming capabilities* of MATHCAD are somewhat limited (see Chapter 10), it permits *procedural programming*, which suffices to write small programs as we will see in the course of this book.

  The programming of more complex algorithms, as you find in many *electronic books*, is solved in MATHCAD through the inclusion of programs written in the C *programming language*.

  ◆

If computer algebra fails for the exact (symbolic) calculation, you can use the methods for numerical calculation integrated in the systems. Thus, MATHCAD and the other systems can be used to realize the major techniques required for the *solution* of *mathematical problems:*

* manipulation and exact calculation of expressions
* use of formulae and equations from various mathematical disciplines
* use of numerical methods
* graphical representations

  ◆

To summarize, the use of *computer algebra methods* and *numerical methods* to solve a mathematical problem on the *computer* can be characterized as follows:

- *Computer algebra* provides the following *advantages*:
  - * *Input* in the *form* of *formulae* for the *problem* to be solved.
  - * The *result* is also returned as *formula*. This methodology is adapted from the manual solution using pencil and paper, and so can be used without requiring extensive programming knowledge.
  - * The use of *exact (symbolic) calculation avoids the occurrence* of *rounding errors.*
- *Computer algebra* has the single (but not insignificant) *disadvantage* that only those problems can be solved for which a *finite solution algorithm* exists.
- To realize a *numerical method* with the computer, you must write a *program* in a *programming language* such as BASIC, C, FORTRAN or PASCAL or make use of available *program libraries.* This requires more detailed knowledge and more effort than the use of *computer algebra systems.*
- *Numerical methods* have the *advantage* in their *universality,* i.e., they can be developed for most problems that arise.
- *Numerical methods* have the *disadvantage* that
  - * *rounding errors* can falsify the result
  - * because the *convergence* of the methods cannot always be assured, *incorrect results* can occur
  - * even in the case of convergence, *termination errors* normally occur.

  Consequently, *numerical methods* normally return only *numerical (approximate) solutions.*
  ♦

Two examples serve to illustrate the *differences* between *computer algebra* and *numerical methods.*

Example 1.3:

a) *Computer algebra* does not approximate the *real numbers*

$$\sqrt{2} \text{ and } \pi$$

after being input with a *decimal number*

$$\sqrt{2} \approx 1.414214 \text{ and } \pi \approx 3.141593$$

as required for numerical methods, but records it *symbolically,* thus, for example, the subsequent calculation for

$(\sqrt{2})^2$

yields the exact value 2.

b) The solution of the simple *linear equation system*

a·x + y = 1 , x + b·y = 0

that contains two *parameters* a and b also shows a typical *difference* between *computer algebra* and *numerical methods*. The *computer algebra* has the advantage that the *solution* is found *depending* on a and b, whereas *numerical methods* require the numerical values for a and b. The *computer algebra systems* return the *solution (formula solution)*

$$x = \frac{b}{-1 + a \cdot b} \ , \ y = \frac{-1}{-1 + a \cdot b}$$

The user must recognize that only the inequality a·b ≠ 1 is required for the parameters a and b, because otherwise no solution exists.

◆

Because in future the *use* of *mathematics* will be mainly realized using *computers*, the use of *computer algebra systems* will continually increase in importance.

Consequently, the efforts in the *further development* of *computer mathematics* concentrate on *further developing* and combining the methods of *computer algebra* and *numerical methods* in *computer algebra systems:*

*   In order to obtain effective finite solution algorithms for ever larger classes of problems, the computer algebra methods represent a focal point of current research.

*   The new versions of the systems include more effective, newly developed numerical methods that you can use should the exact calculation using computer algebra fail.

☞

However, as with the other computer algebra systems, you *cannot expect wonders* from MATHCAD:

*   *Exact solutions* are possible only for those problems for which the mathematical theory provides a *finite solution algorithm*.

*   Although (modern) *standard methods* provided for the *numerical solution* frequently offer acceptable approximations, they do not necessarily converge.

However, MATHCAD and the other computer algebra systems remove the need for many tedious calculations, and can be augmented with user-written programs. MATHCAD also permits the performed calculations to be made available in printable form. This has enabled MATHCAD to be devel-

oped into an *effective tool* for the *solution of mathematical problems* in *engineering* and *science*.
♦

1.3 Development of MATHCAD

MATHCAD was *originally* just a *system* for *numerical calculations*. However, the later versions under WINDOWS (from version 3) have a *license* from MAPLE for *exact (symbolic) calculations* using *computer algebra*, i.e., they contain a *minimum variant* of the *symbolic processor* from MAPLE.
MATHCAD contains a *minimum variant* of the *symbolic processor* from MAPLE for *exact (symbolic) calculation* that can be used to perform significant *mathematical operations exactly (symbolically)*, such as

* *manipulation* and *calculation* of *expressions*
* *matrix calculations*
* *solution* of *equations*
* *calculation* of *sums (series)* and *products*
* *differentiations*
* *Taylor expansions*
* *integration* (for suitably simple functions)
* *calculations* from *probability theory* and *statistics*.
♦

Although this book is based on *Version 8 Professional* from 1998 under WINDOWS 95/98 for IBM-*compatible personal computers* (PCs), MATHCAD also exists for APPLE computers and computers that run under the UNIX operating system. The operation is similar under the various operating systems.
♦

The *current Version 8 Professional* of MATHCAD has the following *improved operation* and *new functional characteristics* compared with the previous version 7:

* *Extension* of the *mathematical functions* and *improvements* to the *computing speed*. Consequently, *functions* for *optimization* (maximization/minimization) that have been added for the first time can be used to solve problems subject to constraints.
* *Increased dimensional problems* with up to one hundred variables can be solved.
* *Improved 3D graphical displays*.

* A *number* of *errors* from the previous versions have been *corrected*.
* *Improved integration* with *other systems* such as EXCEL, LOTUS, MATH-LAB and WORD.
* *Improved user interface*.
* *Improvements* in the *word processing*.
* *Support* of all important *standards*, such as SI units, full *compatibility* with WINDOWS 95/98/NT and OLE2.
* The MATHCAD *worksheets* can be saved in the following *formats*:
  - MATHCAD 7 and MATHCAD 6 format
  - HTML format
  - RTF format.

The higher-performance *Version 8 Professional* rather than the *standard version 8* is envisaged for those users who

* require a larger selection of mathematical functionality
* want to write their own programs
* wish to use other systems, such as EXCEL and MATLAB, in MATHCAD.
◆

MATHCAD is developed and marketed by MATHSOFT, a company that was founded in USA in 1984.
MATHSOFT is actively further developing MATHCAD. The *versions*
1 (1986)
2 (1988)
3 (1991)
4 (1993)
5 (1994)
6 (1995)
7 (1997)
8 (1998)
have been produced since 1986.
A budget version with the name MATHCAD 99 is available with a reduced function repertoire (corresponding to the full version 3.1). Cheaper student versions are also available for versions 6, 7 and 8.

Because the user interface for Version 8 described in the book has changed only slightly from Version 7, most of the problems discussed in the book can also be calculated using Version 7. The same applies for Versions 5 and 6. The author's previous books [2,3,4] contain additional details for these versions. However, all earlier versions have reduced capabilities and have fewer functions than those supplied with Version 8.
◆

1.4 Comparison of MATHCAD with other Systems

The author's books [1,3,4] provide an overview of the functioning, main operational areas and history of *computer algebra systems,* introduce the most important *systems* AXIOM, DERIVE, MACSYMA, MAPLE, MATHEMATICA, MATHCAD, MATLAB and MuPAD for PCs, and compare them in solving various mathematical problems.

The October 1994 issue of the English computer publication *PC Magazine* contains further *comparisons.* This magazine placed MATHCAD 5.0 PLUS behind MAPLE in the general evaluation and behind MATHEMATICA in the user-friendliness, in each case in second position.

Although the other systems have also been improved in the meantime, in the *author's opinion* MATHCAD must now be at *first position* in the *user friendliness,* because no other system provides the *layout capabilities* of the *worksheet.*

♦

The author also considers that the new MATHCAD in its *computational capabilities* represents a *comparable system* to AXIOM, DERIVE, MACSYMA, MAPLE, MATHEMATICA, MATLAB and MuPAD:

* MATHCAD in its *capabilities* with regard to *exact (symbolic) calculation* is somewhat inferior to the *universal computer algebra systems* AXIOM, MACSYMA, MAPLE, MATHEMATICA and MuPAD

* MATHCAD with regard to the *integrated numerical methods* is superior to the other systems.

   ♦

To *summarize,* MATHCAD is useful for those persons who,

* in addition to exact calculations, often need to solve problems that can only be computed using approximations

* place value on an easily readable representation of the performed calculations

* require calculations with measurement units.

As the result of these characteristics and other properties discussed in the book, MATHCAD has developed to be the *favoured system* for *engineers* and *scientists.*

♦

2 Installation of MATHCAD

To *install* MATHCAD 8 *Professional*, you require a PC with at least a *Pentium processor* that has 16 MB (better 32 MB) RAM and approximately 80 MB free space of the hard disk (for a complete installation). Because MATHCAD is supplied on CD-ROM, you need an appropriate drive for the installation.

2.1 Program Installation

The *installation* of MATHCAD is performed *menu-controlled* using WINDOWS 95/98 in the *following steps:*

I. *Start* the SETUP.EXE *file* from the *program CD-ROM.*

II. *Enter* the *serial number* when requested.

III. You can choose between a *complete* or *minimum installation* and a *user*-defined installation. The complete installation recommended for the novice user installs all program files for MATHCAD, *MathConnex* and the files for **Resource Center** *electronic book.*

IV. A *name* for the *directory* on the *hard disk* can be selected into which the *files* from MATHCAD are *copied.*

V. A *name* for the WINDOWS *program group* can be selected.

VI. Finally, all *files are copied from the CD-ROM* to the hard disk in the *selected directory.*

After a *successful installation*, the *activation* of the *menu sequence*

Start ⇒ Programs

opens the WINDOWS *program window* that now contains the new entry (*program group*) with the name selected for the installation. After clicking this name, the *program group* appears in which you can *activate the following:*

* *Mathcad 8 Professional* the MATHCAD system

* *MathConnex* a supplementary program to connect MATHCAD work-sheets
* *View Release Notes* to display important information for the new version of 8 MATHCAD.
    ♦

The *MATHCAD installation CD-ROM* also contains the *following* items:

* The ADOBE ACROBAT READER 3.0 (*file* AR32E30.EXE) *program* in the DOC *directory* that is required to *read* the *MATHCAD user's manual* and a *guide* for the *MathConnex*.
* The *MATHCAD user's manual* and a *guide* for the *MathConnex* in the DOC *directory*.
* The *Internet Explorer* from MICROSOFT in the IE directory that you can install if this has not been done previously during the installation of WINDOWS. Even when the computer does not have an *Internet connection*, you require *Explorer* to work with the **Resource Center**.
* The **Practical Statistics** *electronic book* that you can use to solve statistics problems (see Section 27.4). The supplied SETUP.EXE file must be used to separately install this book.
    ♦

2.2   MATHCAD Files

After the *installation*, the MATHCAD files are contained on the hard disk in the *directory* selected for the installation:

* This directory contains MCAD.EXE (*executable program file*) used to start MATHCAD.
* This directory also has a number of *subdirectories*.
    * The MAPLE subdirectory contains a *minimum variant* of the *symbolic processor* from MAPLE used to perform *exact* (*symbolic*) *calculations*.
    * QSHEET *examples* and *samples* for the programming and solution of mathematical problems (*files* with the .MCD *extension*).
    * TEMPLATE *templates* for worksheets (*files* with the .MCT *extension*).

2.3 Help System

The MATHCAD *help system* has been further extended and completed so
that the user receives answers or help for all questions and problems that
may occur.
The MATHCAD *help window*

Mathcad Help

is *opened* using one of the *following activities:*

- Click the

  button in the standard toolbar.
- Press the F1-key
- *Activate* the *menu sequence*

  Help ⇒ Mathcad Help

In the opened MATHCAD *help window,* you can use

- **Contents**

  to *display detailed information* for the *specific areas*
- **Index**

  to *display explanations* for all *terms* and *designations relevant* for
  MATHCAD
- **Search**

  to *display information relating* to a *search expression.*

The **Resource Center** *integrated electronic book* that provides additional
aids and guides for the use of MATHCAD is *opened* in one of the following
two ways:

* Use the *menu sequence*

  Help ⇒ Resource Center

* Click the

  button in the standard toolbar (see Figure 2.1)

  ♦

The **Resource Center** that consists of a collection of extracts from electronic books *contains* the *following books*:

  * **Overview**
    provides an *overview* of Version 8 for MATHCAD.

* **Getting Started**

  provides a step-by-step *introduction* to the operation of MATHCAD.

* **Advanced Topics**

  provides an introduction to the *new features* contained in *Version 8* of MATHCAD, for example, the solution of differential equations and optimization problems.

* **Quick Sheets**

  permit *access* to *existing* MATHCAD *worksheets* to solve a range of *mathematical problems*.

* **Reference Tables**

  The *reference tables* contain *important formulae* and *constants* from the areas of *mathematics, engineering* and *science*.

* **Mathcad in Action**

  provides an *introduction* to the *engineering problems* for which MATHCAD provides solutions.

* **Web Library**

  permits *access* to the *MATHCAD WWW library* provided you have an *Internet connection* and the Microsoft *Internet Explorer* has been installed on your computer. This library contains *electronic books* and interesting *worksheets* that you can *load*.

* **Collaboratory**

  If you have an *Internet connection* and have installed *Internet Explorer*, you can use this to participate in the *Online Forum* of MATHCAD users.

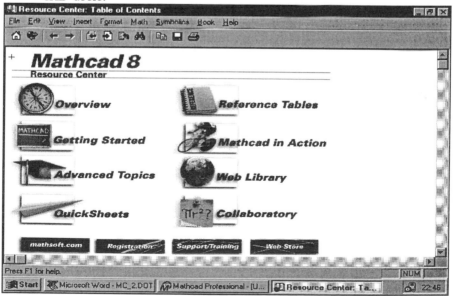

Figure 2.1. Resource Center

The *window* that appears for the **Resource Center** contains its own *standard toolbar* (see Figure 2.1) whose buttons/icons are explained when you place the mouse pointer on the appropriate icon. You can use these *symbols* to *scroll, search* in the book and *copy* and *print* interesting parts.
♦

MATHCAD also provides *additional capabilities* to *obtain help* for any difficulties. We discuss some of the interesting items in the following:

* After the *activation* of the *menu sequence*

  Help ⇒ Tip of the Day...

  a *dialogue box* appears with a *useful tip* for MATHCAD.

* If you place the *mouse pointer* on a *button/icon* of the *standard toolbar, formatting toolbar* or *math toolbar*, its *meaning* is displayed.

* Notes are output in the *status bar* for those *operations* performed. If you place the mouse pointer on a menu, a short explanation for the associated *submenus* (see Section 3.1) appears in the *status bar*.

* If you place the *cursor/editing lines* on a *command, function* or *error message* and then press the ⒡-key, the *help* is *displayed* in the work window.

* You can obtain a *help* for the *submenus* in the *menu bar* and the *buttons/icons* in the *standard toolbar* and *formatting toolbar* as *follows:*

  I. *Press* the ⇧⒡ *key combination* to change the *cursor* into a *question mark*.

  II. *Click* with this *question mark* on a *submenu* or *icon* to *display* the associated *help*.

  III. *Press* the ⒺⓢⒸ *key* to change *question mark* back to the normal *cursor*.

Because the *help functions* from MATHCAD are very *complex*, we recommend that you *experiment* with the help to gain experience.
♦

2.4 AXUM

AXUM is another *program system* marketed by MATHSOFT for the

* *data analysis*
* *creation of graphics*

and can be integrated into the MATHCAD operation.

Because the AXUM manual contained in the DOC directory on the *MATHCAD installation CD-ROM* contains *detailed notes*, we do not need to provide any additional explanations for this topic.

3 MATHCAD User Interface

The *user interface* of MATHCAD *Version 8* has changed only slightly compared with the *previous version 7*. We discuss the main *characteristics* in the following section. In the course of the book you will learn further special characteristics of the user interface used to solve specific problems.

If you have already worked with *WINDOWS programs*, you will not have any difficulties with the MATHCAD *user interface*, because it has the *same structure* in

* *menu bar*

* *standard toolbar*

* *formatting toolbar*

* *worksheet*

* *status bar*

These are augmented with the

* *math toolbar*

that is frequently used to work with MATHCAD, because it is required for all *mathematical operations*, to *create graphics* and for *programming*.

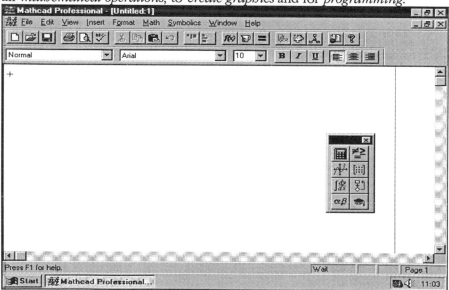

Figure 3.1. MATHCAD 8 Professional user interface

☞

After starting MATHCAD 8 *Professional* under WINDOWS, the *user interface* shown in Figure 3.1 is displayed, in which the *menu bar, standard toolbar, formatting toolbar, math toolbar* and *status bar* can be seen.

♦

The *following sections* of this chapter provide a detailed *description* of the individual *parts* of the *user interface.*

3.1 Menu Bar

The *menu bar* at the upper edge of the user interface contains the *following menus:*

File - Edit - View - Insert - Format - Math - Symbolics - Window - Help

The individual *menus* contain the *following submenus;* three dots following the menu name indicate that a *dialogue box* appears in which the required settings can be made:

- **File**

    File contains the *file operations* typical for WINDOWS programs (open, close, save, print, etc.).

- **Edit**

    Edit contains the *editing operations* typical for WINDOWS programs (cut, copy, paste, delete, find, replace, check spelling, etc.).

- **View**

    View is used to

    * *Display* and *remove* the *standard toolbar, formatting toolbar, math toolbar* and *status bar*
    * Click on **Animate...** to *create animated graphics*
    * Click on **Zoom...** to *increase the size.*

- **Insert**

    The following can be activated here:

    * **Graph**

        Open graphic windows (see Chapter 18)

    * **Matrix...**

        Insert a *matrix* (see Section 15.1)

    * **Function...**

        Insert an *built-in (predefined) function* (see Section 17.1)

* **Unit...**

  Insert measurement units (see Chapter 11)

* **Picture**

  Insert pictures (see Section 18.6)

* **Math Region**

  Switch to *math mode* (see Section 4.2)

* **Text Region**

  Switch to *text mode* (see Section 4.1)

* **Page Break**

  Issue a *page break*

* **Hyperlink**

  Create *hyperlinks* between *MATHCAD worksheets* and *templates* (see Section 4.3.3)

* **Reference...**

  Make a *reference* to other *MATHCAD worksheets* (see Section 4.3.3)

* **Component...**

  Exchange data between *MATHCAD worksheets* and other applications (see Section 9.3)

* **Object...**

  Insert an *object* in a *MATHCAD worksheet* (see Section 4.3.4).

- **Format**

  Format numbers, equations, text, graphics.

- **Math**

  * **Calculate**
  * **Calculate Worksheet**
  * **Automatic Calculation**

  MATHCAD menu items here *control* the *calculations* (see Section 6.3)

  * **Optimization**

    Optimize the *interaction* between the *symbolic* and the *numerical calculation* (see Section 6.3.5)

  * **Options...**

    The *dialogue box* that appears for

    – **Built–In Variables...**

      changes the *built-in (predefined) variables,* such as the *starting value* **ORIGIN** for the *indexing* or the *accuracy* **TOL** used for numerical calculations

    – **Calculation**

sets the *automatic calculation* and *optimization* of expression

    – **Unit System**

    sets a *unit system*

    – **Dimensions**

    sets dimension names.

- **Symbolics**

This menu contains all *submenus* for the *exact (symbolic) calculation* that we explain and use starting at Chapter 12.

- **Window**

arranges the *worksheet.* If *several worksheets* are open,

    * **Cascade**

    * **Tile Horizontal**

    * **Tile Vertical**

can be used to *arrange* these.

- **Help**

contains the *help functions* for MATHCAD that are described in Section 2.3.

The integrated *help functions* in MATHCAD (see Section 2.3) provide further *explanations* for the *menus* and *submenus.*
The main part of the book also provides a detailed discussion of the important menus and submenus.

♦

3.2 Standard Toolbar

The *standard toolbar* located *below* the *menu bar* contains a number of *buttons/icons* known from other WINDOWS *programs*:

* *Open File*

* *Save File*

* *Print*

* *Cut*

* *Copy*

* *Paste*

and other *MATHCAD buttons/icons* that we will discuss in the course of the book.

Because MATHCAD *names* the *buttons/icons* when you place the *mouse pointer* on a *button/icon*, we can omit a detailed explanation here.

◆

The

View ⇒ Toolbars ⇒ Standard

menu sequence can be used to *display* or *remove* the *standard toolbar.*

◆

3.3 Formatting Toolbar

A *formatting toolbar* located below the *standard toolbar* is used to *set* the *fonts* and *character forms*; because this is known from word processing, we can omit an explanation here.

The

View ⇒ Toolbars ⇒ Formatting

menu sequence can be used to *display* or *remove* the *formatting toolbar.*

◆

3.4 Math Toolbar

The *math toolbar* is used most often when working with MATHCAD, because it is required

*   for all *mathematical operations*
*   to *create graphics*
*   for *programming.*

The

View ⇒ Toolbars ⇒ Math

menu sequence can be used to *display* or *remove* the *math toolbar*

♦

The MATHCAD *math toolbar* contains the *buttons/icons* for the eight *operator palettes* that can be opened with a mouse click on these buttons. These *palette buttons/icons* (with *palette name*) follow:

1. *Arithmetic Toolbar*

2. *Evaluation and Boolean Toolbar*

3. *Graph Toolbar*

4. *Vector and Matrix Toolbar*

5. *Calculus Toolbar*

6. *Programming Toolbar*

7. *Greek Symbol Toolbar*

8. *Symbolic Keyword Toolbar*

If you do not know the *name* of a *palette button/icon*, you can *display* it by placing the *mouse pointer* on the appropriate *button/icon*. The menu sequence

View ⇒ Toolbars ⇒...

can also be used to display the operator palettes.

♦

☞

The eight *operator palettes* of the *math toolbar* contain *buttons/icons* for

* *mathematical symbols/operators*
  with buttons for differentiation, limits, integrals, summations, products, roots, matrices, etc.

* the *creation* of *graphic windows*

* *letters* from the *Greek alphabet*

* *programming operators*

* *keywords*

and, with few exceptions, are immediately understandable. The main part of the book contains a detailed description of the buttons/icons required for the individual calculations.

If, however, you do not know the *meaning* of a *button/icon* from an *operator palette*, you can display it by placing the *mouse pointer* on the corresponding *button/icon*.

♦

We show in the *following section* the *eight operator palettes* in the same sequence as the palette buttons/icons just discussed:

1. *Arithmetic Toolbar*

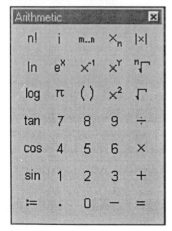

2. *Evaluation and Boolean Toolbar*

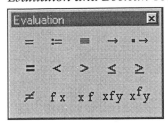

3. *Graph Toolbar*

4. *Vector and Matrix Toolbar*

5. *Calculus Toolbar*

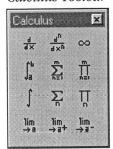

6. *Programming Toolbar*

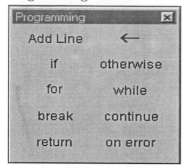

7. *Greek Symbol Toolbar*

8. *Symbolic Keyword Toolbar*

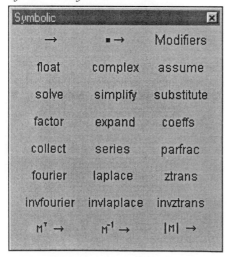

In this book, we will always use the specified number to indicate the operator palettes just discussed.

♦

If you do not explicitly close them, the *opened operator palettes* remain in the *worksheet.* This means that it is theoretically possible to have *all operator palettes open.* However, because only limited space remains when several operator palettes are open, we recommend that you keep *open* only those *palettes required* often for the current calculations.

♦

The *operators* and *symbols* of the *operator palettes* are used for performing both *exact* and *numerical calculations* and will be discussed in detail in the course of the book for the solution of individual problems.

♦

3.5 Worksheet

The *worksheet* (or *document*)

- is appended to the formatting bar and is framed below by the status bar
- occupies the *largest part* of the *user interface* and is used for the *main work* with MATHCAD:
  * *input mathematical expressions, formulae* and *equations*
  * *perform calculations*
  * *create graphics*
  * *input text.*
- can be arranged as a *worksheet* that is *characterized* with a *collection* of
  * *math regions*
    to *perform* all *calculations* and the *input* of *expressions, formulae* and *equations,*
  * *graphic regions*
    for the *representation* of *2D* and *3D graphs,*
  * *text regions*
    for the *input* of *text.*

☞

The MATHCAD *worksheet* can be formatted into a form that is *ready for printing*.

Although the *division* of the *worksheet* into *math, graphic* and *text regions* can also be found in other *computer algebra systems*, these do not provide such comprehensive *layout capabilities* as MATHCAD offers (see Chapter 4):

- *Expressions, formulae, equations, graphics* and *text* can be added at *any position* of the *worksheet*.
  If they are already in the worksheet, they can be *moved*.

- The *mathematical symbolism* used by MATHCAD corresponds to the *mathematical standard*, i.e., thanks to the comprehensive *operator palettes, mathematical expressions* can be created in a *form* that is *ready for printing*.
  Consequently, you can use MATHCAD by itself to create *documents* that *contain calculations*, i.e., avoid the use of a word processing system.

- All common *capabilities* of *word processing systems*, such as *cut, copy, change* the *font type* and *size, spell check*, etc., are *integrated* in MATHCAD.

We use the *example* of the *equation solution* in *Figure 3.2* to demonstrate a *possible arrangement* of the *worksheet. Figure 4.1* provides *another example*.

♦

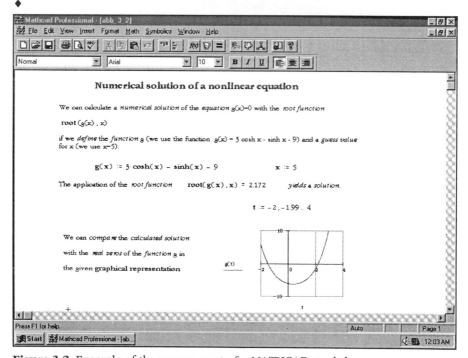

Figure 3.2. Example of the arrangement of a MATHCAD worksheet

☞

Even when professional *word processing systems* under WINDOWS, such as WORD or WORDPERFECT, are used for text processing, you can use MATHCAD to perform any required *calculations*. In this case, because MATHCAD uses the usual *mathematical symbolism*, you can use the clipboard to pass the complete calculation to the text processing.

This book, which was produced using WORD for WINDOWS (Version 6.0c) and in which all calculations and graphics were taken directly from MATHCAD, provides an example of this recommendation.

♦

3.6 Status Bar

The *status bar* known from many WINDOWS programs that is located under the worksheet provides the following *information* from MATHCAD on

*   the *current page number* of the opened *worksheet*

*   the *operations* just *performed*

*   the *math mode* (e.g., *auto* in the *automatic mode*)

*   *help functions*

*   etc.

☞

The menu sequence

View ⟹ Status Bar

can be used to display/remove the status bar.

♦

4 MATHCAD Worksheet

The MATHCAD *work area* can be arranged like a *worksheet* (or *document*) that is characterized by having a *collection* of

* *text regions* (*text fields*)

  for the *input* of *text*

* *math regions* (*calculation fields*)

  to *perform* all *calculations* and the *input* of *expressions*, *formulae* and *equations*

* *graphic regions*

  to *represent 2D* and *3D graphics.*

These *regions* can be *inserted* in the *worksheet* at the *position marked* by the *cursor.* These capabilities are referred to as being *input* in

* *text mode*
* *math mode* (*formula mode*)
* *graphic mode*

Before we investigate the text and math regions, we first discuss several *general characteristics* of the *worksheet.*

We start with the various *forms* of the *cursor* in MATHCAD that are required for input and correction:

* **Crosshair +**

  The *crosshair* appears at the *start* of MATHCAD or when you click with the *mouse* at *any empty space* in the *worksheet.*

  You can use it to *determine* the *position* in the worksheet where the *input* is to be made in *text, math* or *graphic mode.* This means that you can perform the following tasks at the *position marked* by the *crosshair:*

  * *open* a *text region* (*text field*) where the *text input starts*, as described in Section 4.1

  * *open* a *math region* (*calculation field*) where you enter *mathematical expressions*, as described in Section 4.2

  * *open* a *graphic region* (*graphic window*), as described in Chapter 18.

* **Insertion point** (vertical line) |

The *insertion point* appears in the *text field* when you *switch* to *text mode*. It is already known from *word processing systems* and is used by MATHCAD

* to *identify* the *current position* in the *text*
* to *enter* or *delete numbers* or *characters*.

- **Editing lines**

Editing lines appear in the *calculation field* when you *switch* to *math mode* and are used

* to *mark* individual *digits, constants* or *variables* for the *input*, the *correction* or the *symbolic calculation* (see Example 4.1) and have here one of the *forms*

  ̲x̲| or |x

  i.e.., it can be set in front or behind the term
* to *mark* complete *expressions* for the *input* (see Example 4.1), *copying* or for the *symbolic* or *numerical calculation* and have here the *form*

  expression|

You *create* the *editing lines* with a *mouse click* on the expression and/or by pressing the

⊏⊐ or ⬇⬆⬅➡ *keys*.

Because such *editing lines* are not used in other computer algebra systems, we suggest that you practice on some exercises, such as the following Example 4.1.

Example 4.1:

a) If the *expression*

$$x^2 + \sin(x) + 1$$

is to be *symbolically differentiated* or *integrated* with, regard to x, the *variable* x must be *marked* once with *editing lines* before invoking the corresponding *menu sequence*, i.e.,

$$x^2 + \sin(\underline{x}) + 1 \quad\text{or}\quad x^2 + \sin(|x) + 1$$

or

$$\underline{x}^2 + \sin(x) + 1 \quad\text{or}\quad |x^2 + \sin(x) + 1$$

i.e., the *editing lines* can be placed *before* or *behind* the *variables*.

b) We wish to *enter* the *expression*

$$\frac{x+1}{x-1} + 2^x + 1$$

We *start* with the *input* of x+1, which *yields*

$x + 1|$

Now press ☐-*key* to *mark* the *complete expression*, i.e.,

$\underline{x + 1|}$

We *now* enter the *division symbol* / and *then* x–1, which *yields*

$$\frac{x + 1}{x - 1|}$$

To *add*

2^x

press the ☐-*key twice* to *mark* the *complete expression* with *editing lines*, i.e.

$$\frac{x + 1|}{x - 1|}$$

You can now *enter*

+ 2^x

which *yields*

$$\frac{x + 1}{x - 1} + 2^{x|}$$

To also *add*

1

the ☐-*key* must be pressed to *mark* 2^x with *editing lines*, i.e.

$$\frac{x + 1}{x - 1} + \underline{2^{x|}}$$

You can now *enter* +1

The *editing lines* are used for the *input* of *expressions* to *build* the *expression*, i.e., to return to the required level of the expression.
Rather than using the ☐-*key*, which works best, you can also try using the mouse click or the ⬇⬆⬅➡-*keys* for the *input* of *editing lines*.

    ♦

We will now discuss *other general properties* of the *worksheet:*

• The necessary *division* of the worksheet into *text, math* and *graphic regions differentiates* MATHCAD from *word processing systems.*

• If the *text, math* and *graphic regions* overlap in a *worksheet,* they can be *separated* using the *menu sequence*

Format ⇒ Separate Regions

- The *worksheet* contains *expressions, formulae, equations, graphics* and *text* that can be *moved* with the following steps:

  I. A *mouse click* causes them to be surrounded with a *selection rectangle*.

  II. A *small hand* appears when the *mouse pointer* is *placed* on the *border* of the *selection rectangle*.

  III. They can then be *moved* while keeping the *mouse button pressed*.

- The *formatting toolbar* can be used to assign *fonts* to *text* and *math regions*.
  The **Format** *menu* permits *other layout capabilities*. Because these capabilities are easy to test, we do not provide any further description here.

- MATHCAD has a number of *templates* (*files* with the .MCT *extension*), in which *formats, fonts,* etc. for *text* and *calculations* are defined and assist the user in laying out the worksheet. These *templates* are contained in the TEMPLATE *subdirectory* for MATHCAD.
  Because these *templates* have similar characteristics as templates known from *word processing systems*, we do not provide any further details.

  ◆

We have already shown in Figure 3.2 in Chapter 3 an *example* for *laying out* a *worksheet* with *text, math* and *graphic regions*; we now show in Figure 4.1 a further *worksheet* that we *layout* using the *manipulation* of *expressions* (Chapter 13) and *save* as *file* with the name TRANSFOR.MCD.
In the shown *worksheet*, MATHCAD displays a *short comment* for each calculation. You can achieve this by clicking

Show Comments

in the *dialogue box* that appears with the

Symbolics ⇒ Evaluation Style...

menu sequence. You can also *set* in this *dialogue box* whether a computed *result* is to be displayed *next* to or *below* the expression.

◆

In the *following sections* of this chapter, we provide a detailed discussion of the layout of *text* and *math regions*, whereas the display of *graphic regions* is discussed in Chapter 18.

◆

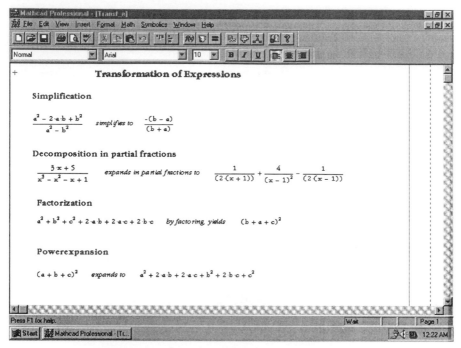

Figure 4.1. Example for the layout of a MATHCAD worksheet

4.1 Text Arrangement

You are automatically in *math mode* at the *start* of MATHCAD, i.e., you can *enter mathematical expressions* in the *worksheet* at the location marked by the cursor.

☞

To permit *text* to be *entered* at the location marked with the cursor in the *worksheet*, you must *switch* into *text mode*. This can be done in the *following ways*:

- *Enter* the *quote character* ['] from the *keyboard*.
- *Activate* the *menu sequence*

  Insert ⇒ Text Region

- If you have *already entered characters* in *math mode*, *press* the [＿＿＿]-*key*.

  ♦

You can recognize the *text mode* at the *text field,* in which the *insertion point* | is located and is *framed* by a *rectangle* (*selection rectangle*). During the *text input,* the *text field* is continually *extended* and the *insert bar* is positioned after the character last entered.

♦

The ⏎-key cannot be used to end the *text input.* Rather, this causes only a *new line* in the text.

You *leave* the *text mode* in one of the *following ways:*

* *click* with the *mouse* outside the text,
* *enter* the Ctrl-⬆-⏎ *-key combination.*

    ♦

MATHCAD also *realizes* the most important *functions known* from *word processing systems* under WINDOWS. You can find these in

* the **File, Edit, View, Insert, Format** *menus* of the *menu bar*
* known (standardized) *buttons/icons* of the *standard toolbar* and *formatting toolbar.*

    ♦

Because the *text processing functions* in MATHCAD have the same meaning as for word processing systems, we discuss in the following only some *important* or *MATHCAD specific* functions:

* *Fonts* and *character sizes* can be set in the *formatting toolbar.*
* You can enter an *equation* or *formula* in the *text:*

  I. *First* activate the *menu sequence* to *create* a *math region*

    Insert ⇒ Math Region

  II. You can *then enter* the required *equation/formula.*

  III. *Finally, click* again in the *text* to *return* to *text mode.*

  If the *equation/formula* to be inserted is already located in the *worksheet,* you can *copy* it to the required text location as usual.

  If the *equation/formula* has only an *illustrative character,* i.e., it is not to be used for calculation, it can be *deactivated* by clicking on **Disable Evaluation** in the *dialogue box* for **Calculation** that appears with the

  Format ⇒ Properties...

  menu sequence.

* *Text* contained in the worksheet can be

  * *deleted*

by marking while keeping the mouse key pressed or surrounding it with a *selection rectangle* and then pressing the *Delete-key* or clicking it with the known *cut button/icon*

from the *standard toolbar*

* *corrected*
  by placing the cursor (insertion point) at the appropriate location and then corrected

* *moved*
  with a *mouse click* (pressed mouse button) to surround the appropriate *text* with a *selection rectangle*, and then place the mouse pointer at the border of the rectangle until a hand appears, and finally move the text while keeping the mouse button pressed.

      ◆

Despite its *text processing functions*, MATHCAD cannot compete with professional *word processing systems* such as WORD or WORDPERFECT:

* If *calculations predominate* in a document, you can also use MATHCAD to write the associated text.

* If *text predominates* in a document, you should use a word processing system, such as WORD for WINDOWS, and use the clipboard to pass the calculations made with MATHCAD to the text.

This last method is always required if you want to produce magazines or books. This book is an example of this. It was created with WORD for WINDOWS (Version 6.0c) and the included calculations and graphics from MATHCAD.

◆

4.2 Layout of Calculations

If the *cursor* at a *free position* of the *worksheet* has the *form* of the + *crosshair*, you can start with the *input* of *expressions*, *formulae* and *equations*, i.e., to switch to the *math mode*.

You can recognize the *math mode* at the *calculation field* after entering the first character through the presence of the *editing lines* and the *enclosing rectangle* (*selection rectangle*). The *calculation field* is continually *extended*

during the *input* of a *mathematical expression* and the *editing lines* are positioned after the last input character (see Example 4.1b).

♦

Various

*   *mathematical operators*
*   *mathematical symbols*
*   *Greek letters*

from the *operator palettes* of the *math toolbar* are available by mouse-click for the *input* of *expressions, formulae* and *equations.* The *operators* and *symbols* may appear with *placeholders* should values be required. After entering the appropriate values in the *placeholders* and closing the complete expression with *editing lines*, the *exact (symbolic) calculation* or the *numerical calculation* with the set *accuracy* (see Chapter 6) can be initiated in one of the *following ways*:

*   *Activate* the **Symbolics** *menu*
*   *Enter* the
    *   *symbolic equal sign* →
    *   *numerical equal sign* =

♦

The following actions can be performed on a *math region* in the *worksheet*:

*   *delete* and *move*

    This is done the same way as for text (see Section 4.1)

*   *correct*

    Mathematical expressions can be corrected in many ways:

    *   You can *correct Individual characters* as follows:

        With a mouse click set the *editing lines* before or after the character to be corrected. Then use the *Delete-key* or ⬅-*key* to delete the character and to insert a new character.

    *   You can *insert* a *mathematical operator* as follows:

        With a mouse click set the *editing lines* at the appropriate position and then enter the operator.

    *   You can *delete* a *mathematical operator* as follows:

        With a mouse click set the *editing lines* before or after the operator to be deleted. Then use the *Delete-key* or ⬅-*key* to delete the operator.

Because we have discussed only the *most important correction capabilities*, we suggest that the reader does a few exercises before he or she starts working with MATHCAD so that he or she can acquire a standard repertoire of operations used for correction work.

♦

The *math regions* in a *worksheet* for MATHCAD are *processed* from *left to right* and *from top to bottom*. You must take account of this when you use *defined quantities* (functions, variables) (see Sections 8, 10.3, 17.2.3). They can only be used for *calculations* when these are performed to the right or below the *assignment*. MATHCAD *shows inverted* those quantities that are *not* yet *defined* when they are used, i.e., framed with a black box.

♦

4.3 Worksheets Management

Now that we have already described in the previous sections the main properties of MATHCAD worksheets, we discuss in this section some *global properties*, such as *open*, *save* and *print*, the *layout, hyperlinks*, the *insertion* of *objects* and describe the integrated *MathConnex*.

4.3.1 Open, Save and Print

As for all WINDOWS programs, the *open*, *save* and *print* of *files* belong to the *standard operations*.

MATHCAD can perform the following operations on a *file*:

* *Open*
  using the usual *menu sequence* for WINDOWS programs

  File ⇒ Open...   (**File ⇒ New...** for a new file)

  where a *differentiation* is made *between*

  * *worksheets* (file extension .MCD)
  * *templates* (file extension .MCT)
  * *electronic books* (file extension .HBK)
* *Save*
  using the usual *menu sequence* for WINDOWS programs

  File ⇒ Save As...

  to diskette or hard disk with the *following extensions*:

  * .MCD
    represents the *standard extension* for *worksheets* and is used for MATHCAD *documents*

* .MCT
  this is used to save the *worksheet* as *template*
* .RTF
  this is used to save the *worksheet* in a *form* (*rich text format*) that can also be *read* by *word processing systems.*
- *Print*
  using the usual *menu sequence* for WINDOWS programs
  File ⇒ Print...

4.3.2   Layout

We use the term *layout* to describe the usual *arrangement* of the *worksheet* for the print out. This means the *definition* of *page borders, page breaks, header lines* and *footer lines.* Similar for word processing systems, this is done in MATHCAD as *follows:*
Using the *menu sequence*

- **File → Page Setup...**

  the
  * *page size*
  * *page borders*
  * *page orientation*

  can be set in the *dialogue box* that appears.

- **Insert → Page Break**

  can insert a *manual page break.*
  Otherwise MATHCAD *automatically* adds a *page break* after the size specified for a page.

- **Format ⇒ Headers/Footers...**

  can be used to enter the *page size* and the *text* for the *heading line* and *footing line* in the *dialogue box* that appears.

4.3.3   References and Hyperlinks

MATHCAD permits *two types* of *connections* between its *worksheets:*

- *Reference*

  This can be used to *access another worksheet* without opening it, i.e., you can *use* the *expressions, formulae* and *equations* present in this worksheet without being required to enter or copy them.

- *Hyperlink*

You can *add* these in *worksheets* to *open* them from *other worksheets.*

We discuss in the following the *creation* and the principal properties of *references* and *hyperlinks:*

- *Reference*

  The

  Insert ⇒ Reference...

  menu sequence can be used to *insert* it at *any position* of the *worksheet.* MATHCAD inserts the

  > ➡ Reference:A:\test_1.mcd

  symbol when you specify the *path* of the MATHCAD *worksheet* that is to reference the connection in the *dialogue box* that appears. In this case the *reference* is made to the test_1.mcd *worksheet* (file) which is on diskette in the drive A. This reference permits the *use* of all *calculations* and *results* from test_1.mcd without requiring to open test_1.mcd.

- *Hyperlink*

  The

  Insert ⇒ Hyperlink...

  menu sequence can be used to *create* it at *any position* of the *worksheet* when you specify the *path* of the MATHCAD *worksheet* that is to reference the connection in the *dialogue box* that appears. This *menu sequence* can only be *used* when previously in the MATHCAD *worksheet*

  * a *text item was marked*
  * the *insert bar was placed* in a *text region*
  * the *editing lines were placed* in a *math region*
  * a *graphic region* was clicked.

  You can recognize a set *hyperlink* by the *hand* that appears when you place the mouse pointer at the appropriate position.

  A *double-click* with the *mouse on a set hyperlink* opens the associated *worksheet.*

The *advantage* of *references* and *hyperlinks* is obvious. You can *connect* several *worksheets* with each other and so *use* them *concurrently.*

We suggest that the reader experiments with them and uses the integrated help should any problems occur.

♦

4.3.4 Insertion of Objects

The *insertion* of *objects* is performed in MATHCAD in a similar manner as for the usual word processing systems in the *following ways:*

* using the *clipboard*
* using the
  dialogue box that appears after the

  Insert ⇒ Object...

  menu sequence.

4.3.5 MathConnex

The MathConnex *system* integrated in *MATHCAD 8 Professional* permits extended connections between MATHCAD worksheets and other program systems.

Because the DOC *directory* on the *MATHCAD program CD-ROM* (see Section 2.1) contains a *guide* for MathConnex, we do not provide further details here

◆

5 Electronic Books

As with all *computer algebra systems*, MATHCAD has the following *structure* (see Section 1.1):

- *User interface*

  for the *interactive work* between *user* and *computer.*

- *Kernel*

  loaded for *every use* of the *system.* This contains the programs for the *solution* of *basic problems.*

- *Supplementary programs*

  used to *extend* the system and only need to be *loaded/opened* on *demand.*

  MATHCAD provides the following three types of *supplementary programs:*

  * *Electronic Books*

    they consist of a *collection* of *worksheets* for the individual *areas* from mathematics, engineering, science and economics and are created in the form of a book, i.e., they contain the basic *formulae, calculation methods* and *facts* with *explanations,* and are divided into chapters.

  * *Extension Packs*

    They extend and supplement the built-in (predefined) functions integrated in MATHCAD.

  * *Electronic Libraries*

    These contain three electronic books for technical areas.

Electronic books and *extension packs* solve advanced or complex problems. MATHCAD has more than 50 *electronic books* and several *extension packs* and *electronic libraries* (see Section 5.2) for the solution of problems from the areas of mathematics, engineering, science and economics that have been prepared by specialists of the appropriate areas.

Consequently, you should first search in the provided *electronic books* and *extension packs* if you cannot find a realization in MATHCAD for a problem to be solved.

♦

A further extensive area for the creation of electronic books is for the preparation of education material using MATHCAD. MATHCAD's electronic books for *education* and *training*, marketed under the name *Education Library,* can be used here as basis.

♦

5.1 Properties, Structure and Handling

The *electronic books* for MATHCAD

- *extend* MATHCAD, because they represent a *collection* of *MATHCAD worksheets* that covers problems that cannot be solved using the menus/commands/functions integrated in MATHCAD
- are *books* in the metaphorical sense, because

  * you can use the appropriate buttons/icons of the standard toolbar to *scroll* in a *book*
  * they have a *table of contents* (see Figures 5.1 and 5.2), and in which you can use a *mouse click* to *open* individual *chapters*
  * they have a *search function* that you can use to *search* for arbitrary *terms*

  Furthermore, you can

  * add *your own comments* and *additions*
  * *extend* existing *formulae* or *modify* them to your *own requirements*

  and *save* the *book* in this changed or extended form. Note here that the changes are not saved when the book is closed. However, you can save the changed book in the usual way in the required directory or, in more recent books, *save* using the *menu bar* by activating the *menu sequence*

  Book ⇒ Save Section   or   **Book ⇒ Save All Changes**

  If you wish to reread a changed book that has been saved, invoke the *menu sequence*

  Book ⇒ View Edited Section

- *contain equations, formulae, functions, constants, tables, explanations* (*explanatory text*), *graphics* and *calculation methods* for *many areas,* which with the usual method with copy and insert button can be *accepted* into your own *MATHCAD worksheet* using the clipboard and vice versa. This provides you with *online access* while working with MATHCAD and so avoids the tedious input of calculation and text passages and the long-winded searching in books and tables.

- *are continually extended* and created for new areas.
- *have* the *advantage* that they are written in the *form* of a *training manual,* i.e., in the language of the user.
- *are contained* in *files* with the .HBK *extension after* the *installation.*
- *are not supplied* with MATHCAD, but must be *purchased separately* and consequently also *installed separately.*
- *are provided* by both MATHSOFT and other companies.

An *electronic book* can be *opened using* one of the *menu sequences*

* **File ⇒ Open...**
* **Help ⇒ Open Book...**

by entering in the displayed *dialogue box* the *path* of the *file* with the .HBK *extension* that contains the sought book.

After *opening* an *electronic book,* a *window* (*title page*) appears with its own *standard toolbar* (see Figure 5.1). Some of the *uses* of these *buttons/icons*

* *Search*
  for terms
* *Scroll*
  in the book
* *Copy*
  selected areas
* *Print*
  sections

As with all modern WINDOWS programs, the *meaning* of the individual *buttons/icons* is *displayed* when you place the mouse pointer on the appropriate button/icon.

♦

These *electronic books* are *used interactively,* i.e., parameters, constants and variables can be changed for special calculations and MATHCAD then calculates in automatic mode the associated result.

For example, you can *enter other coefficients* a, b and c in the *section* shown in *Figure 5.3* for the *solution* of *quadratic equations* from the **Quick Sheets** *electronic book* (see Figure 5.2) and MATHCAD *calculates* the corresponding *results.*

However, the specified *notes* for working with *electronic books* only apply for *more recent books.* Not all these aids are currently available for older books. In this case, you must the usual method to load into the MATHCAD worksheet the individual documents of the book whose files have the .MCD extension.

♦

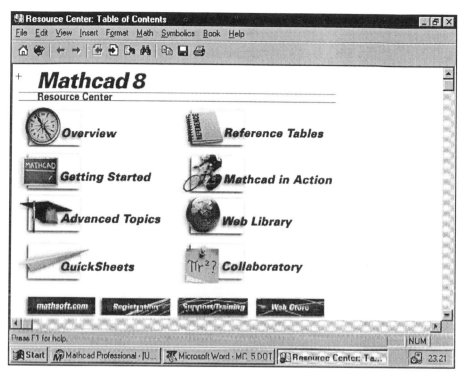

Figure 5.1. Title page of the **Resource Center** electronic book

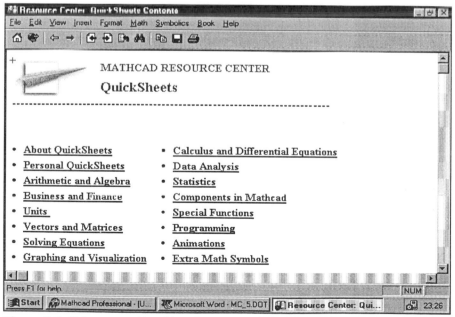

Figure 5.2. Title page with table of contents of the **QuickSheets** electronic book from the **Resource Center**

You can get a first impression of the *electronic books* for MATHCAD with the *integrated*

Resource Center

that is *opened* using the *menu sequence*

Help ⇒ Resource Center

The *title page* is shown in *Figure 5.1*.

It contains more than 600 worksheets from several *electronic books* and uses examples to supply *help* for many problems:

* **Overview**

    shows an *overview* of MATHCAD Version 8.

* **Getting Started**

    provides a step-by-step *introduction* to working with MATHCAD.

* **Advanced Topics**

    shows an overview of the *new features* in MATHCAD Version 8, such as for the solution of differential equations and optimization problems.

* **Quick Sheets**

    permits the *access* to *complete* MATHCAD *worksheets* for the solution of a range of *mathematical problems*.

* **Reference Tables**

    The *reference tables* contain *important formulae* and *constants* from the areas of *mathematics, engineering* and *science*.

* **Mathcad in Action**

    shows an *overview* in the *technical disciplines* provided for the MATHCAD solutions.

* **Web Library**

    permits the *access* to the MATHCAD WWW *library*, in case you have an *Internet connection* and the Microsoft's *Internet Explorer* has been installed on the computer. This *library* contains *electronic books* and interesting *worksheets* that you can *download*.

* **Collaboratory**

    If you have an *Internet connection* and have installed *Internet Explorer*, you can participate here on the *Online Forum* for MATHCAD users.

    ♦

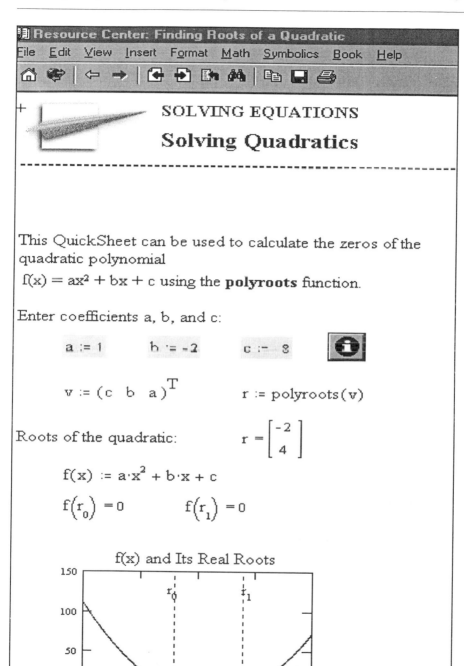

This QuickSheet can be used to calculate the zeros of the quadratic polynomial

$f(x) = ax^2 + bx + c$ using the **polyroots** function.

Enter coefficients a, b, and c:

$a := 1 \qquad b := -2 \qquad c := 8$

$v := (c \quad b \quad a)^T \qquad r := \text{polyroots}(v)$

Roots of the quadratic: $\qquad r = \begin{bmatrix} -2 \\ 4 \end{bmatrix}$

$f(x) := a \cdot x^2 + b \cdot x + c$

$f(r_0) = 0 \qquad\qquad f(r_1) = 0$

Figure 5.3. Extract from the *Solving Equations* chapter of the **QuickSheets** electronic book

5.2 Books Available, Extension Packs and Libraries

More than 50 electronic books are currently available for the areas of *electrical engineering, mechanical engineering, civil engineering, mathematics, economics* and *natural science.*
For *mathematics,* these are the *electronic books* for

* *higher (advanced) mathematics*
* *differential equations*
* *numerical methods*
* *financial mathematics*
* *statistics.*

The *electronic books* are not part of the supplied MATHCAD and must be purchased separately.

♦

A listing of the *important*

* *Electronic books:*
  * **Standard Handbook of Engineering Calculus (Machine Design and Analysis and Metalworking)**

    This book contains more than 125 applications from the areas of machinery design and calculation and for metal processing; calculations for milling, stamping, drilling; gearbox calculations; tension and expansion calculations; power transmission, etc.

  * **Standard Handbook of Engineering Calculus (Electrical and Electronics Engineering)**

    This book contains more than 70 practical examples from the areas of motors, power current engineering, electronic components and circuits, electrical devices, optics, electronic switching analysis, calculation of transformers and amplifiers, etc.

  * **CRC Material Science and Engineering**

    This book contains material constants (density, melting point, crystal structures,...), ceramics and polymer calculations, data tables for composite materials, metals and oxides, etc.

  * **Tables from the CRC Handbook of Chemistry and Physics**

    This book contains formulae, graphics and more than 80 tables for chemistry and physics.

* **Theory and Problems of Electric Circuits**

  This book contains circuit concepts, complex frequency calculations, Fourier analysis and Laplace transforms, etc.

* **Mathcad Electrical Power Systems Engineering**

  This book contains calculations for energy conversion, protection of energy systems, simulation and protection of three-phase motors, etc.

* **Topics in Mathcad: Electrical Engineering**

  This book contains calculations for electromagnetism, circuit analysis, signal processing, and filter design, etc.

* **Building Thermal Analysis**

  This book contains calculations for thermal transmission in walls, pipes and buildings, etc.

* **Formulas for Stress and Strain**

  This book contains tables with material constants, calculations of elasticity modules, tensions, bending stresses, expansions, stabilities, etc.

* **Astronomical Formulas**

  This book contains astronomical and physical constants, facts about astronomical phenomena, stars and star systems, etc.

* **Topics in Mathcad: Advanced Math**

  This book provides solutions for differential equations, computes eigenvalues, performs conformal mappings and matrix operations, etc.

* **Topics in Mathcad: Differential Equations**

  This book provides a range of numerical methods for the solution of initial and boundary value problems for ordinary and partial differential equations that occur in physics.

* **Finite Element Beginnings**

  This book contains an introduction to the theory and use of the methods of finite elements.

* **Topics in Mathcad: Numerical Methods**

  This book contains advanced numerical methods for the solution of boundary value problems and eigenvalue problems, partial differential equations, integral equations and elliptic integrals.

* **Mathcad Selections from Numerical Recipes Function Pack**

  This function packet contains 140 numerical methods, etc. for interpolation, optimization, integration and solution of algebraic equations and differential equations.

* **Explorations with Mathcad: Exploring Numerical Recipes**

  This book uses more than 140 examples to show the method of operation of numerical algorithms.

* **Exploration with Mathcad: Exploring Statistics**

  Guide for statistical methods. It handles parametric and non-parametric hypothesis testing.

* **Topics in Mathcad: Statistics**

  This book contains facts related to combinatorics, time series analysis, Kendall rank correlation coefficients and Kolmogorov-Smirnov test, etc.

* **Mathcad Treasury of Statistics, Volume I : Hypothesis Testing**

  This book contains parametric and non-parametric hypothesis tests, variance analyses, ranking value methods according to Wilcoxon, ranking summation tests according to Mann-Whitney and the Student, chi-square, and F distribution.

* **Mathcad Treasury of Statistics, Volume II : Data Analysis**

  This book contains point estimates, variance analyses, multiple regressions, time series analyses, the methods of minimum squares and the calculation of confidence intervals, etc.

* **Personal Finance**

  This book contains graphical methods for the determination of value reductions, methods for the analysis of capital investments and for the refinancing of installment payments, facts on credit costs and mortgages.

* **Mathcad Education Library: Algebra I and II**

  These books provide solutions for the basic problems of algebra.

* **Mathcad Education Library: Calculus**

  This book contains exercises for the basic mathematical computational methods.

* **McGraw Hill's Financial Analyst**

  This book provides a guide through the financial world.

* **Mathcad 8 Treasury**

  This book provides a guide on the use of MATHCAD 8.

* **The MathSoft Electronic Book Sampler**

  This book provides a cross-section through the electronic books for MATHCAD.

* **Queuing Theory**

  This book discusses the theory of queuing systems.

* **Engineering Calculations**

  This book contains 70 examples from the standard manual Electrical Engineering and Calculations.

- *Extension Packs:*
  * **Signal Processing Extension Pack**
  * **Image Processing Extension Pack**
  * **Steam Tables Extension Pack**
  * **Numerical Recipes Extension Pack**
  * **MATHCAD Expert Solver**
  * **Wavelets Extension Pack**
- *Electronic Libraries:*
  * **Electrical Engineering Library**
  * **Civil Engineering Library**
  * **Mechanical Engineering Library**

6 Exact and Numerical Calculations

To *solve mathematical problems,* the major part of the book (Chapters 12-27) *proceeds as follows:*

- *First* we explain the *menus/commands/functions* of MATHCAD for the *exact (symbolic) solution.*

- Because the exact solution is not always successful, we *then* discuss the *numerical (approximate) solution* using the *numerical functions* integrated in MATHCAD. This method returns a finite *decimal number* (floating point number) as *approximation* for the *result.*

We have chosen this *methodology* for *calculations,* because we *first* recommend the *exact calculation* (if possible). The *numerical calculation* is used only when this fails.

♦

The *processing* of the *exact (symbolic)* or *numerical (approximate) solution* of a given *problem* takes place in MATHCAD in the *following steps:*

I. First use the

    Symbolics ⇒ Evaluation Style...

    menu sequence to display a *dialogue box* in which by clicking

    Show Comments

    you *request* MATHCAD to *display* a short *comment* for each *calculation.* You can also *set* in this *dialogue box* whether the *result* is to be shown *next to* or *below* the *expression to be calculated.*

II. Before you can start with *calculations,* you must first *enter* the *problem to be calculated* in the *worksheet.* The following methods are available:

    - With *mouse click* from the *operator palettes* of the *math toolbar* various

      * *mathematical operators*
      * *mathematical symbols*
      * *Greek letters*

      If necessary, the *operators* and *symbols* appear with *placeholders* for any required values.

- *Commands/functions*

  can be *entered* in one of the *following ways*:

  - using the *keyboard*
  - by *copying*
  - by *clicking* the *button* for *built-in/predefined functions*

    in the *standard toolbar* above the displayed *dialogue box*,

  - by *activating* the *menu sequence*

  Insert ⇒ Function...

  above the displayed *dialogue box*.

  Most *functions* require an *expression* in the *argument* (parenthesized).

  If you want to calculate or execute *several expressions/functions* successively, it is possible to enter them all and then initiate the calculation after the last one has been entered. We discuss this method in Section 6.3.

III. *After making* the *input*, you can *initiate* the *calculation* in one of the *following ways*:

- Input the

  - *symbolic equal sign* → (for *exact calculation*) with the key combination Ctrl-$\boxed{.}$
  - *numerical equal sign* = (for *numerical calculation*) with the keybord

  and then press the $\boxed{\leftarrow}$-key (in *automatic mode* – see Section 6.3.1), after *marking* the expression with *editing lines* (see Chapter 4).

- *Activate* the *menu sequences* from the *menu bar*, where the

  Symbolics

  menu is used.

  A *menu sequence* is the *successive execution* of *menus* and *submenus* that we write in the form

  menu_1 ⇒ menu_2 ⇒ ... ⇒ menu_n

  where the ⇒ *arrow* represents a *mouse click*, which is also used to end the *complete menu sequence*.

  If three points appear after a *submenu*, this means that a *dialogue box* appears requiring completion.

  Most *menu sequences* require that the *expression* to be calculated or a *variable* of the expression be marked with *editing lines* (see Chapter 4).

MATHCAD allows *both specified calculation types* for *some problems*, i.e., you can use both a *symbolic equal sign* → and a *menu sequence* for the *exact calculation*.

Because of the simple handling, we suggest that you use the *symbolic equal sign* → for *exact calculations* and the *numerical equal sign* = for *numerical calculations*, provided that the problems to be solved permit this.

The exact *procedure* used for the *solution* of given *problems* is explained in the appropriate chapters.

♦

Use the Ctrl-⏎ *key combination* if a *line break* is required in the *input* of *commands/functions* or *expressions*.

♦

When working with MATHCAD, you frequently need to *reuse* the returned *results* in subsequent calculations. There are various possibilities for the required *assignments* (*solution assignments*). Notes and examples are provided in the appropriate chapters, for example in

* Chapter 16 for the solution of equations

* Chapter 19 for the result of a differentiation

* Chapter 20 for the result of an integration.

♦

If MATHCAD does *not* find *any solution* for the *calculation* of a problem, it can indicate this in various ways:

I.   It *issues* a *message* that no solution was found.

II. The *calculate command* is *returned unchanged*.

III. The *calculation* is *not completed* within a reasonable time.

IV. It *issues* a *note* that the *result* is *MAPLE-specific* and whether it is to be *saved* in the *clipboard*. However, no result was calculated in the majority of these cases. You can see this when you display the contents of the clipboard in the worksheet.

If, in case III., you want to *terminate* the *calculation*, this is done in MATHCAD by pressing the Esc -key.

♦

MATHCAD permits *two forms* for all *calculations*:

* *automatic mode*

* *manual mode*

that we discuss in Sections 6.3.1 and 6.3.2. ♦

In the next two Sections 6.1 and 6.2, we provide a detailed description of
performing exact and *numerical calculations* using MATHCAD. The major
part of the book (Chapters 12-27) provides further *details.*
♦

6.1 Exact Calculation Using Computer Algebra

MATHCAD uses a minimum variant of the *symbolic processors* from MAPLE
to *perform exact (symbolic) calculations;* this processor is *loaded automatically* at the start of MATHCAD.

MATHCAD provides *two methods* for the *exact (symbolic) calculation* of the
expressions/functions contained in the worksheet once you have *marked*
the *expression/function* or a *variable* with *editing lines:*

I.   *Use* a *menu sequences* of the

   Symbolics

   menu from the *menu bar.*

II. *Input* of the *symbolic equal sign* → (see Section 6.1.1) and then press the
   ⏎-key.

Each of these *methods*

*   *initiates* the *operation* of the *symbolic processor* adopted from MAPLE

*   is *described* in *detail* for the *solution* of *individual problems* in the
   course of the book.
   ♦

Because the *use* of the *symbolic equal sign* → (that can also be used in
combination with so-called *keywords*) is not provided for other systems, we
provide a detailed discussion in the following two Sections 6.1.1 and 6.1.2.
♦

If you use *decimal numbers* in the *expressions* for the *exact calculation,* the
result is also returned as a *decimal number.*
♦

If you want to display an *exact result* as *decimal number (floating point
number),* you must enclose the expression with editing lines and activate
the *menu sequence*

Symbolics ⇒ **Evaluate** ⇒ **Floating Point...**

You can enter the required *precision* (maximum 4000 decimal places) in the displayed

Floating Point Evaluation

dialogue box.

♦

6.1.1   Symbolic Equal Sign

The *use* of the *symbolic equal sign* → provides a number of *advantages* for *processing exact (symbolic) calculations*:
The *symbolic equal sign*

*   can *avoid* the need to activate the corresponding *submenu* from the

    Symbolics

    menu, as will be shown for specific calculations in the course of the book. The additional *use* of *keywords* serves the same purpose (see Section 6.1.2).

*   can be used for the *exact calculation* of *function values* (see Example 6.1f).

*   also *returns* a *decimal approximation* for a mathematical expression if you enter in *decimal form* all the *numbers* used in the expression (see Example 6.1g).

Two methods are provided to *activate* the *symbolic equal sign* →:

I. *Click* the

button in the *operator palette no. 2* or *8* or *enter* the *key combination*

Ctrl-⬚

after having entered the *expression* to be *calculated* in the *worksheet* and *marking* it with *editing lines.* You can also first click on the button and then enter the expression in the displayed placeholder.

II. *Click* the

button in the *operator palette no. 8.*
In contrast to I., the symbolic equal sign here has two placeholders in which the expression and a keyword are entered.

♦

☞

If you perform one of the *following activities after entering* the *symbolic equal sign* →:

*   *mouse click* outside the expression
*   *press the* ⏎-key

the *exact* (*symbolic*) *calculation* of the *expression* is initiated, provided the *automatic calculation* (*automatic mode*) was previously activated from the

Math ⇒ Automatic Calculation

menu sequence. You recognize it with the *word* **Auto** in the *status bar.*
If the *automatic mode* has been switched off (*manual mode*), then

*   *press* the F9-key

to initiate the *calculation* (see Section 6.3.2).
♦

We will learn the *first uses* of the *symbolic equal sign* → in *Example 6.1.* The main part of the book (Chapters 12-27) contains other examples.

Example 6.1:

We use the *symbolic equal sign* to solve the following problems:

a)
$$\int_1^2 x^x \, dx \;\rightarrow\; \int_1^2 x^x \, dx$$

MATHCAD *cannot exactly calculate* this *definite integral* and will output it unchanged after the symbolic equal sign is entered.

b)
$$\int_0^5 x^6 \cdot e^x \, dx \;\rightarrow\; 6745 \cdot \exp(5) - 720$$

The *symbolic equal sign* can *exactly calculate* this *definite integral.*

c)
$$\frac{d^2}{dx^2} \frac{1}{x^4 + 1} \;\rightarrow\; \frac{32}{(x^4 + 1)^3} \cdot x^6 - \frac{12}{(x^4 + 1)^2} \cdot x^2$$

The *symbolic equal sign* performs this *differentiation exactly.*

d)
$$\prod_{k=1}^n \frac{1}{k + 4} \;\rightarrow\; \frac{24}{\Gamma(n + 5)}$$

The *symbolic equal sign* calculates this *product exactly.*

e) The use of the *symbolic equal sign* permits *defined functions*

   e1) to be *differentiated exactly*:

   $$f(x) := \sin(x) + \ln(x) + x + 1$$

   $$\frac{d}{dx}f(x) \to \cos(x) + \frac{1}{x} + 1$$

   e2) to be *integrated exactly*:

   $$f(x) := \sin(x) + \ln(x) + x + 1$$

   $$\int f(x)\, dx \to -\cos(x) + x\cdot\ln(x) + \frac{1}{2}\cdot x^2$$

   e3) to be used for the *calculation* of *limits*:

   $$f(x) := \frac{2\cdot x + \sin(x)}{x + 3\cdot\ln(x + 1)}$$

   $$\lim_{x \to 0} f(x) \to \frac{3}{4}$$

f) We calculate the *value* of the sin x *function* at the point $\frac{\pi}{3}$ *exactly* and *numerically*.

   exact calculation using the symbolic equal sign:

   $$\sin\left(\frac{\pi}{3}\right) \to \frac{1}{2}\cdot\sqrt{3}$$

   numerical calculation using the numerical equal sign:

   $$\sin\left(\frac{\pi}{3}\right) = 0.866 \quad \blacksquare$$

g) The *symbolic equal sign* does *not change real numbers* in *symbolic form*, such as

   $$\sqrt{2} \to \sqrt{2}$$

   The same applies for *mathematical expressions* that contain real numbers in symbolic form, such as

$$\frac{\sqrt{5} + \ln(7)}{e^3 + \sqrt[3]{2}} \rightarrow \frac{\left(\sqrt{5} + \ln(7)\right)}{\left[\exp(3) + 2^{\left(\frac{1}{3}\right)}\right]}$$

Because the real numbers contained in this expression cannot be further simplified exactly, the symbolic equal sign returns the same expression, although MATHCAD alters the form slightly.

However, if you change the form of the *numbers* in the preceding expressions by specifying them in *decimal notation*, the *symbolic equal sign* returns the following *decimal approximations* (floating point numbers):

$$\sqrt{2.0} \rightarrow 1.4142135623730950488$$

$$\frac{\sqrt{5.0} + \ln(7.0)}{e^{3.0} + \sqrt[3]{2.0}} \rightarrow .19591887566098330253$$

♦

6.1.2 Keywords

Together with the *symbolic equal sign* → , *keywords* provide an effective means of performing *exact calculations without using* the **Symbolics** *menu*.

The use of keywords in the versions 7 and 8 of MATHCAD has changed somewhat from the previous versions. The author's books [2] and [4] describe the use of keywords for the versions 5 and 6 of MATHCAD.

♦

Click the

button to display the *complete palette* of the *symbolic keywords* (*operator palette no. 8*) in the *math toolbar* (see Figure 6.1).

Two capabilities are provided for the *Symbolic Keyword Toolbar* used to *activate* the *keyword*:

I.   *Use* the

button to produce a *symbolic equal sign* with *two placeholders*; the *expression* to be calculated and the associated *keyword* are written in the *first* and *second placeholder* respectively.

II. Click the appropriate *keyword button*; an *expression* in the *following form*

    ■ Keyword →

used for the *keyword* appears in the worksheet at the position indicated by the cursor.

The *expression* to be calculated is usually *entered* in the *placeholder* to the left of the keyword. There are also keywords that have *several placeholders* (see Example 6.2).

Because MATHCAD has already entered the keyword, it is evident from the two procedures that method II. is preferable.

♦

Perform one of the *following activities after* entering one of the previously described *keywords*:

*   *mouse click* outside the expression

*   *press* the ⏎-key

to initiate the *exact* (*symbolic*) *calculation* of *expressions*, if the

Math ⇒ Automatic Calculation

menu sequence has been used previously to activate the *automatic calculation* (*automatic mode*). You can recognize it by the *word* **Auto** in the *status bar*.

If the *automatic mode* has been *switched off* (*manual mode*), then

*   *press* the ⒡9-key

to *initiate* the *calculation* (see Chapter 6.3).

♦

Let us consider *method* of *operation* for *two keywords* used to *transform expressions* (see Chapter 13):

*   *Simplification* of *expressions* (see Section 13.1) by *clicking* the

button for the

simplify

keyword, where the *expression* to be simplified is *entered* in the *placeholder* to the *left* of the *displayed keyword* (see Example 6.2).

*   *Expand* (*multiply/exponentiate*) *expressions* (see Sections 13.4 and 13.5) by *clicking* the

button for the

expand

keyword, where the *expression* to be expanded is *entered* in the *left placeholder* and *expansion variables* (separated by commas*)* are *entered* in the *right placeholder* (see Example 6.2).

Chapter 13 contains further keywords used for the transformation of expressions.

$\blacksquare\to$	$\blacksquare\blacksquare\to$	float	complex
expand	solve	simplify	substitute
collect	series	assume	parfrac
coeffs	factor	fourier	laplace
ztrans	invfourier	invlaplace	invztrans
$M^{\tau}\to$	$M^{-1}\to$	$\lvert M\rvert\to$	Modifiers

Figure 6.1. Palette of symbolic keywords

We will discuss *further keywords* of the *palette* from *Figure 6.1* in the main part of the book.

We suggest that the reader uses the provided Example 6.2 to practice the use of *keywords* in conjunction with the *symbolic equal sign.*

♦

Example 6.2:

The **simplify** and **expand** *keywords* are used in the next two problems to *transform expressions*:

a) The use of the

   simplify

   keyword simplifies expressions with the symbolic equal sign; the expression is written in the left placeholder:

$$\frac{x^4-1}{x+1}\ \textbf{simplify}\ \to\ x^3-x^2+x-1$$

The expression is *not simplified* if you use the *symbolic equal sign without keyword:*

$$\frac{x^4-1}{x+1}\ \to\ \frac{(x^4-1)}{(x+1)}$$

b) The use of the

   expand

keyword **expands** or **multiplies** with the symbolic equal sign; the expression is written in the left placeholder and the variables are written in the right placeholder:

$$(a + b)^3 \; \textbf{expand}, a, b \; \to \; a^3 + 3 \cdot a^2 \cdot b + 3 \cdot a \cdot b^2 + b^3$$

The expression is *not expanded* if you use the *symbolic equal sign with out keyword:*

$$(a + b)^3 \; \to \; (a + b)^3$$

◆

The *symbolic equal sign* used by *itself calculates derivatives, limits, integrals, products* and *sums,* as shown in Examples 6.1a-e.

In contrast, *expressions* with the *symbolic equal sign* can only be *transformed* when you use an appropriate *keyword* (see Example 6.2).

◆

6.2 Numerical Calculations

MATHCAD differentiates between *two possibilities* for *numerical calculations:*

I. Some *mathematical expressions* in the worksheet can be *calculated numerically* immediately, when you perform one of the following operations in *automatic mode* (see Section 6.3.1) after *marking* with *editing lines.*

  * *Enter* the *numerical equal sign* =

  * *Activate* the *menu sequence*

    Symbolics ⇒ Evaluate ⇒ Floating Point...

  This capability can be used for algebraic and transcendental expressions, integrals, sums, products and calculations with matrices, etc.

II. A *numerical function* integrated in MATHCAD must be used for the *numerical calculation* of a mathematical problem. Once the numerical function has been marked with *editing lines,* the *input* of the *numerical equal sign* = *initiates* the *calculation.*

  This capability can be used for the solution of equations and differential equations, for interpolation and regression, etc.

The *solution* of the *problems described* in the course of the book indicates which *method* of *numerical calculation* should be used for a given problem.

♦

Because of the simple handling, the *numerical equal sign* = is recommended for all numerical calculations and can be *input* in the following *two ways*:

* using *keyboard*
* by clicking the

   button in the *operator palette no. 1* or *2*

   ♦

The *numerical equal sign* = should *not* be *confused* with the *symbolic equal sign*

or the *equality operator* (*equality symbol*)

contained in the *operator palette no. 2*.

♦

We saw in Example 6.1g that the *symbolic equal sign* → can also be used for the *numerical calculation* of a mathematical expression. This functions only when all numbers used in the expression are written in decimal notation.

♦

MATHCAD provides *three forms* for the *precision setting* for *numerical calculations*:

* The *number* of *decimal places* (maximum 15) for the *result* (*decimal approximation*) can be *set* as the

   Displayed Precision

   to maximum 15 positions (*default value 3*)
   in the

   Result Format

   dialogue box displayed with the *menu sequence*

Format ⇒ Result...

Decimal must be set here for a *decimal approximation*.

- If you wish to display an *exact result* or a *real number* as *decimal approximation* (floating point number), enclose the expression with editing lines and activate the *menu sequence*

Symbolics ⇒ Evaluate ⇒ Floating Point...

The required *precision* of maximum *4000 decimal positions* can be set in the displayed *dialogue box*

Floating Point Evaluation

- The required precision of a numerical method (default value 0.001) can be *set* for

Convergence Tolerance

in

Built-In Variables

in the

Math Options

dialogue box displayed with the menu sequence

Math ⇒ Options...

The same result is achieved locally by entering the

TOL := .....

assignment in the worksheet.

However, you cannot expect that the specified result has the set precision. You only know that the specified numerical method terminates if the difference of two successive approximations is smaller than **TOL**.

◆

6.3 Control of the Calculation

The *following section describes* a number of *methods* that MATHCAD uses to *control* the *calculations* to be performed.

These include

* *automatic execution* of *calculations* (*automatic mode*)

* *manual execution* of *calculations* (*manual mode*)

* *termination* of *calculations*

* *deactivate calculations*

* *optimization* of the *interaction* between *exact* (*symbolic*) and *numerical* *calculations*.

6.3.1 Automatic Mode

The *automatic mode* is the *standard setting* for MATHCAD. You can recognize its *activation* by the tick at

Automatic Calculation

in the **Math** *menu*.
You can *switch on/off* the *automatic mode* with a *mouse click* here.

You can recognize the *activated automatic mode* by the word **Auto** in the *status bar*. If the *automatic mode has been deactivated*, this is called the *manual mode*.

♦

In *automatic mode*

* every *calculation* is *performed immediately* after entering the symbolic or numerical equal sign.

* the complete *current worksheet is recalculated* if *constants, variables* or *functions* are changed; this applies for *exact calculations* only when the *symbolic equal sign* is used.

♦

If you only want to *view* a read *worksheet*, the *automatic mode* can be disruptive, because you must wait for the calculation of all expressions, equations, etc. contained in the document. In this case, we suggest that you change to *manual mode* (see Section 6.3.2).

♦

6.3.2 Manual Mode

The deactivation of *automatic mode* switches to *manual mode* and is characterized by the following *properties*:

* A *calculation* is performed only when you press the
  (F9)-*key*
  This applies for the complete *current worksheet*.

* If *variables* and *functions* are *changed*, all *calculations* based on them *remain unchanged* in the worksheet, unless you do not initiate them by *pressing* the (F9)-*key*.

☞

The *manual mode* is *preferable* when you

*   *view* a *worksheet*

*   only wish to investigate the *effects* of *changes* for just a few expressions in the current *worksheet*.

The *manual mode* has the *advantage* that you do not need to wait for MATHCAD to perform the calculations after the *input* of *expressions, formulae* and *equations* or for *scrolling*.

♦

6.3.3   Termination of Calculations

If, for some reason, you want to *interrupt* or *terminate* the running *calculations* for MATHCAD, then *press* the

(Esc)-*key*

and click on OK in the displayed

Interrupt processing

dialogue box. MATHCAD then issues a *message* that the *calculation* was *interrupted*.

If you want to *continue* this *interrupted calculation*, you must click on the appropriate expression and then perform one of the *following activities*:

*   *activate* the **Math** ⇒ **Calculate** *menu sequence*

*   *press* the (F9)-*key*.

6.3.4   Deactivate Expressions

If you want to *deactivate* the *calculation* of *individual expressions, formulae* or *equations* in a *worksheet* even though you are in *automatic mode* (see Section 6.3.1), we suggest the *following method* once the expression to be deactivated has been marked with editing lines: After *activation* of the

Format ⇒ Properties...

menu sequence, click on the

Disable Evaluation

in

Calculation Options

in the displayed

Properties *dialogue box* for **Calculation**.

MATHCAD *indicates* the *deactivation* with a small black rectangle between the corresponding expression.

♦

If you want to *undo* a *deactivation*, i.e., you want to *reactivate* the *calculation* of *expressions, formulae* or *equations*, perform the same procedure as for the deactivation, except that the *marking* for

Disable Evaluation

must be removed for

Calculation Options

The deactivation rectangle now disappears.

♦

6.3.5   Optimization

Although MATHCAD normally performs exact and numerical calculations independent from each other, you can request MATHCAD to first attempt a *symbolic calculation* before performing a *numerical calculation*. MATHCAD refers to this as *optimization* of *calculations*.

You can *switch on* or *off* the *optimization* of the *calculations* performed by MATHCAD by *activating* the *menu sequence*

Math ⇒ Optimization

It is *switched on* when a *tick* appears next to

Optimization

♦

The *enabled optimization* in MATHCAD is *characterized* by the *following properties:*

- It attempts to perform a *symbolic calculation before* a *numerical calculation*, if the *expression* to be *calculated* is in an *assignment* (*assignment statement*) (see Example 6.3).

- An *exact calculation* was *successful*, if, after the activation, an *asterisk* appears next to the *expression* to be calculated.

- A *double-clicking* of the *asterisk* displays a

  Optimized Result

  dialogue box that contains the *exact result* of the *calculation*.

Example 6.3:

An *optimization* of the *calculation* is used for the following examples:

a) We *calculate numerically* an *integral* contained in an assignment statement with

a1) *deactivated optimization:*

$$a := \int_{0}^{1} \sqrt{1 - x^2} \, dx \qquad a = 0.785208669629317$$

a2) *activated optimization:*

$$a := \int_{0}^{1} \sqrt{1 - x^2} \, dx \ {}_* \qquad a = 0.785208669629317$$

The *asterisk* displayed next to the entered integral indicates that an
exact solution was *found* for the *integral*. The result for a at the side
is a *decimal approximation* of the *exact result*
$a = \pi/4$
that is displayed by a *double-click* on the *asterisk* in the

Optimized Result

dialogue box.

b) We *calculate numerically* an *algebraic expression* contained in an assignment statement with

b1) *deactivated optimization:*

$$a := \frac{\sqrt{64} + \sin\left(\dfrac{\pi}{3}\right) + 5^2}{\cos\left(\dfrac{\pi}{3}\right) + \sqrt[3]{27}} \qquad a = 9.676007258224127$$

b2) *activated optimization:*

$$a := \frac{\sqrt{64} + \sin\left(\dfrac{\pi}{3}\right) + 5^2}{\cos\left(\dfrac{\pi}{3}\right) + \sqrt[3]{27}} \qquad {}_* \qquad a = 9.676007258224127$$

The *asterisk* displayed next to the entered integral indicates that an
exact value was *found* for the *expression*. The result for a at the side
is a *decimal approximation* of the *exact result*
$$\frac{66}{7} + \frac{1}{7}\sqrt{3}$$

that is displayed by a *double-click* on the *asterisk* in the
Optimized Result
dialogue box.

c) We *calculate numerically* an *algebraic expression* contained in an assignment statement with *activated optimization:*

$$b := \frac{\sqrt{5} + \ln(2)}{\sin\left(\frac{\pi}{3}\right) + \sqrt[3]{2}} \qquad\qquad b = 1.377840515686663$$

No asterisk appears here, although

$$\sin\left(\frac{\pi}{3}\right) = \frac{1}{2} \cdot \sqrt{3}$$

can be transformed. This may be because this transformation also returns
a real number, with the consequence the complete expression cannot be
further simplified.
However, the input of the *symbolic equal sign* returns the *simplified result*

$$b \rightarrow \frac{\left(\sqrt{5} + \ln(2)\right)}{\left(\frac{1}{2} \cdot \sqrt{3} + \sqrt[3]{2}\right)}$$

◆

6.4 Error Messages

Errors can arise for *calculations* in MATHCAD that

*   were caused by the user (e.g. division by zero)

*   occur in MATHCAD functions.

If an *error* occurs, the corresponding *expression* is marked with an *error
message* and the part of the expression in which the error occurred displayed in a different colour (red).

You can obtain *help* for the displayed *error message* by *pressing* the
F1 -*key.*

◆

7 Numbers

In this chapter we discuss in detail the number types that are of particular interest for working with MATHCAD. MATHCAD

* *can process* both *real* and *complex numbers;* we consider their possible representations in the next two Sections 7.1 and 7.2

* *interprets* every *designation* that *starts* with a *digit* as being a *number*

* *can process* the *numerical values* of some *constants;* we discuss the important ones in Section 7.3.

The *number format* for all *number types* can be *set* in the

Result Format

dialogue box that is displayed from the

Format ⇒ Result...

menu sequence. Because we will see this *dialogue box* shown in Figure 7.1 many times during the course of the book, we will not discuss it here.

♦

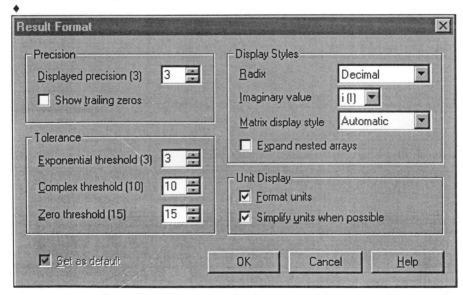

Figure 7.1. Result Format dialogue box for the setting of the number format

7.1 Real Numbers

Real numbers can be *shown* in MATHCAD in the *following* ways:

- *exact* as *symbol,* such as

    * e (*Euler's number*)

    * π (*Pi*)

    * $\sqrt{2}$

    * $\dfrac{\sqrt{5} + \ln(2)}{\sin\left(\dfrac{\pi}{3}\right) + \sqrt[3]{2}}$

        (*algebraic expression* with *real numbers* in *symbolic notation*)

- *numerical approximated* in

    * *decimal notation*

    * *binary notation*

    * *octal notation*

    * *hexadecimal notation*

    The associated *setting* is *made* in the

    Result Format

    dialogue box as

    * **Decimal**

    * **Binary**

    * **Octal**

    * **Hexadecimal**

    We use the *decimal notation* in *numerical calculations* for *real numbers* in this book.

Because the computer does not permit an unlimited number of digits for the *decimal notation* of *real numbers, approximate values* can occur here. MATHCAD permits the use of maximum *4000 decimal digits* for the approximation.

The *required number* of *decimal digits* can be *set* and calculated in the

Floating Point Evaluation

dialogue box displayed from the

Symbolics $\Rightarrow$ Evaluate $\Rightarrow$ Floating Point...

menu sequence.

♦

The *following settings* for *decimal approximation* of *real numbers* can be made in the *dialogue box* **Result Format**:

* **Displayed precision**

    Maximum 15 decimal digits (*default value=3*) can be *set* here for the *precision* to be *used* with the *numerical equal sign* = (*numerical calculation*).

* **Exponential threshold**

    An *integer* n in the interval 0 to 15 can be entered here (*default value=3*) that causes MATHCAD to represent *real numbers* greater than 10^n or smaller than 10^{-n} in *exponential form*.

* **Zero threshold**

    An *integer* n in the interval 0 to 307 can be entered here (*default value=15*) that causes MATHCAD to represent *real numbers* less than 10^{-n} as zero.

* **Show trailing zeros**

    If this is activated, every *number* will be *represented* with the number of *decimal digits set* for

    Displayed precision

    For example, if precision 6 has been set, 3 will be displayed as 3.000000.

    ◆

Chapter 12 contains information on the *basic arithmetic operations* for *real numbers*.

◆

7.2 Complex Numbers

A *complex number* z must be *entered* and *represented* in MATHCAD in one of the *forms*

* z := a + bi

* z := a + bj

i.e., without the multiplication sign between the imaginary part b and the imaginary unit i or j, where

* the *real part* a is calculated using

    Re (z)

* the *imaginary part* b is calculated using

  Im (z)

* the *argument* $\varphi = \arctan \dfrac{b}{a}$ is calculated using

  arg (z)

* the *absolute value* (*modulus*)

  $|z|$

  is formed using the *button*

  of the *operator palette no. 1 or 4.*

* the *conjugate complex number* $\bar{z} = a - bi$ is formed for the complex number z by *framing* it with *editing lines* and *pressing* the

  key.

The *imaginary unit*

$\sqrt{-1}$

can be *input* as i or j. Note that no multiplication sign and no blank may appear between the imaginary part b and the imaginary unit i.
The *setting* can be made in the

Result Format

dialogue box whether MATHCAD is to display the *imaginary unit* in the *worksheet* as i or j.
♦

MATHCAD can perform the *operations* for complex numbers either *exactly* or *numerically,* i.e., by entering the symbolic or numerical equal sign (see Example 7.1).
♦

If a calculation operation with *complex numbers* returns *several results,* MATHCAD normally outputs the *principal value* (see Example 7.1c).
♦
MATHCAD performs without difficulty the *basic arithmetic operations* and the above specified *operations* for *complex numbers; see* Example 7.1.

Example 7.1:

We define two *complex numbers*

$z_1 := 2 + 3i$ $z_2 := 1 - 5i$

and perform here *operations* using the *numerical and symbolic equal sign*:

a)

Real part $\quad$ $Re(z_1) = 2 \quad Re(z_2) = 1$

$\quad\qquad Re(z_1) \rightarrow 2 \quad Re(z_2) \rightarrow 1$

Imaginary part $\quad Im(z_1) = 3 \quad Im(z_2) = -5$

$\qquad\qquad Im(z_1) \rightarrow 3 \quad Im(z_2) \rightarrow -5$

Argument $\qquad arg(z_1) = 0.983 \qquad arg(z_2) = -1.373$

$\qquad\qquad arg(z_1) \rightarrow atan\left(\dfrac{3}{2}\right) \quad arg(z_2) \rightarrow -atan(5)$

Absolute value $\qquad |z_1| = 3.606 \quad |z_2| = 5.099$

$\qquad\qquad |z_1| \rightarrow \sqrt{13} \quad |z_2| = \sqrt{26}$

Conjugate $\qquad \bar{z}_1 = 2 - 3i \quad \bar{z}_2 = 1 + 5i$

$\qquad\qquad \bar{z}_1 \rightarrow 2 - 3i \quad \bar{z}_2 \rightarrow 1 + 5i$

b) We now perform the *addition, subtraction, multiplication* and *division:*

$z_1 + z_2 = 3 - 2i \qquad z_1 - z_2 = 1 + 8i$

$z_1 + z_2 \rightarrow 3 - 2i \qquad z_1 - z_2 \rightarrow 1 + 8i$

$z_1 \cdot z_2 = 17 - 7i \qquad \dfrac{z_1}{z_2} = -0.5 + 0.5i$

$z_1 \cdot z_2 \rightarrow 17 - 7i \qquad \dfrac{z_1}{z_2} \rightarrow \dfrac{-1}{2} + \dfrac{1}{2} \cdot i$

c) If a *calculation operation* with *complex numbers* returns *several results*, the *principal value* is the usual output:

$$\sqrt{z_1} = 1.674 + 0.896i \quad \sqrt{z_2} = 1.746 - 1.432i$$

$$\sqrt{z_1} \;\to\; \sqrt{\frac{1}{2}\cdot\sqrt{13}+1} + i\cdot\sqrt{\frac{1}{2}\cdot\sqrt{13}-1}$$

$$\sqrt{z_2} \;\to\; \sqrt{\frac{1}{2}\cdot\sqrt{26}+\frac{1}{2}} - i\cdot\sqrt{\frac{1}{2}\cdot\sqrt{26}-\frac{1}{2}}$$

The *calculation* of the *n th root* using the *root operator* from *operator palette no.1* forms an exception, as shown in the following example:
The *result* for the *cubic root* of −1 is returned *appropriate* for the *form:*
Whereas the *principal value* is *calculated* for the *form*

$$(-1)^{\frac{1}{3}} = 0.5 + 0.866\,i$$

the *use* of *n-th root operator* from the *operator palette no.1*

$$\sqrt[3]{-1} = -1$$

returns the *real result* −1. This is because the *n-th root operator* has the *property* that it always *returns* a *real result* (if available).
♦

7.3 Built-In Constants

Some of the *constants* known to MATHCAD follow (the *designations* for the *input* are parenthesized):

* $\pi = 3.14159...$ (π from the *operator palette no. 1*)

* *Euler's number* $e = 2.718281...$ (e from the *keyboard*)

* *imaginary unit* $i = \sqrt{-1}$ (1i from the *keyboard*)

* *infinity* ∞ (∞ from the *operator palette no. 5*)

* *percentage character* %=0.01   (% from the *keyboard*)

Note when you *use* the *imaginary unit* i, MATHCAD can only recognize this when a number (without multiplication sign) prefixes it. A j can be written in place of i for the *imaginary unit* (see Section 7.2).

◆

MATHCAD uses the 10^{307} *numerical value* to represent *infinity* ∞ in *numerical calculations*.

◆

The *designations* for *built-in constants* (*predefined constants*) *are reserved* in MATHCAD and should not be used for other quantities (variables or functions), because this would lose the predefined values.

◆

8 Variables

MATHCAD provides a number of means of *representing variables*. A differentiation is made between the *following variable forms*:

* *built-in variables*
* *simple variables*
* *indexed variables* with *literal index* (*literal subscript*)
* *indexed variables* with *array index* (*array subscript*)

The next section discusses the main *properties* of the *variables* used by MATHCAD.

8.1 Built-In Variables

MATHCAD has a number of *built-in* (*predefined*) *constants* and *variables*, for which we have already constants in Section 7.3. The *designations* for *built-in constants* and *variables* are *reserved* in MATHCAD and should not be used for other quantities (e.g., functions), because otherwise they would no longer be available.

Whereas *fixed values* are *assigned* to the *built-in* (*predefined*) *constants*, such as

e, π, i, %

(see Section 7.3), values other than the *standard values* used by MATHCAD can be assigned to the *built-in* (*predefined*) *variables*.

♦

We discuss the important *built-in variables* in the following (the *standard values* used by MATHCAD are parenthesized):

* **TOL** (=0.001)

    This specifies the *precision used* by MATHCAD for *numerical calculations* (see Section 6.2); 0.001 is used as *standard value*.

- **ORIGIN** (=0)

  This specifies the *index* (*array index*) of the first element (*starting index*) for *vectors* and *matrices*; MATHCAD uses 0 as *standard value* for the start index. Note that calculations with matrices and vectors usually start with the *array index* 1 (see Chapter 15), consequently you must assign the value 1 to the **ORIGIN**.

- **PRNCOLWIDTH** (=8)

  This determines the *column width* used for *writing* with the **WRITEPRN** *function* (see Chapter 9).

- **PRNPRECISION** (=4)

  This determines the *number* of *digits* to be output when the **WRITEPRN** *function* is used for *writing* (see Chapter 9).

If you want to use *different values* for the *built-in variables* from the *standard values* used by MATHCAD, you can set these *globally* for the *complete worksheet* for

Built-In Variables

in the

Math Options

dialogue box displayed from the

Math ⇒ Options...

Menu sequence.
The *assignment operator* := can be used to assign only *locally* different values for the built-in variables, for example

ORIGIN := 1

defines the local starting value 1 for the indexing.

♦

8.2 Simple and Indexed Variables

Variables play a large role for all mathematical calculations; variables can be either simple or indexed. MATHCAD takes this into consideration and can represent all variable types.
In MATHCAD, *names* (*designations*) of *variables* are *represented* for

- *simple variables*

  as *combination* of *letters* (even Greek), *numbers* and certain *special characters*, such as underscore _ , percentage %, etc., for example,
  x, y, x1, y2, ab3, x_3

- *indexed variables*

  in *indexed form*, such as x_1, y_n, z_a, $a_{i,k}$

MATHCAD differentiates between uppercase and lowercase. Every variable name must start with a letter.

Note that MATHCAD does *not distinguish* between the *names* of *variables* and *functions* (see Section 17.2.3).

Consequently, *ensure* when you specify *variable names* that you do not use any *names* of *built-in functions* or *built-in constants* in MATHCAD, because otherwise these will no longer be available.

♦

Depending on the usage, MATHCAD provides *two possibilities* for the *representation* of *indexed variables*:

I.   If you want to interpret a variable x_i as *component* of a *vector* **x**, you must create this using the

button from the *operator palette no. 1* or *4* by entering x and the index (array index) i in the displayed *placeholders*

■
▪

and so produce

x_i

II. If you are only interested in **x** *variables* with *subscripted index* i, you obtain this by typing a point ⬚ after entering the x. The subsequent input of i now appears subscripted and you obtain

x_i

In contrast to the array index, this type of index is designated as a literal index.

Because a blank appears between variables and index for the *literal index* and the literal index has the same size as the variable, whereas the index is shown smaller than the variable for the *array index*, MATHCAD can *recognize* the *difference* between the two types of *indexed variables*.

♦

The *assignment operators* := and ≡ can be used to assign *numbers* or *constants* to the variables (see Section 10.3), namely the *assignment operator*

- :=

  is *created* by

  *   *entering* the *colon* from the *keyboard*
  *   *clicking the*

button in *operator palette no. 1*

• ≡

is *created* by *clicking* the

button in *operator palette no. 2*

♦

The two *assignment operators differ* in that

* :=

defines the *local assignment*, whereas

* ≡

defines the *global assignment*

As with programming languages, these two different *assignment operators* can be used to define *local* and *global variables*.

MATHCAD during the processing of the *worksheet* first analyses all *global variables* from top left to bottom right. Only then are the *local variables* used for the calculation of the expressions.

♦

Note the following for *assignments:*

* No numerical values may have been previously assigned to the variables used for *symbolic calculations* (e.g., differentiation).
  If a numerical value has already been assigned to the variable x and you want to reuse this as symbolic variable, e.g., for the differentiation of a function f(x), you can do this with a *redefinition*

  x := x

* All *variables* used in *numerical calculations* must have previously been *assigned numerical values*.
  Undefined variables are shown here in a different colour and a short *error message* indicates the non-definition.

  ♦

8.3 Range Variables

MATHCAD provides so-called *range variables* that can be assigned *several values* from an *interval*.

Range variables are *defined* in the *form*

v := a, a + Δv.. b

where the *two periods* must be *entered* in one of the *following ways*

* by clicking the

   button from *operator palette* no.*1* or *4*

* by *entering* a *semicolon* from the keyboard

Instead of the *local assignment operator* := , you can also use the *global assignment operator* ≡ to *define range variables.*

♦

A *range variable* v with this definition accepts all *values* between a (initial value) and b (end value) with the *increment* (*step size*) Δv.

If the *increment* (*step size*) Δv is omitted, i.e., you *define* a *range variable* v in the *form*

v .– a .. b

v *accepts* the *values* between a and b with the *increment/decrement* (*step size*) 1, i.e., for

* a < b

   then i = a , a+1 , a+2 , ... , b

* a > b

   then i = a , a–1 , a–2 , ... , b

♦

Note the following for *range variables*:

* Only *simple variables* can be used, i.e., indexed variables are not permitted here.

* MATHCAD can only *assign* a *maximum* of *50 values* to a *range variable.*

♦

In MATHCAD, *range variables* and thus also their initial values, increments/decrements and end values, can be *any real numbers.* We discuss in Chapters 14 and 18 non-integer *range variables* for range sums and range products or for the graphical display of functions.

♦

You can *display* the *values assigned* to a range *variable* as a *value table* (output table) if you enter the *numerical equal sign* (see Example 8.1a). Note that MATHCAD represents only the first 50 values in this table. You must use additional range variables if you require more values.

♦

Although *range variables* can assume several values, they cannot be used as vectors (see Chapter 15). You can only interpret these as *lists*. Example 8.1c3 illustrates how *vectors* can be created using range variables.

♦

Range variables are required for purposes such as

* *graphical display* of functions (see Chapter 18)
* the *formation* of *loops/iterations* in *programming* (see Section 10.5)
* *calculation* of *sums* and *products* (see Chapter 14)
* *calculation* of *function values* (see Example 8.1b)
* *definition* of *vectors* and *matrices* (see Example 8.1c).

♦

We now use several characteristic examples to illustrate the use of range variables.

Example 8.1:

a) We define *range variables* u and v in the intervals

   [1.2,2.1] and [-3,5]

   with *increment*

   0.1 and 1, respectively

   and enter the *numerical equal sign* = to output the *calculated values* as a *value table* (*output table*):

u := 1.2 , 1.3 .. 2.1   v := -3 .. 5

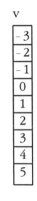

u
1.2
1.3
1.4
1.5
1.6
1.7
1.8
1.9
2
2.1

v
-3
-2
-1
0
1
2
3
4
5

b) We calculate the sin x *function* for the *values*

$$x = 1, 2, 3, \ldots , 7$$

by defining x as *range variable* with *increment 1*:

$$x := 1 .. 7$$

x   sin(x)

1	0.841
2	0.909
3	0.141
4	-0.757
5	-0.959
6	-0.279
7	0.657

The input of the *numerical equal sign* = after x and sin(x) returns the *value table* (*output table*) for the *defined range variables* x and the *value table* (*output table*) for the associated *function* values for sin x.

c) We use range variables to create vectors and matrices (see Section 15.1):

  c1) If you use the **ORIGIN** *built-in variable* (see Section 8.1) to set the *start index* to 1, you can use a *range variable* j to *create* a *column vector* **x**. In our example, the j-th component of the vector x is calculated from j+1:

$$j := 1 .. 9 \qquad\qquad x_j := j+1$$

$$x = \begin{pmatrix} 2 \\ 3 \\ 4 \\ 5 \\ 6 \\ 7 \\ 8 \\ 9 \\ 10 \end{pmatrix}$$

  c2) If you use the **ORIGIN** *built-in variable* (see Section 8.1) to set the *start index* to 1, you can use two *range variables* i and k to *create* a *matrix* **A**. In our example, the element of the i-th row and the k-th column of the matrix A is calculated from i+k:

$$i := 1 .. 2 \qquad\qquad k := 1 .. 3 \qquad A_{ik} := i+k$$

$$A = \begin{pmatrix} 2 & 3 & 4 \\ 3 & 4 & 5 \end{pmatrix}$$

c3) As shown in the following example, you can also use *range variables* to *assign* a specified *number table* (*input table*) to a *vector.*

If, for example, you want to *assign* the five values (*number table*)

4 , 6 , 2 , 9 , 1

to the *vector* **x**, *define* a *range variable* i with the five index values 1, 2, 3, 4, 5:

i := 1 .. 5

You can then enter the *following assignment statement* for x with the array index i

x_i :=

and then enter the individual *numbers* of the *number table* with *commas;* MATHCAD displays:

x_i :=

4
6
2
9
1

If you now enter the designation of the *vector* **x** with numerical equal sign =, the specified vector is displayed if you used the **ORIGIN** *built-in variable* (see Section 8.1) to set the *starting index* to 1:

$$x = \begin{pmatrix} 4 \\ 6 \\ 2 \\ 9 \\ 1 \end{pmatrix}$$

♦

8.4 Strings

In addition to numbers, constants, and variables, MATHCAD can also process *strings/string expressions.*

MATHCAD considers a *string* to be a *finite sequence* of *characters* present
on the *keyboard. ASCII characters* are also permitted in strings.

♦

Strings cannot occur by themselves in MATHCAD. They can

*   be *assigned* to *variables*
*   *entered* as *elements* of a *matrix*
*   *entered* as *arguments* of a *function*

i.e., used only at those places where there is an empty *placeholder.*

♦

Strings are entered in an empty placeholder with the following steps:

I.   Click the placeholder
II. Enter the double quote ["] from the keyboard
III. Enter the required characters from the keyboard.

♦

Proceed as *follows* to *enter ASCII character* within a string:

I.   Press the [Alt]-key
II. Enter a 0 (zero)
III. Enter the appropriate ASCII code

♦

Strings can be *used* in a number of operations, such as for the

*   *creation* of user-defined *error messages*
*   *programming*

and can be processed using *string functions* (see Section 17.1.3).

♦

Example 8.2:

As an exercise in the use of strings, we *define* a *vector* **v** whose *components*
consist of *strings* and then consider the components:

$$v := \begin{bmatrix} \text{"one"} \\ \text{"two"} \\ \text{"three"} \end{bmatrix}$$

$v_1 = \text{"one"}$ $v_2 = \text{"two"}$ $v_3 = \text{"three"}$

♦

9 Data Management

This chapter discusses the (*data management*) in MATHCAD. These *data* are typically stored in *files*.

The *data management* in MATHCAD differentiates between

* *input/import/read* of *files* from hard disk or some other data media

* *output/export/write* of *files* to hard disk or some other data media

* *exchange* of *files* with *other systems*.

MATHCAD can *input, output* or *exchange files* in a number of *file formats*, for example, for

* AXUM

* EXCEL

    files with the .XLS *extension*

* MATLAB

    files with the .MAT *extension*

* ASCII editors

    files with the .DAT, .PRN, .TXT *extension*

In general, the

Insert ⇒ Component...

menu sequence is used to *control* the *input, output* and *exchange* of *files*, where the following settings are made in the *Component Wizard* that appear in the

* *first page*

    the *type* of the associated *components*.

* *second page*

    whether *input* (reading) or *output* (writing) is to be performed.

* *third page*

    the *file format* and the *path* of the *file*.

The sections of this chapter discuss both this *general form* and other special capabilities for the *input* and *output* of data. ♦

The *following section* is concerned with the *management* (input and output) of *ASCII files* that principally consist of numbers (*number files*).

Other components are managed similarly. For example, EXCEL can be used to exchange numbers in the form of matrices (see Example 9.2d).

MATHCAD also has *input* and *output functions* (*read* and *write functions*), which can also be designated as *file access functions*, for

*   *input* (*read*)

*   *output* (*write*)

of *data* (*numbers*) from or in *unstructured/structured ASCII files*.

♦

Unstructured and *structured ASCII files differ* in the following ways:

*   *unstructured files*

    are *characterized* by the *numbers* being *arranged* successively and *separated* with one of the *delimiters/separators*

    *   *blank*

    *   *comma*

    *   *tabulator*

    *   *line-feed*

    Unstructured files normally are *characterized* by having the .DAT *extension*. However, other extensions can be used starting with version 7 of MATHCAD.

*   *Structured files*

    differ from unstructured files only by the *arrangement* of the *numbers*. The numbers must be arranged in *structured form* (*matrix form* with *rows* and *columns*), i.e., each row must have the same number of numbers that are separated by *delimiters*.

    With the exception of the *line-feed* delimiter, which is required here to identify the rows, the same delimiters can be used as for unstructured files.

    Structured files are normally *identified* with the .PRN *extension*. However, other extensions can be used starting with version 7 of MATHCAD.

Strictly speaking, *unstructured files* represent a *special case* of structured files, so that input/output functions for structured files can also be used for unstructured files. Unstructured files can be considered to be matrices of type (1,n) or (n,1), i.e., as row or column vectors.

♦

9.1 Data Input

As already mentioned at the start of this chapter, MATHCAD can read a range of file formats. However, we will restrict ourselves in the following section to the important case of *ASCII files*. Other file formats are read in a similar manner.

Although MATHCAD can *read* both *unstructured* and *structured files* in *ASCII format, number files* usually occur in *practical applications*.

Obviously, when you *read files*, you must know from where MATHCAD is to *read* the *required file*. If nothing is specified, MATHCAD searches for the file in the *working directory*. This is the directory from where the current MATHCAD worksheet was loaded or most recently stored.

If the *file* is located in *another directory*, you must tell MATHCAD the *path as* we see in the following section.

♦

Two possibilities are provided for *reading*:

I. Use of the *menu sequence*

   Insert ⇒ Component...

   where in the displayed *Component Wizard*

   * *click on*

     File Read or Write

     in the *first page*

   * *click on*

     Read from a file

     in the *second page*

   * *enter* the *file format*
     at

     File Format
     (for *ASCII files:* **text files**)

     and the *path* of the *file* (e.g., A:\data.prn)
     in the *third page*

   Enter in the free placeholder in the displayed *symbol* the name for the variable/matrix to which the read file is to be assigned (see Example 9.1a).

II. *Use* of *input functions* (*read functions*):

   * **READ** (*data*)

     reads a *number* from the unstructured/structured *data* file that can be assigned to a variable. Examples 9.1c and d illustrate the use.

* **READPRN** (*data*)

> *reads* the structured *data* file into a matrix. A row or column of *data* is assigned to each row or column of the matrix. This function can also be used to read an unstructured file, which then appears as a row or column vector (see Example 9.1c2).

The complete *path* of the *file* must be specified as a *string* for the *input functions* for *data*.

If, for example, the structured file *data.prn* is to be read from the *diskette* in *drive A* and be assigned to a *matrix* **B**, specify

B := READPRN (″ A:\ *data.prn* ″)

   ◆

Both read capabilities are initiated with a mouse click outside the expression or by pressing the ⏎-key.

◆

We recommend that you use the *input functions* to *read ASCII files*. Note that the function name must always be written in *uppercase*.

◆

Ensure for *reading* that the read values are displayed as a *scrollable output table* if the file has more than nine rows or columns (see Examples 9.1a and b) when the *display* with *indexes* (standard setting from MATHCAD) is required.

◆

The following *Example 9.1* demonstrates the *use* of the *input functions*. Because the *matrices* used here are introduced first in Chapter 15, consult that chapter if you have any questions.

Example 9.1:

In the following examples, the *start values* used for the *indexing* are immediately visible.

a) The *diskette* in *drive A* of the computer *contains* the *structured* ASCII file *data.prn* that has the following form (*column form:* 19 lines, 2 columns):

1 20
2 21
3 22
4 23
5 24
6 25
7 26
8 27
9 28

10 29
11 30
12 31
13 32
14 33
15 34
16 35
17 36
18 37
19 38

The file is to be *read* and *assigned* to a *Matrix* **B**. The two discussed *capabilities* can be used as follows:

- Use of the *menu sequence*

  Insert ⇒ Component...

  where the following must be entered in the individual pages of the displayed *Component Wizard*:

  * a file is to be *read*

  * the *file format* for *ASCII files:*

    Text Files

  * the *path* A:\data.prn

  The name of the *matrix* **B** to which the read file is to be assigned must be entered in the free placeholder in the *Symbol* that then appears. Thus, the following appears on the screen:

  B :=

  A:\data.prn

- Use of the *input function:*

  B := READPRN ("A:\data.prn")

Both methods are initiated with a mouse click outside the expression or by pressing the ⏎-key and yield the following result for the *matrix* **B**:

B =

	0	1
0	1	20
1	2	21
2	3	22
3	4	23
4	5	24
5	6	25
6	7	26
7	8	27
8	9	28
9	10	29
10	11	30
11	12	31
12	13	32
13	14	33
14	15	34

∎

b) The *diskette* in *drive* A of the computer *contains* the *structured* ASCII file
   data.prn of the following form (*row form*: 2 rows, 19 columns):

   1 2 3 4 5 6 7 8 9 10 11 12 13 14 15 16 17 18 19
   20 21 22 23 24 25 26 27 28 29 30 31 32 33 34 35 36 37 38

   The file is to be *read* and *assigned* to a *matrix* **B**. The procedure is the
   same as for Example a.
   MATHCAD returns the following for the read *matrix* **B**:

B =

	0	1	2	3	4	5	6	7	8	9	10	11
0	1	2	3	4	5	6	7	8	9	10	11	12
1	20	21	22	23	24	25	26	27	28	29	30	31

∎

Both Examples a and b for reading show the previously explained effect,
namely, for the *representation* with *display* of the *indexes* (standard setting
of MATHCAD) not all read values are displayed directly on the screen. A
scrollable output table is displayed. You can view all read values by clicking
with the mouse on the displayed representation. This permits the *scrolling*
of the *file*. The following *screenshot* shows this for the *file read* in Example
b:

7	8	9	10	11	12	13	14	15	16	17	18
8	9	10	11	12	13	14	15	16	17	18	19
27	28	29	30	**31**	32	33	34	35	36	37	38

$B =$ (with row indices 0 and 1)

You obtain a *display* of the *read file* on the screen in *matrix form* (i.e., *without Indices*) as follows:

After activation of the *menu sequence*

Format → Result...

set the

Matrix

option for the

Matrix display style

in the displayed

Result Format

dialogue box.

All values are shown directly for the *display in matrix form*, as you can see for Example b:

$$B = \begin{bmatrix} 1 & 2 & 3 & 4 & 5 & 6 & 7 & 8 & 9 & 10 & 11 & 12 & 13 & 14 & 15 & 16 & 17 & 18 & 19 \\ 20 & 21 & 22 & 23 & 24 & 25 & 26 & 27 & 28 & 29 & 30 & 31 & 32 & 33 & 34 & 35 & 36 & 37 & 38 \end{bmatrix} .$$

c) When the *unstructured* ASCII file *data.dat*

$$9, 7, 8, 5, 4, 6, 7, 8, 3, 2, 3, 4, 5, 6, 7, 1, 2, 3, 4, 5, 6, 7, 6, 5, 4, 3, 2$$

contained on the diskette in drive A is used.

c1) The **READ** *function* returns

$a := \textbf{READ} \ ("A:\backslash data.dat")$

$a = 9$

and

$i := 19 .. 27$

$j := 1 .. 10$

$a_i := \textbf{READ} \ ("A:\backslash data.dat")$

$b_j := \textbf{READ} \ ("A:\backslash data.dat")$

a_i

9
7
8
5
4
6
7
8
3

b_j

9
7
8
5
4
6
7
8
3
2

Reading the a_i shows that the read is always made from the start of the file, even when you use other indices. You must read all values up to that required and then select the required value, if only individual values are required from the file.

c2) The **READPRN** *function* returns

C := READPRN ("A:\data.dat")

the result in the *form*

C = (9 7 8 5 4 6 7 8 3 2 3 4 5 6 7 1 2 3 4 5 6 7 6 5 4 3 2)

i.e., as *matrix* **C** with one row (row vector) whose elements are called using

$C_{1,k}$

d) When the *structured* ASCII file *data.txt*

2 4

6 8

contained on the diskette in drive A is used.

d1) The **READ** *function* returns

a := **READ** ("A:\data.txt")

a = 2

and

i := 1 .. 3

a_i := **READ** ("A:\data.txt")

$$a = \begin{pmatrix} 2 \\ 4 \\ 6 \end{pmatrix}$$

d2)The **READPRN** *function* returns

$$\mathbf{B} := \mathbf{READPRN} \; (\; ''\text{A:}\backslash\text{data.txt}'' \;)$$

$$B := \begin{pmatrix} 2 & 4 \\ 6 & 8 \end{pmatrix}$$

♦

9.2 Data Output

As previously mentioned at the start of this chapter, although MATHCAD can write a range of file formats, we concentrate in the following section on the most important case of *ASCII files*:

MATHCAD can *write matrices* with *elements* in *ASCII format* in *structured* and *unstructured files*.

Other file formats are written similarly, as we show in Example 9.2d for an EXCEL file.

Obviously, you must know when you *write files* where MATHCAD is to *write* the *required file*. Unless something else is specified, MATHCAD writes the file into the *working directory*. This is the directory from which the current MATHCAD worksheet was most recently loaded or saved.

If the *file* is to be located in *another directory*, you must inform MATHCAD of the *path*, as we see in the following.

♦

Two methods are available for *writing*:

I. Using the *menu sequence*

Insert ⇒ Component...

where in the displayed *Component Wizard*

* *click*

File Read or Write

in the *first page*

* *click*

Write to a file

in the *second page*

* *enter*
the *file format*
(for *ASCII files*: **Formatted Text**)
for

File Format

and the *path* of the *file* (e.g., A:\data.prn)
in the *third page*

The name of the variables/matrix to be written is entered in the free placeholder in the *symbol* that then appears (see Example 9.2a).

II. *Use* of *output functions* (*write functions*):

* **WRITE** (*data*)

   writes a *number* in the new unstructured file *data*.

* **APPEND** (*data*)

   adds a *number to* the existing unstructured file *data*.

* **WRITEPRN** (*data*)

   writes a matrix in the structured file *data*, i.e., a row or column of the matrix is assigned to each row or column of *data*.

* **APPENDPRN** (*data*)

   adds a *matrix to* the existing structured file *data*, i.e., matrix and file must have the same number of columns. The *data* file then also contains the rows of the matrix.

The complete *path* for the file *data* must be specified as *string* for the *write functions*.

If, for example, the matrix **B** contained in the worksheet is to be written in the structured file *data.prn* on the *diskette* in *drive A*, then specify

WRITEPRN (*"A:\ data.prn"*) := **B**

Both write methods are initiated with a mouse click outside the expression or by pressing the ⏎-key.

◆

using the

Math ⇒ Options...

menu sequence
the

PRN File Settings

with the *predefined variables* (*built-in variables*)

PRNPRECISION

and

PRNCOLWIDTH

for the *output function*

WRITEPRN

can be used to specify the

* *accuracy* (default value 4)
* *column width* (default value 8)

in the *dialogue box* displayed for **Built-In Variables**.

♦

We recommend that you use the *output functions* for *writing*. Note that these must always be written in *uppercase*.

♦

The *use* of *output functions* is demonstrated in the following *example*. Because the *matrices* used here are introduced first in Chapter 15, consult that chapter if you have any questions.

Example 9.2:

a) We have set the value 1 as start value for the indexing of the *matrix* **B** used in this exercise,
   i.e., **ORIGIN:=1**.

   We *write* the *matrix* **B** contained in the worksheet

$$\mathbf{B} := \begin{pmatrix} 1 & 2 & 3 & 4 \\ 5 & 6 & 7 & 8 \\ 9 & 10 & 11 & 12 \end{pmatrix}$$

using the

* *menu sequence*

   Insert ⇒ Component...

   where the following must be entered in the individual pages of the *Component Wizard* that appears:

   * a *file* is to be *written*
   * the *file format*

      Formatted Text

   * the *path* A:*data.prn*

   The name of the *matrix* **B** to be *written* must be entered in the free placeholder in the displayed *symbol*:

A:\data.prn

B

* *output function*

   WRITEPRN ("A:*data.prn*") := **B**

on the *diskette* in *drive* A as structured ASCII file *data.prn*, by closing
both methods using a mouse click outside the expression or pressing the
⏎-key.

The *matrix* **B** is then present in the following form in the *data.prn* file
on the *diskette:*

1	2	3	4
5	6	7	8
9	10	11	12

b) Use the **APPENDPRN** *output function* to *append* the *matrix* **C** contained
in the worksheet

$$\mathbf{C} := \begin{pmatrix} 4 & 3 & 2 & 1 \\ 8 & 7 & 6 & 5 \end{pmatrix}$$

to the *data.prn* file on diskette in drive A created in Example a:

APPENDPRN ("A:*data.prn*") := **C**

The *data.prn* file then has the following form:

1	2	3	4
5	6	7	8
9	10	11	12
4	3	2	1
8	7	6	5

c) We use

WRITE ("A:*data.dat*") := 3

to *write* the number 3 in the new unstructured file *data.dat*
and use

APPEND ("A:*data.dat*") := 4

to append the number 4 to this file.

d) We *write* the *matrix* **B** contained in the worksheet as the *data.xls EXCEL*
file on the *diskette* in drive A:

$$\mathbf{B} := \begin{pmatrix} 1 & 2 & 3 & 4 \\ 5 & 6 & 7 & 8 \\ 9 & 10 & 11 & 12 \end{pmatrix}$$

This is done using the *menu sequence:*

Insert ⟹ **Component...**

where the following must be entered in the individual pages of the displayed *Component Wizard*:

* a *file* is to be *written*
* the *file format*

   Excel
* the *path* A:\ *data.xls*

The name of the *matrix* **B** to be *written* must be entered in the free placeholder in the displayed *symbol*:

A:\data.xls

B

Click with the mouse outside the expression or press the ⏎-key to initiate the write.

♦

9.3 Data Exchange

We have already learnt a form of the *data exchange* for MATHCAD in the previous two sections:

input and *output* of *files.*

However, this is not the only form for data exchange starting with the Version 7 of MATHCAD. The data exchange is now possible with other *program systems,* such as AXUM, EXCEL and MATLAB. This data exchange is also performed using the *Component Wizard* used in the Sections 9.1 and 9.2. Please refer to the manual and the integrated help for further details.

10 Programming

If MATHCAD does not supply any functions to solve a problem, it provides users with *programming capabilities* that they can use to *create* their *own programs.*

Some *computer algebra systems* such as AXIOM, MACSYMA, MAPLE and MATHEMATICA, permit *programming* in the usual *programming styles:*

- *procedural*
- *recursive* or *rule-based*
- *functional*
- *object-oriented.*

These systems can also be described as being *programming languages.*

Although MATHCAD does not provide such comprehensive programming capabilities, it masters the *procedural programming* with the *basic facilities:*

- *assignments*
- *branches/decisions* (*conditional statements*)
- *loops.*

Simple *recursive programming* tasks can also be realized in MATHCAD (see Example 10.3c).

♦

MATHCAD has a programming feature that all programs must be written in the form of *function subroutines/subprograms/procedures.* This well-known program type returns a number, a vector, a matrix or a string as the result.

♦

In addition, the *programming* in *computer algebra systems,* and thus also in MATHCAD, can make use of the functions/commands integrated in the system. This represents a large advantage compared with programming using conventional programming languages such as BASIC, C, FORTRAN and PASCAL.

♦

☞

Since *Version 6,* MATHCAD provides a *programming palette* that *simplifies* the *programming,* because it contains the programming operators for assignments, branches and loops that can be added simply with a click of the mouse.

This *programming palette (operator palette no. 6)* is shown in Figure 10.1. The individual *operators* of this palette are *discussed* in the course of this chapter.

MATHCAD also permits the input of statements from the keyboard.

♦

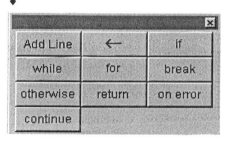

Figure 10.1. Programming palette

In the following sections we *discuss* the components of the *programming capabilities* in MATHCAD and provide hints for the creation of your own programs. We consider

- *first* (Sections 10.1 and 10.2)
  comparison operators, logical operators and the *definition* of *custom operators*

- *then* (Sections 10.3 to 10.5)
  the three *fundamentals* of *procedural programming*
  * *assignments*
  * *branches*
  * *loops*

- *finally* (Section 10.6)
  the *creation* of simple *programs.*

☞

The examples contained in the following sections are designed to motivate the readers and to make it possible for them to write programs in MATHCAD to solve any tasks that might arise in the course of their work.

♦

10.1 Boolean Operators and Logical Operators

Comparison operators, also known as *Boolean operators*, and *logical operators* are required to form *logical expressions* used for the *programming* of *branches* (see Section 10.4).

MATHCAD provides the *following Boolean operators*:

*   *equal*

    =

*   *less than*

    <

*   *greater than*

    >

*   *less than or equal*

    ≤

*   *greater than or equal*

    ≥

*   *unequal*

    ≠

that all can be added by mouse-click from *operator palette no. 2*.

The two operators < and > can also be entered from the keyboard.

The *equal operator* = created using the

button from *operator palette no. 2* cannot be confused with the *numerical equal sign* = (see Section 6.2) that is entered from the keyboard or using operator palette no. 1 or 2. Because the equality operator is displayed with thick lines, the two operators differ optically.

♦

All *Boolean operators* can be used for *real numbers* and *strings* (see Section 8.4).

Only the two = and ≠ operators are appropriate for use with *complex numbers*.

♦

The *Boolean operators* and the two *logical operators*:

*   *logical* AND

*   *logical* OR

(entered in MATHCAD using the multiplication sign * or plus sign +) can be used to form *logical expressions*.

In contrast to algebraic expressions (see Chapter 13), logical expressions can assume only the two *values* 0 (*false*) or 1 (*true*).

Example 10.1:

a) The following expressions are examples of logical expressions:

$$x = y \qquad x \leq y \qquad x \neq y \qquad (a \geq b) + (c \leq d) \qquad (a < b) * (c > b)$$

b) The *logical expression*

   b1) $u := (1 < 2) + (3 < 2)$

      with the *logical* OR returns the value 1 (true).

   b2) $v := (1 < 2) * (3 < 2)$

      with the *logical* AND returns the value 0 (false).

    ◆

10.2 Definition of Operators

MATHCAD permits you to define your own *operators*, which are designated as *user-defined operators* or *custom operators*. They have similar characteristics as functions (see Section 17.2.3). *Defined operators* differ from *defined functions* in the following ways:

- They can be activated after their definition with one of the *operators*

from the *operator palette no. 2* (see Example 10.2); these operators are designated as

  * *prefix*

  * *postfix*

  * *infix*

  * *tree*

operators. You can also use the *custom operators* simply by entering it without using the operator palette no. 2

- *Symbols* are frequently used for the operators.

☞

Because the *symbols* often used for operators are not present on the keyboard, you can insert these by *copying* from the **QuickSheet**

Extra Math Symbols

of the

Resource Center

using the clipboard.

This **QuickSheet** is opened using the *menu sequence*

Help ⇒ Resource Center ⇒ QuickSheets ⇒ Extra Math Symbols

♦

MATHCAD also provides the capability to copy *defined operators* into the

My Operators

personal quicksheet using the

Help ⇒ Resource Center ⇒ QuickSheets ⇒ Personal QuickSheets ⇒

My Operators

menu sequence.

♦

The following Example 10.2 demonstrates the use of custom operators for three simple problems. We suggest that you also try these three examples so that you get a feel for the definition and use of such operators.

Example 10.2:

a) We use the

＋

operator to define the *division of two numbers* x and y, where this operator is copied from the **Quicksheet**

Extra Math Symbols

＋ (x,y) := x/y

The use of the ＋ *operator* to solve a concrete problem (e.g., 1/2) is performed by clicking the button for the *infix operator*

and then *completing* the three displayed *placeholders* as follows

1 ＋ 2

The final *input* of the *symbolic* or *numerical equal sign* returns the *result:*

1 ＋ 2 → 1/2   or   1 ＋ 2 = 0.5

The following input is also possible (without using the infix operator):

＋ (1,2) → 1/2   or   ＋ (1,2) = 0.5

b) Names can also be used in place of symbols for a newly defined operator; this is shown in the following *operator*

Plus

for the *addition of two numbers* x and y:

Plus (x,y) := x + y

This operator is input as usual in math mode from the keyboard.
The use of the **Plus** *operator* to solve a concrete problem (e.g., 1+2) is performed by clicking the button for the *infix operator*

and then *completing* the three displayed placeholders as follows

1 **Plus** 2

The final *input* of the *symbolic* or *numerical equal sign* returns the *result:*

1 **Plus** 2 → 3   or   1 **Plus** 2 = 3

The following input is also possible (without using the infix operator):

Plus (1,2) → 3   or   **Plus** (1,2) = 3

c) We use the

operator to define the *root* of a *number* x, this operator is copied from the **Quicksheet**

Extra Math Symbols

$\sqrt{}$ (x) := $\sqrt{x}$

The use of the $\sqrt{}$ *operator* to solve a concrete problem (e.g., square root of 2) is performed by clicking the button for the *prefix operator*

and then *completing* the two displayed *placeholders* as follows

$\sqrt{}$ 2

Here only the final *input* of the *numerical equal sign* is appropriate, because the result is a real number that often can only be approximated with a finite decimal number, e.g.,

$\sqrt{}$ 2 = 1.414213562373095

The following input is also possible (without using the prefix operator):

$\sqrt{}$ (2) = 1.414213562373095

♦

10.3 Assignments

Assignments play a dominant role in working with MATHCAD. We have already often used them in the previous chapters. They are required for the *assignment* of *values* to *variables* and for the *definition* of *functions* and *operators*.

As in other programming languages, MATHCAD differentiates between *local* and *global assignments*, for which we discussed the following operators in Chapter 8:

- *Local assignments* are realized in MATHCAD using the

  :=

  sign, which is created with *one* of the *following operations*:

  * click on the button for the *assignment operator*

  in the *operator palette no. 1* or *2*.

  * *enter* the *colon* from the *keyboard*.

- *Global assignments* are realized using the

  ≡

  sign, which is created by clicking the button

  for the *assignment operator* in *operator palette no. 2*.

Local assignments within *function subroutines* can only be realized using the button

for the *assignment operator* from *operator palette no. 6* (*programming palette*), as illustrated in the following examples.

♦

As for programming languages, *local* and *global assignments* can be used to define *local* and *global variables*.

MATHCAD during the *processing* of a *worksheet first* analyses all *global variables* starting at the top left and then proceeding down right. Only then are the *local variables* used for the calculation of existing expressions.

♦

10.4   Branches

Branches (*conditional statements*) are normally formed in programming
languages with the **if** *statement* and return various results depending on the
expressions involved. Together with arithmetical and transcendental expressions, logical expressions are normally used here.

We have already discussed *logical expressions* in Section 10.1 (Example
10.1).

♦

MATHCAD uses both the **if** *statement* and the **until** *statement* for the programming of *branches*:

*   The **if** statement (see Example 10.3) can be entered and used in two
    ways:

    *   *input* from the *keyboard* in the *form*

        if (*expr, res1, res2*)

        The result *res1* is output here if the expression *expr* is non-zero (for
        arithmetical and transcendental expressions) or *true* (for logical ex-
        pressions), otherwise the result *res2*.

    *   *input* by *clicking* the button for the **if**-operator

        in the *operator palette no. 6* (*programming palette*).

*   The **until** *statement* (see Example 10.4) has the following form:

    until (*expr* , *w*)

    Here the *value* of *w* is calculated until the *expr* expression assumes a
    negative value. If *expr* is a *logical expression,* ensure that MATHCAD sets
    1 for *true* and 0 for *false*. Consequently, you must deduct a value be-
    tween 0 and 1 for *logical expressions* in order that the *w* expression is
    no longer calculated for *false*.

Problems occur in MATHCAD for the **if** and **until** *statements* if you use assignments rather than expressions for the *res1*, *res2* and *w*, as shown in Example 10.5a.

♦

In the following Examples 10.3 and 10.4, respectively, we illustrate the details of the use of the **if** statement and the **until** statement for specific tasks.

Example 10.3:

a) We *define* the following continuous *function* of two variables

$$z = f(x,y) = \begin{cases} x^2 + y^2 & \text{if} & x^2 + y^2 \le 1 \\ 1 & \text{if} & 1 < x^2 + y^2 \le 4 \\ \sqrt{x^2 + y^2} - 1 & \text{if} & 4 < x^2 + y^2 \end{cases}$$

to demonstrate the use of the **if** statement. This *function* is *defined* in MATHCAD *either*

*   with

$$f(x,y) := \mathbf{if}\left(x^2 + y^2 \le 1, x^2 + y^2, \mathbf{if}\left(x^2 + y^2 \le 4, 1, \sqrt{x^2 + y^2} - 1\right)\right)$$

where the **if** statement is entered from the keyboard in *nested form*

or

*   with

$$f(x,y) := \begin{vmatrix} x^2 + y^2 & \text{if} & x^2 + y^2 \le 1 \\ 1 & \text{if} & 1 < x^2 + y^2 \le 4 \\ \sqrt{x^2 + y^2} - 1 & \text{otherwise} \end{vmatrix}$$

using the button for the *operator*

twice to add lines and the button for the **if**-operator

from *operator palette no. 6 (programming palette)*

b) The number of the positive components *pos_comp* of a column vector **a** contained in the worksheet can be calculated with the **if** statement and the *summation operator* (see Chapter 14) in the following manner, when the **ORIGIN:=1** has been used to set the initial indexing to 1:

$$\text{pos\_comp} := \sum_{i=1}^{\text{last}(a)} \text{if}\,(a_i \ge 0, 1, 0)$$

or

$$\text{pos\_comp} := \sum_{i=1}^{\text{rows}(a)} \text{if}\,(a_i \ge 0, 1, 0)$$

The **last** and **rows** *functions* used return the *number* of the *components* of the *vector* **a** (see Section 15.2).

c) We use the **if** statement to create a simple *recursive program* to *calculate* the *factorial* of a positive integer n:

n! = n · (n–1) · (n–2) · ... · 1

The "Error" *message* (as string) is to be output if a negative integer was entered for n.

We create the associated *recursive function* both without and with the use of the programming palette:

c1) A first variant of the *function subroutine* can be obtained by entering a nested **if** statement from the keyboard in the *form*

fact(n) := **if** (n < 0 , "Error" , **if** (n = 0 , 1 , n*fact(n–1)))

c2) A second variant of the *function subroutine* can be obtained in the *form:*

$$\text{fak}(n) := \begin{vmatrix} 1 & \text{if} & n=0 \\ n \cdot \text{fak}(n-1) & \text{if} & n>0 \\ \text{"error""} & \text{otherwise} \end{vmatrix}$$

by using the button for the *operator*

twice to add lines and the button for the **if** operator

and the button for the **otherwise** operator

from the *operator palette no. 6 (programming palette).*

♦

10.5 Loops

Loops are used to execute a sequence of statements and are formed in the programming languages with **for** or **while**.

MATHCAD provides the following methods for *forming loops*:

* *Use* of *range variables* (see Example 10.4):

    These *loops* with a *specified number* of *passes* begin in MATHCAD with a *range assignment* for the *loop index* (*loop counter/control variable*) i

    i := m .. n

formed using the *assignment* := and the *button*

from the *operator palette no. 1* or *4*.

MATHCAD designates a *variable* i defined like the *loop index* as *range variable* (see Section 8.3).

The numbers m and n determine the *range* for the *loop index* i and specify the *initial value* (*starting value*) and *end value*, where 1 is used as increment/decrement (step size), i.e., for

*   m < n

    then i = m , m+1 , m+2 , ... , n

*   m > n

    then i = m , m−1 , m−2 , ... , n

If you require a *step size* Δi for the *loop index* i that is not 1, then write

i := m , m+Δi .. n

Because the *loop indexes* in MATHCAD can be defined as *range variables*, the initial values, the step sizes (increments/decrements) and end values can be *any real numbers*. We use such *loop indexes* in the following Examples 10.4b and 10.5b.

♦

The *statements/functions* to be executed in the loop that normally depend on the loop index i follow it in the definition of the loop index.

If you wish to *nest several loops*, you must define the individual range variables in succession. For example, you write for a double loop with the range variables i and k

i := m .. n   k := s .. r

• The button for the **for** operator (see Examples 10.4c and d1)

from *operator palette no. 6* (*programming palette*) can be used for **for** *loops* with a *given number* of *passes*. A mouse-click on this button displays the following at the position marked by the cursor in the worksheet:

    for ∎∈∎

    ∎

The *loop index* and its *range of values*, and the *statements/functions* to be executed are entered here in the placeholders *after* and *under* **for** respectively. If you wish to nest several loops, you must activate the **for** operator the appropriate number of times.

• Using the button for the **while** operator

| while |

from the *operator palette no. 6* (*programming palette*). This permits the
formation of **while** loops without a specified number of passes (iterative
loops), such as those required for iteration procedures (see Example
10.4d2).

☞

MATHCAD requires loops both for calculations (iterations) and for the
graphical display of functions and data.

♦

The following Example 10.4 shows the exact *procedure* for the *use* of *loops*.

Example 10.4:

a) *Graphical displays* of *functions*

$z=f(x,y)$

of two variables can only be realized in MATHCAD when a *matrix* **M**
was previously calculated that contains as elements the function values
$f(x,y)$ in the given (x,y) values.
This matrix **M** can be calculated using a *nested loop* with *range variables*
i and k, such as:

$N := 12$

$i := 1 .. N \qquad k := 1 .. N$

$x_i := -3 + 0.5 \cdot i \qquad y_k := -3 + 0.5 \cdot k$

$M_{i,k} := f\left(x_i, y_k\right)$

The *function values* of a previously defined function $f(x,y)$ are calculated
here in 144 equidistant points over the square

$-3 \le x \le 3 , -3 \le y \le 3$

and stored in the *matrix* **M**.
This computed *matrix* **M** can be used to draw the function $f(x,y)$ in a
graphical window (3D graph) as described in Section 18.2.

b) Let us now use the example of *function value calculations* to discuss the
handling of *non-integer loop indexes:*
Here we use the *simple function*

$f(x) := x^2 + 1$

that we want to calculate at the points $x = 1.1, 1.3, 1.5, ..., 2.9$ (i.e., with
the *increment* 0.2).
MATHCAD achieves this with the *range variables*

$x := 1.1 , 1.3 .. 2.9$

The *input* of the *numerical equal sign* = after f(x) and x returns the required function values for the values of the defined range variable x:

f(x)		x
2.21		1.1
2.69		1.3
3.25		1.5
3.89		1.7
4.61		1.9
5.41		2.1
6.29		2.3
7.25		2.5
8.29		2.7
9.41		2.9

MATHCAD can display graphically the function values f(x) calculated
using the range variables x (see Section 18.1):

$$x := 1.1, 1.3 .. 2.9$$

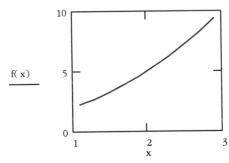

c) If the *elements* of a *matrix* **A** are to be calculated using a specified rule,
this can be performed using *nested loops*.
This we demonstrate in two different ways in the following example, in
which we calculate the elements of a matrix **A** of the type (5,6) as the
sum of the row and column numbers:

- *Nested loops* using *range variables:*

$$i := 1..5 \quad k := 1..6$$

$$A_{i,k} := i + k$$

- *Nested loops* using the **for** *loops:*
  we use here the following *buttons*

from the *operator palette no. 6 (programming palette)*

$$A := \quad \begin{vmatrix} \text{for } i \in 1..5 \\ \quad \text{for } k \in 1..6 \\ \quad\quad A_{i,k} \leftarrow i + k \\ A \end{vmatrix}$$

Each of the two loops used here creates the *matrix*

$$A = \begin{pmatrix} 2 & 3 & 4 & 5 & 6 & 7 \\ 3 & 4 & 5 & 6 & 7 & 8 \\ 4 & 5 & 6 & 7 & 8 & 9 \\ 5 & 6 & 7 & 8 & 9 & 10 \\ 6 & 7 & 8 & 9 & 10 & 11 \end{pmatrix}$$

d) We perform the *calculation* of the *square root* $\sqrt{a}$ of a positive number a using the known convergent *iteration method*

$$x_1 = a$$

$$x_{i+1} = \frac{1}{2} \cdot \left(x_i + \frac{a}{x_i} \right) \quad , \quad i = 1, 2, \ldots$$

This can be realized in MATHCAD using one of the following *loops*; in particular, we calculate the square root of 2 (i.e., a=2):

d1)*Loops* using a *given number* N of *passes:*

- Using *range variables:*

  $$a := 2 \quad N := 10$$
  $$x_1 := a \quad i := 1..N$$

  $$x_{i+1} := \frac{1}{2} \cdot \left(x_i + \frac{a}{x_i} \right)$$

  $$x_{N+1} = 1.414213562373095$$

- You can write the following *function subroutine* **qroot**:

  $$\text{qroot}(a, N) := \quad \begin{vmatrix} x \leftarrow a \\ \text{for } i \in 1..N \\ \quad x \leftarrow \frac{1}{2} \cdot \left(x + \frac{a}{x} \right) \end{vmatrix}$$

by using the *buttons*

from the *operator palette no. 6* (*programming palette*).

The loop with 10 passes produces the following result for the square root of 2:

qroot (2, 10) = 1.414213562373095

We have given a *fixed number* N of *passes* (iterations) in the previous two *loops*. However, because you do not know in advance when the required accuracy is achieved, this has limited practicability for iteration algorithms.

Consequently, we consider the following loops with a variable number of passes.

♦

d2)*Loops* with a *variable number* of *passes:*

A *variable number* of *loop passes* can be realized in MATHCAD by ending the calculation with an *accuracy test* (*error test*) using the relative or absolute *error* of two successive results.

For our simple iteration method for the *square root calculation,* we use the effective *error estimations*

$$\left| \; x_i^2 - a \; \right| < \varepsilon$$

for the *absolute error*
or

$$\left| \frac{x_i^2 - a}{a} \right| < \delta$$

for the *relative error.*

In the following examples we perform an *iteration* with MATHCAD in which we end the calculation when the required *accuracy* (*absolute error* ε) is attained:

• *Loop* using *range variables* and **until**:

The following program variant must contain a *fixed number* N of *loop passes.* However, the calculation is terminated if the given accuracy is attained beforehand.

$a := 2 \quad \varepsilon := 10^{-15} \quad N := 10$

$i := 1 .. N$

$$x_1 := a$$

$$x_{i+1} := \mathbf{until}\left(\left|x_i^2 - a\right| - \varepsilon, \ \frac{1}{2} \cdot \left(x_i + \frac{a}{x_i}\right)\right)$$

$$x = \begin{pmatrix} 2 \\ 1.5 \\ 1.416666666666667 \\ 1.41421568627451 \\ 1.41421356237469 \\ 1.414213562373095 \\ 0 \end{pmatrix}$$

The vector **x** of the iteration values shows that the penultimate component supplies the result and that the required accuracy is achieved with less than N iteration steps when the last component of the vector is 0:

$$x_{\mathrm{last}(x) - 1} = 1.414213562373095$$

- *Loop* using **while**:

  We use here the buttons

  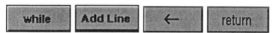

  from the *operator palette no. 6* (*programming palette*).

  Because the loop is exited when the accuracy is attained, you do not need to specify a fixed number of loop passes here. We specify *two variants* for a *function subroutine*:

  * Use of the *tolerance* **TOL** that is a MATHCAD *built-in variable*:

  $$\mathrm{qroot}(a) := \begin{vmatrix} x \leftarrow a \\ \mathrm{while} \ \left|x^2 - a\right| > \mathrm{TOL} \\ \quad x \leftarrow \frac{1}{2} \cdot \left(x + \frac{a}{x}\right) \end{vmatrix}$$

  $$\mathrm{qroot}(2) = 1.414213562373095$$

  The **TOL** *tolerance* specified in the loop can be defined in one of the following ways:

  - using the *menu sequence*

    Math $\Rightarrow$ Options... $\Rightarrow$ Built-In Variables

    – by an *assignment* of the form

       TOL := ......

      before calling **qroot**.

  * Use of a *tolerance* (absolute error) ε that is entered with the function call:

$$\text{qroot}(a, \varepsilon) := \begin{vmatrix} \text{return "number} < 0\text{"} & \text{if} & a < 0 \\ x \leftarrow a \\ \text{while } \left| x^2 - a \right| > \varepsilon \\ \quad x \leftarrow \frac{1}{2} \cdot \left(x + \frac{a}{x} \right) \end{vmatrix}$$

$$\text{qroot}(2, 10^{-10}) = 1.41421356237469$$

$$\text{qroot}(-2, 10^{-10}) = \text{"number} < 0\text{"}$$

This program also uses the **return** statement to permit the output of an *error message* if a negative argument is inadvertently specified for the square root.

♦

10.6 Programming Examples

The *programming languages* integrated in the AXIOM, MACSYMA, MAPLE and MATHEMATICA *computer algebra systems* permit the creation of programs having a comparable quality to those produced in conventional programming languages such as BASIC, C, FORTRAN and PASCAL.
Although MATHCAD has somewhat reduced programming capability, you can still use the advantage of all computer algebra systems:
all integrated commands/functions can be used for the programming.

MATHCAD allows you to save the created *programs* as *worksheets*.
As previously discussed in Chapter 4, text and graphics can be used to supplement the program elements.
You can combine these worksheets as your own user *electronic books* for a specific area. You can use an ASCII editor to view the files from the professional electronic books to get ideas for their layout.
However, these *self-produced electronic books* differ somewhat from the *professional books* that have a better user interface and can be included in

subprograms for conventional programming languages to realize complex algorithms.

♦

Note when you create *programs* that *MATHCAD processes* them from *left* to *right* and from *top* to *bottom*. This must be taken into consideration in the use of defined variables and functions. They can only be used to the right of or below the definition.

♦

In the following sections also use the *statements*

on error, **break** and **return**

that are realized using the *buttons*

from *operator palette no. 6 (programming palette)*.
With the exception of the **continue** *statement*, that is activated with the button

we have now used all operators of the *programming palette*. This instruction is used to terminate the current calculation in a loop, where, however, the complete loop is not terminated, but rather continued with the next loop index.

♦

In Example 10.6 in Section 10.6.2, we demonstrate the creation of our own *MATHCAD programs*.
We would like here to encourage the reader to write his or her own *electronic books*, although naturally these do not achieve the form of the purchased books. However, this should not discourage the MATHCAD users from attempting to produce their *own* simple *electronic books*.
For example, MATHCAD can be used to create an electronic book from the notes, thesis and degree work produced during university study. Similarly, employee's investigations and computations, and lecturer's notes can also be recorded in this form.

10.6.1 Error Handling

Because the programs shown in this book and the algorithms they use are not complex, the error determination for such programs does not play a large role. However, much attention should be devoted to the *error determination* for larger programs, because unlike *syntax errors*, MATHCAD, obviously cannot recognize errors in a program (*programming errors*).

You can recognize such *programming errors* in the following ways:

* *incorrect results* are returned

* MATHCAD indicates a *division by zero*

* the *calculation does not end.*

Starting with version 7, MATHCAD provides two additional *statements*

return and **on error**

that are realized using the *buttons*

from *operator palette no. 6* (*programming palette*). We discuss the **return** *statement* in Examples 10.4d2 and 10.6.

The **on error** *statement* operates as follows: the following expression

∎ on error ∎

with two placeholders appears in the worksheet after clicking the button in the programming palette.

Enter the *expression* to be evaluated in the right-hand placeholder. If an *error* occurs here, the expression entered in the left-hand placeholder is evaluated.

We leave it as an exercise to the reader to find a use for this type of error determination.

10.6.2   Examples

We first consider in Example 10.5 additional uses for branches and loops. The problems handled here should serve as encouragement for the reader

* to experiment with the programming capabilities

* to master any problems that occur.

Example 10.5:

a) The *number* of *components* for **a** that are less than or equal to those for **b** is to be determined for two *vectors* **a** and **b** (which have the same dimension):

The two *vectors* **a** and **b** whose components are contained in the structured files *data_a* and *data_b* on diskette or hard disk are first *read.* Section 9.1 describes how to perform the read:

a := READPRN (data_a)   b := READPRN (data_b)

$$a = \begin{pmatrix} 3 \\ 4 \\ 5 \\ 7 \\ 1 \\ 2 \\ 4 \\ 8 \\ 9 \\ 3 \end{pmatrix} \qquad b = \begin{pmatrix} 1 \\ 2 \\ 4 \\ 5 \\ 7 \\ 8 \\ 9 \\ 2 \\ 9 \\ 6 \end{pmatrix}$$

The *number* of *components* for **a** that are less than or equal to those for **b**:

* is *obtained* using the **if** statement

$$\text{number} := \sum_{i=1}^{\text{rows(a)}} \text{if}\,(a_i \le b_i, 1, 0)$$

number = 5

or

$$\text{number} := \sum_{i=1}^{\text{last(a)}} \text{if}\,(a_i \le b_i, 1, 0)$$

number = 5

when the indexing starts at 1, i.e., you set

ORIGIN := 1

* is *not obtained* if you attempted to use *assignments* as arguments in the **if** statement, as the following example shows:

number := 0   i := 1 .. rows(a)

if $(a_i \le b_i, \text{number} := \text{number} + 1, \text{number})$

number = 0

* is *not obtained* if you used the following program variant:

number := 0   i := 1 .. rows(a)

number := number + if $(a_i \le b_i, 1, 0)$

*   is *obtained* if the last two *errors* are changed using the following *trick* (the use of *number* as a vector with one component):

$$number_1 := 0 \qquad i := 1..\,rows(a)$$

$$number_1 := if\,(\,a_i \le b_i\,, number_1 + 1, number_1\,)$$

$$number_1 = 5$$

or

$$number_1 := 0 \qquad i := 1..\,rows(a)$$

$$number_1 := number_1 + if\,(\,a_i \le b_i\,, 1, 0\,)$$

$$number_1 = 5$$

*   is *obtained* using the *buttons*

from *operator palette no. 6* (*programming palette*) as follows using a *function subroutine* **number**:

$$number(a, b) := \quad \begin{array}{|l} number \leftarrow 0 \\[4pt] for \ \ i \in \ 1..\,rows(a) \\[4pt] \quad number \leftarrow number + 1 \ \ if \ \ a_i \le b_i \\[4pt] number \end{array}$$

$$number(a,b) := 5$$

b) In the following simple example, we use a logical expression in the **if** statement that contains the logical OR to determine the number of values from 0, 0.1, 0.2, ... , 1 that are less than or equal to 0.4 or greater than or equal to 0.7.
We provide two solutions for the problem:

*   using *non-integer range variables* and the button for *summation operator*

from *operator palette no. 5*:

$$x := 0, 0.1 .. 1$$

$$v := \sum_x if(\,(x \le 0.4) + (x \ge 0.7)\,, 1, 0\,)$$

$$v = 9$$

* using the *buttons*

from *operator palette no. 6 (programming palette)*

$$v := \begin{vmatrix} v \leftarrow 0 \\ \quad \text{for } x \in 0, 0.1 \ldots 1 \\ \qquad v \leftarrow v + 1 \quad \text{if } (x \leq 0.4) + (x \geq 0.7) \\ v \end{vmatrix}$$

$v = 9$

♦

Finally, we consider three programs to solve simple problems for probability theory and numerical analysis. You can save these programs in the specified form and then read them again for subsequent calculations. These examples also use the *statements* **break** and **return** that are realized using the *buttons*

from *operator palette no. 6 (programming palette)*.

Example 10.6:

a) In this example we create a program (electronic book) to calculate *probabilities* and *distribution functions* for discrete probability distributions as discussed in Section 26.4. Although MATHCAD supports both the

dbinom (k , n , p)   *(Binomial distribution)*

dhypergeom (k , M , N–M , n)   *(Hypergeometric distribution)*

dpois (k , λ)   *(Poisson distribution)*

probabilities, we calculate these again for the purpose of an exercise:

Discrete distribution functions

*The calculation of discrete distribution functions requires the **binomial coefficients**:*

$$\text{Binomial } (a , k) := \text{if} \left(k = 0 , 1 , \frac{\prod\limits_{i=0}^{k-1}(a - i)}{k!} \right)$$

*The probability for the **binomial distribution** is calculated from (p – probability , n – number of experiments):*

$$PB(n, p, k) := \text{Binomial}(n, k) \cdot p^{k} \cdot (1-p)^{n-k}$$

This yields the distribution function

$$\text{FB}(x,n,p) := \text{if}\left(x \le 0, 0, \text{if}\left(\text{floor}(x) \le n, \sum_{k=0}^{\text{floor}(x)} \text{PB}(n,p,k), 1 \right) \right)$$

*The probability for the **hypergeometric distribution** is calculated from (N – total number of elements, from which M have the required property, n – number of experiments):*

$$\text{PH}(N,M,n,k) := \text{Binomial}(M,k) \cdot \frac{\text{Binomial}(N-M, n-k)}{\text{Binomial}(N,n)}$$

This yields the distribution function

$$\text{FH}(x,N,M,n) := \text{if}\left(x \le 0, 0, \text{if}\left((\text{floor}(x) > n) + (\text{floor}(x) > M), 1, \sum_{k=0}^{\text{floor}(x)} \text{PH}(N,M,n,k) \right) \right)$$

*The probability for the **Poisson distribution** is calculated from (λ – expected value):*

$$\text{PP}(\lambda, k) := \lambda^k \cdot \frac{e^{-\lambda}}{k!}$$

This yields the distribution function

$$\text{FP}(x, \lambda) := \text{if}\left(x \le 0, 0, \sum_{k=0}^{\text{floor}(x)} \text{PP}(\lambda, k) \right)$$

It is desirable to store the specified program as a worksheet, e.g., with the name DISCDIST.MCD. You can re-read this worksheet for subsequent MATHCAD work sessions and make use of the functions defined in it.

For example, we can solve the following problem with the created program:

We wish to determine the number of fish living in a lake. M fish are caught, marked and then returned to the lake. A new batch of n fish are then caught. The number k of marked fish in this batch can be used to estimate the total number of fish in the lake.

You must know that this problem can be solved using the *hypergeometric distribution*.

You can now use the *probability* PH(N,M,n,k) of the *hypergeometric distribution* defined in our worksheet DISCDIST.MCD to calculate the probabilities for a series of N values. Using the *maximum-likelihood estimate*, you take those N values with the highest probability as estimate for the total number of fish. This was 600 fish for the assumed numbers (M=60, n=100 and k=10):

PH(500, 60, 100, 10) = 0.11322

PH(550, 60, 100, 10) = 0.13711

PH(570, 60, 100, 10) = 0.14192

PH(580, 60, 100, 10) = 0.14335

PH(590, 60, 100, 10) = 0.14416

PH(595, 60, 100, 10) = 0.14436

PH(600, 60, 100, 10) = 0.14441

PH(610, 60, 100, 10) = 0.14412

PH(620, 60, 100, 10) = 0.14335

b) We write a *function subroutine* **newton** for the known *Newton iteration method* to determine a *real zero* for a given differentiable *function* f(x) of a real variable x.

$$x^{k+1} = x^k - \frac{f(x^k)}{f'(x^k)} \qquad k = 1, 2, \dots$$

The following *termination criteria* (*stopping rules*) offers itself for the convergence:

* The *absolute error* of two successively calculated values is less than ε, i.e.,

$$\left| x^{k+1} - x^k \right| = \left| \frac{f(x^k)}{f'(x^k)} \right| < \varepsilon$$

* The *absolute value* of the *function* f(x) is less than ε, i.e.,

$$\left| f(x^k) \right| < \varepsilon$$

However, the algorithm does not necessarily converge, even when the *starting value*

$$x^1$$

is near to the sought zero. We take this into consideration in the versions b4 and b5 of our program variants by specifying the *number* N of *iterations* in advance.

The *function* f(x) must have been defined in a *function definition* (see Section 17.2.3) for each of the following specified variants for a function subroutine. However, only the name f of the function can appear as argument in the procedures. The other quantities in the argument for **newton** have the following meanings:

* s

  starting value for the iterations (*initial estimate*)

* ε

  accuracy

b1)A simple program variant has the following form:

$$
\text{newton}(s, f, \varepsilon) := \left| \begin{array}{l} x \leftarrow s \\[1ex] \text{while} \quad | f(x) > \varepsilon \\[1ex] \qquad x \leftarrow x - \dfrac{f(x)}{\dfrac{d}{dx}f(x)} \end{array} \right.
$$

This variant has the following *disadvantages:*

I. The calculation does not end if the algorithm is non-convergent.

II. A division by zero can occur if the derivative of the function is zero at a calculated point.

b2)The following program variant corrects disadvantage II. from b1:

$$
\text{newton}(s, f, \varepsilon) := \left| \begin{array}{l} x \leftarrow s \\[1ex] \text{while} \quad \left(| f(x) | > \varepsilon \right) \cdot \left(\left| \dfrac{d}{dx}f(x) \right| > \varepsilon \right) \\[1ex] \qquad x \leftarrow x - \dfrac{f(x)}{\dfrac{d}{dx}f(x)} \end{array} \right.
$$

b3)Compared with variant b2, the **return** *statement* is also used to *issue* a *message* if the computed derivative is zero:

$$
\text{newton}(s, f, \varepsilon) := \left| \begin{array}{l} x \leftarrow s \\[1ex] \text{while} \quad | f(x) | > \varepsilon \\[1ex] \left| \begin{array}{l} \text{return } "f'(x)=0" \quad \text{if} \quad \left| \dfrac{d}{dx}f(x) \right| < \varepsilon \\[1ex] x \leftarrow x - \dfrac{f(x)}{\dfrac{d}{dx}f(x)} \end{array} \right. \end{array} \right.
$$

b4)The following program variant corrects both disadvantages from b1; the given number N of iterations is realized using a loop.

The **return** *statement* is also used to *issue* a *message* if the computed derivative is zero:

$$\text{newton}(s, f, \varepsilon, \underline{N}) := \left| \begin{array}{l} x \leftarrow s \\[2mm] \text{for } i \in 1.. N \\[2mm] \quad \text{while } \left| f(x) \right| > \varepsilon \\[2mm] \qquad \left| \begin{array}{l} \text{return } "f'(x)=0" \quad \text{if } \left| \dfrac{d}{dx}f(x) \right| < \varepsilon \\[4mm] x \leftarrow x - \dfrac{f(x)}{\dfrac{d}{dx}f(x)} \end{array} \right. \end{array} \right.$$

b5)The following program variant also differs from b4 by using the **return** *statement* to issue a *message* if the given number N of iterations is exceeded.

Furthermore, the **break** *statement* is used in place of a **for** loop to realize a given number of iterations:

$$\text{newton}(s, f, \varepsilon, N) := \left| \begin{array}{l} x \leftarrow s \\[2mm] i \leftarrow 0 \\[2mm] \text{while } \left| f(x) \right| > \varepsilon \\[2mm] \quad \left| \begin{array}{l} \text{return } "f'(x)=0" \quad \text{if } \left| \dfrac{d}{dx}f(x) \right| < \varepsilon \\[4mm] i \leftarrow i + 1 \\[2mm] \text{break if } i > N \\[2mm] x \leftarrow x - \dfrac{f(x)}{\dfrac{d}{dx}f(x)} \end{array} \right. \\[8mm] \text{return } "i>N" \quad \text{if } i > \underline{N} \end{array} \right.$$

We use the last program variant b5 to determine the single real zero of the polynomial function

$$f(x) := x^7 + x + 1$$

You can obtain an *initial approximation* (initial estimate) for the *Newton´s method* from the following graph

$x := -1, -0.99 .. 1$

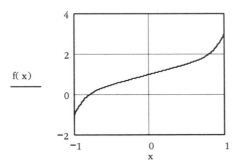

$$f(x)$$

The program is used as follows. If we take as

starting value s=0,

number of *iterations* N=100,

tolerance ε=0.0001

and obtain

newton (0 , f , 0.0001 , 100) = −0.796544857980085

c) Write a *function subroutine* **Max** (**A**) that calculates the *maximum element* of a *given matrix* **A**. We use a *nested loop* (see Section 10.5) and the *matrix functions* **cols** and **rows** to determine the number of columns and rows for the *matrix* **A** and set the starting index for the indexing to 1, i.e., **ORIGIN** := 1 :

$$\text{Max}(A) := \begin{array}{|l} \text{Max} \leftarrow A_{1,1} \\ \text{for } i \in 1 .. \text{cols}(A) \\ \quad \text{for } k \in 1 .. \text{rows}(A) \\ \qquad \text{Max} \leftarrow A_{i,k} \quad \text{if } \text{Max} \leq A_{i,k} \\ \text{Max} \end{array}$$

The function call **Max** (**A**) is used to calculate the maximum element for the following *matrix* **A**:

$$A := \begin{bmatrix} -3 & 1 & 3 \\ 2 & -5 & -6 \\ -7 & 3 & 4 \end{bmatrix} \qquad \text{Max}(A) = 4$$

This function subroutine has only been written as example, because MATHCAD provides an equivalent *matrix function* **max** (**A**) (see Section 15.2).♦

11 Dimensions and Units of Measure

MATHCAD has the advantage compared with other computer algebra systems in that all computations can be performed with *measurement units*. This is of particular importance for *engineers* and *scientists*.
Although MATHCAD recognizes all common (*measurement*) *unit systems:*
SI, MKS, CGS and US
the *SI unit system* with the following *basic dimensions*

* *mass*
* *length*
* *time*
* *charge*
* *temperature*
* *luminosity*
* *substance*

is used prevalently in Europe.

The

Math ⇒ Options...

menu sequence displays a *dialogue box*

Math Options

in which

* **Unit System**

  the required *unit system*

* **Dimensions**

  the *dimensions* can be *set* for the selected unit system.

♦

The

Insert ⇒ Unit...

menu sequence displays a *dialogue box* (see Figure 11.1)

Insert Unit

in which the *measurement units* that *belong* to the individual *dimensions* can be *extracted* and then *inserted* with a mouse click in the worksheet at the cursor position.

♦

If you want to *assign* a *measurement unit* to a *number,* merely multiply the number with the unit of measure. You can enter the designation of the measurement unit directly from the keyboard or with the menu sequence just described.

♦

If you click on (activate) the

Display dimensions

control box for

Dimensions

in the

Math Options

dialogue box
the results for all computations with measurement units are not displayed with the unit of measure but with the dimension name.

If, instead of the basic dimensions, you want to use *other dimensions* such as *number of items, currency,* you can now change the *dimension name* in the corresponding *text field* of the *dialogue box.*

♦

You can *convert* the *measurement unit* of the computation result into *a different measurement unit* by *clicking* with the *mouse* on the *result.* The *following cases* can occur:

- If a *unit placeholder* appears to the right of the result, you can enter the *new measurement unit* in one of the following ways:
  * *Input* with the *keyboard* in the *unit placeholder.*
  * *Insert* with a *double mouse click* on the *unit placeholder* using the
    Insert Unit
    dialogue box that appears
- If *no unit placeholder* appears, you can enter the *new measurement unit* in one of the following ways:
  * Use the keyboard to *delete* the *old unit of measure* and *enter* the *new measurement unit* in the placeholder that appears.
  * *Double mouse click* on the old unit of measure in the
    Insert Unit

dialogue box that appears

For both methods, subsequently pressing the ⏎ key or a mouse click (outside) computes the *result* in the *new measurement unit* (see Example 11.1b).

MATHCAD automatically sets the *units placeholder* for the numeric computation of expressions. It disappears only when a mouse click is made external to expression.

♦

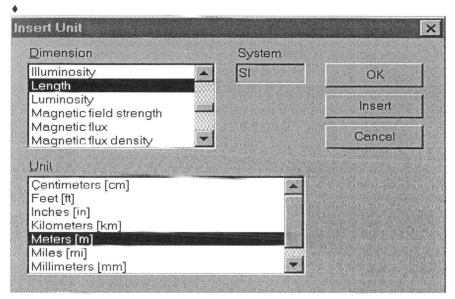

Figure 11.1. Insert Unit dialogue box

We will now use several examples to show the capabilities of MATHCAD for computation using measurement units.

Example 11.1:

a) MATHCAD recognizes the attempt to *add units* that are *not compatible*:

$$25 \cdot \text{kg} + 7 \cdot \text{s} =$$

> The units in this expression do not match.

and *no result will be computed*.

b) The result for the *addition of compatible measurement units* will be output in these measurement units:

b1)We add grams

$$25 \cdot \text{gm} + 250 \cdot \text{gm} = 0.275 \bullet \text{kg} \quad \blacksquare$$

b2)We add kilograms and grams

$$20 \cdot kg + 12 \cdot gm = 20.012 \cdot kg \quad \blacksquare$$

The placeholder at the right of the result is the *units placeholder* for the conversion of the measurement unit.

If you want to *convert* the results shown in kilogram *kg* into grams *gm*, enter the designation *gm* for grams in the *units placeholder* or double-click on the units placeholder to display the

Insert Unit

dialogue box (see Figure 11.1) from which you can obtain and insert the *gm* designation used for gram.

A mouse click outside the result returns the *result* in *grams*, for example for b1):

$$25 \cdot gm + 250 \cdot gm = 275 \cdot gm$$

c) We perform a computation with two measurement units, *length* and *time*

$$a := \frac{3 \cdot cm + 23 \cdot mm}{4 \cdot s + 2 \cdot min} \qquad a = 4.274 \cdot 10^{-4} \cdot m \cdot s^{-1} \quad \blacksquare$$

The placeholder shown at the right of the result is the *units placeholder* for the conversion of the measurement units.

♦

12 Basic Arithmetic Operations

Obviously, performing the basic arithmetic operations does not belong to the main uses of MATHCAD. You can continue to use an electronic calculator for this purpose. However, because these operations are often required during a MATHCAD work session, we provide a short discussion here.

MATHCAD uses the following *operation symbols* for the *basic arithmetic operations*

- + (*addition*)
- − (*subtraction*)
- * (*multiplication*)
- / (*division*)
- ∧ (*power*)
- ! (*factorial*)

that can be entered from the keyboard, where MATCHCAD converts the *multiplication, division* and *power symbols* to the usual representation as *multiplication point, division bar* and *exponent* immediately after input.

In addition to input from the keyboard, the following *operators* can be used to enter *power* and *root*:

* *power* with the *power button*

* *square root* with the *root button*

* *n-th roots* with the *root button*

from *operator palette no. 1.*

♦

An expression formed from real numbers and these operation symbols is designated as being a *number expression.*

The usual *priorities* apply for processing the operators in a number expression, i.e.,

* exponentiation is performed *first*,
* *then* multiplication (division), and
* *finally* addition (subtraction).

If you are not sure of the sequence with which the operations are performed, it is better to use additional parentheses.

♦

MATHCAD provides the *following ways* of calculating *number expressions*:

* *exact* (*symbolic*)
* *numerical*

Chapter 6 has already provided a detailed description of both these capabilities.

♦

MATHCAD and the other *computer algebra systems* are already an *improvement over an electronic calculator* for the *basic arithmetic operations*, because these only compute numerically, whereas the systems can compute both numerically and exactly.

♦

13 Transformation of Expressions

The term *expressions* refers to both *algebraic* and *transcendental expressions*. The *transformation/manipulation* of such *expressions* is *standard* in computer algebra systems and so also part of MATHCAD. In many cases, MATHCAD can quickly and correctly *transform/manipulate* or *calculate* large and complicated *expressions*.

13.1 Introduction

We first provide a short, illustrative description of *algebraic* and *transcendental expressions:*

* We consider an *algebraic expression* to be any combination of *digits* and *letters* (numbers, constants and variables) that are joined by the *basic arithmetic operations* (see Chapter 12)
  * $+$ (*addition*)
  * $-$ (*subtraction*)
  * $*$ (*multiplication*)
  * $/$ (*division*)
  * $\wedge$ (*power*)

Example 13.1:

Some examples of *algebraic expressions:*

a) $(a + b + c + d)^4$

b) $c^2 - 2 \cdot c \cdot d + d^2$

c) $\dfrac{a \cdot x + b}{c \cdot x + x^2} + \dfrac{d}{\sqrt{x}}$

d) $c + \sqrt{\dfrac{d}{a}} + 5^3 + \dfrac{a - 5 \cdot c}{3^2 \cdot d - 25}$

e) $\dfrac{a}{\sqrt[3]{b - x}} + \dfrac{a}{\sqrt[5]{b + x}}$

f) $\dfrac{x^4 - 1}{x^2 + 1}$

$\blacklozenge$

- *Transcendental expressions* are formed in the same way as algebraic expressions but also permit the use of exponential, trigonometric and hyperbolic functions and their inverse functions.

Example 13.2:

Some examples of *transcendental expressions:*

a) $\dfrac{\tan(a+b) + \sqrt{x} + x^2}{\sin x \cdot \sin y + c^2}$
   b) $\dfrac{a^x + 3 + b^3}{\sin x + 1}$
   c) $\dfrac{\log(x+y) + e^x + x \cdot y}{\cos x + a^{2x} + 1}$

◆

Algebraic and *transcendental* expressions can be *transformed/manipulated,* however, only *exact (symbolic) operations* are considered. We *distinguish* between the following *transformations/manipulations:*

- *Algebraic expressions* can be
  * *Simplified* (see Section 13.2).

    Example 13.3:

    a) $\dfrac{x^4 - 1}{x^2 + 1} = x^2 - 1$
       b) $\dfrac{b}{a-b} - \dfrac{a}{a-b} = -1$

    c) $\dfrac{x^3 - 3 \cdot x^2 \cdot y + 3 \cdot x \cdot y^2 - y^3}{x-y} = (x-y)^2$
       d) $\dfrac{b^2 - a^2}{(a+b)^2} = \dfrac{b-a}{a+b}$

    ◆

  * *Decomposed* into *partial fractions* (see Section 13.3).

    Example 13.4:

    a) $\dfrac{5x+1}{x^2 + x - 6} = \dfrac{11}{5} \cdot \dfrac{1}{x-2} + \dfrac{14}{5} \cdot \dfrac{1}{x+3}$

    b) $\dfrac{x^3 + x^2 + x + 2}{x^4 + 3x^2 + 2} = \dfrac{1}{x^2 + 1} + \dfrac{x}{x^2 + 2}$

    c) $\dfrac{x^4 - x^3 - x - 1}{x^3 - x^2} = x + 2 \cdot \dfrac{1}{x} + \dfrac{1}{x^2} - \dfrac{2}{x-1}$

    ◆

  * *Expanded,* i.e., calculation of all powers and products in an expression (see Section 13.4)

    Example 13.5:

    $(a+b+c)^2 = a^2 + b^2 + c^2 + 2 \cdot (a \cdot b + a \cdot c + b \cdot c)$

    ◆

  * *Multiplied* (see Section 13.5).

Example 13.6:

a) $\dfrac{x^2+1}{x^4-1} \cdot \dfrac{x-1}{x+1} = \dfrac{1}{(x+1)^2}$

b) $(x^2+x+1)\cdot(x^3-x^2+1) = x^5+x+1$

◆

* *Factorized* as inverse operation to multiplication (see Sections 13.6 and 16.2).

Example 13.7:

a) $x^5+x+1 = (x^2+x+1)\cdot(x^3-x^2+1)$

b) $a^2+b^2+c^2+2\cdot(a\cdot b+a\cdot c+b\cdot c) = (a+b+c)^2$

c) $x^3-10\cdot x^2+31\cdot x-30 = (x-2)\cdot(x-3)\cdot(x-5)$

◆

* *Reduced* to a *common denominator* (see Section 13.7).

Example 13.8:

a) $\dfrac{1}{x-1}+\dfrac{1}{x+1} = \dfrac{2\cdot x}{(x-1)\cdot(x+1)}$

b) $\dfrac{1}{x^2+1}+\dfrac{x}{x^2+2} = \dfrac{x^3+x^2+x+2}{x^4+3\cdot x^2+2}$

◆

* *Substituted*, i.e., some subexpressions (constants and variables) can be replaced by other expressions (see Section 13.8).

Example 13.9:

You obtain

$$\frac{x^2-1}{a+x} = \frac{(\sin t)^2-1}{a+\sin t}$$

when you replace x by sin t

◆

• Although *transcendental expressions* can also be transformed, practical uses occur mainly in *trigonometric expressions* (see Section 13.9).

☞

The following section discusses the *menu sequences* and *keywords* provided in MATHCAD to perform the *transformations/manipulations*. Note: A concrete *expression* shown as A is assumed to exist in the worksheet. ◆

13.2 Simplification

First mark with *editing lines* the complete *expression* contained in the work-sheet. MATHCAD then provides *three methods* of *simplification*:

I. Use the *menu sequence*

Symbolics ⇒ Simplify

II. Click on the

button in *operator palette no. 8 (keyword palette)* and enter the

simplify

keyword in the free placeholder.
Then press the ⏎ key or *click* with the *mouse* outside the expression to initiate the simplification.

III. Click on the

button for the

simplify

keyword in *operator palette no. 8 (keyword palette)*
Then press the ⏎ key or *click* with the *mouse* outside the expression to initiate the *simplification*.

If you enter a comma after the **simplify** *keyword* for methods II and III, you can use the

Modifiers

button from *operator palette no. 8* and the *equal operator* from operator palette no. 2 to enter an *attribute* (*modifier*) in the displayed placeholder (see Example 13.10), for example

assume = real

simplifies on the assumption that all numbers, constants and variables in the expression are real.

♦
Example 13.10:

Simplify the *expression*

$$\frac{x^4 - 1}{x^2 + 1}$$

using the *three* specified *methods:*

* using *method I.:*

$$\frac{x^4 - 1}{x^2 + 1} \quad simplifies\ to \quad x^2 - 1$$

* using *method II.:*

$$\frac{x^4 - 1}{x^2 + 1} \; \text{simplify} \; \rightarrow x^2 - 1$$

* using *method III.:*

$$\frac{x^4 - 1}{x^2 + 1} \; \text{simplify}, \text{assume} \blacksquare \text{real} \; \rightarrow x^2 - 1$$

Note: We have also used the **real** attribute here.

♦

13.3 Partial Fraction Decomposition

We assume that the A(x) *expression* dependent on the variable x is a *fractional rational expression*, i.e., it is represented as the quotient of two polynomials.

MATHCAD then provides *two methods* to *decompose* an *expression* A(x) into *partial fractions* :

I. Once a *variable* x of the expression A(x) contained in the worksheet has been *marked* with *editing lines*, you can use the *menu sequence*

Symbolics ⇒ Variable ⇒ Convert to Partial Fraction

II. *Click* on the

button for the

convert, parfrac

keyword in *operator palette no. 8* (*keyword palette*) that then displays:

■ convert , parfrac , ■ →

Enter the expression A(x) in the *left-hand placeholder* and the variable x in the *right-hand placeholder.*

Then press the ⏎ key or *click* with the *mouse* outside the expression to initiate the *decomposition.*

Because the *partial fraction decomposition* is closely related to the zeros (see Section 16.2) of the denominator polynomial, it is not surprising that MATHCAD sometimes cannot solve this problem. This is caused by there being no finite algorithm to determine the zeros for polynomials greater than the fourth degree. However, MATHCAD can in many cases also determine integer zeros for denominator polynomials having a higher degree. MATHCAD also has difficulties with the *partial fraction decomposition* when the denominator polynomial has *complex zeros* (see Example 13.11a).

♦

Example 13.11:

a) MATHCAD fails with the simple function

$$\frac{1}{x^4 + 1}$$

that has the *partial fraction decomposition*

$$\frac{1}{2\sqrt{2}} \cdot \frac{x + \sqrt{2}}{x^2 + \sqrt{2} \cdot x + 1} - \frac{1}{2\sqrt{2}} \cdot \frac{x - \sqrt{2}}{x^2 - \sqrt{2} \cdot x + 1}$$

because the denominator polynomial has complex zeros.

In contrast, MATHCAD has no difficulty returning the *following partial fraction decompositions;* we use the two specified methods:

b) *Method I.:*

$$\frac{x^4 - x^3 - x - 1}{x^3 - x^2} \quad expands\ in\ partial\ fractions\ to \quad x + \frac{1}{x^2} + \frac{2}{x} - \frac{2}{(x-1)}$$

c) *Method II.:*

$$\frac{x^3 + x^2 + x + 2}{x^4 + 3 \cdot x^2 + 2} \quad convert,\ parfrac,\ x \quad \rightarrow \quad \frac{1}{(x^2 + 1)} + \frac{x}{(x^2 + 2)}$$

♦

13.4 Expansion

First mark with *editing lines* the complete *expression* contained in the worksheet. MATHCAD then provides *three methods* for *expansion*:

I.   *Use* the *menu sequence*

Symbolics ⇒ Expand

II. *Click* the

button in *operator palette no. 8 (keyword palette)* and enter the

expand

keyword in the free placeholder.

Then press the ⏎ key or *click* with the *mouse* outside the expression to expand.

III. *Click* the

button for the

expand

keyword in *operator palette no. 8 (keyword palette)* and enter the *expansion variable* in the free *right-hand placeholder*.

Then press the ⏎ key or *click* with the *mouse* outside the expression to expand.

This permits the *calculation* of *expressions* of the *following form* without difficulty (n is an integer):

$(a+b+c+...+f)^n$

♦

Example 13.12:

We use all three methods in the following section to expand the expression

$(a+b+c)^2$

* *Method I.:*

$(a+b+c)^2$ *expands to* $a^2 + 2 \cdot a \cdot b + 2 \cdot a \cdot c + b^2 + 2 \cdot b \cdot c + c^2$

* *Method II.:*

$(a+b+c)^2$ expand $\rightarrow a^2 + 2 \cdot a \cdot b + 2 \cdot a \cdot c + b^2 + 2 \cdot b \cdot c + c^2$

* *Method III.:*

$(a+b+c)^2$ expand, a, b, c $\rightarrow a^2 + 2 \cdot a \cdot b + 2 \cdot a \cdot c + b^2 + 2 \cdot b \cdot c + c^2$

♦

13.5 Multiplication

The *expressions* to be *multiplied* are *entered* in the worksheet as *complete expression* in the usual form. Then mark the complete *expression* with *editing lines*. MATHCAD then offers *three methods* of *multiplying* that have the same form as for expansion (see Section 13.4):

I. *Use* the *menu sequence*

    Symbolics ⇒ Expand

II. *Click* the

button in *operator palette no. 8 (keyword palette)* and enter the

expand

keyword in the *free placeholder*.
Then press the ⏎ key or *click* with the *mouse* outside the expression to perform the *multiplication*.

III. *Click* the

button for the

expand

keyword in *operator palette no. 8 (keyword palette)* and enter the *expansion variable* in the free *right-hand placeholder*.
Then press the ⏎ key or *click* with the *mouse* outside the expression to perform the *multiplication*.

Example 13.13:

This example uses all three methods to multiply out the expression

$$(x^3 + x^2 + x + 1) \cdot (x^2 + 1)$$

*   *Method I.:*

    $$(x^3 + x^2 + x + 1) \cdot (x^2 + 1) \, expands \, to \, x^5 + 2 \cdot x^3 + x^4 + 2 \cdot x^2 + x + 1$$

*   *Method II.:*

    $$\left(x^3 + x^2 + x + 1\right) \cdot \left(x^2 + 1\right) \, \text{expand} \, \rightarrow x^5 + 2 \cdot x^3 + x^4 + 2 \cdot x^2 + x + 1$$

*   *Method III.:*

    $$\left(x^3 + x^2 + x + 1\right) \cdot \left(x^2 + 1\right) \, \text{expand, x} \, \rightarrow x^5 + 2 \cdot x^3 + x^4 + 2 \cdot x^2 + x + 1$$

    ◆

13.6 Factorization

First *mark* with *editing lines* the complete *expression* contained in the worksheet. MATHCAD then offers *three methods* for *factorization*:

I. *Use* the *menu sequence*

Symbolics ⇒ Factor

II. *Click* the

button in *operator palette no. 8* (*keyword palette*) and *enter* the

factor

keyword in the *free placeholder*.
Then press the ⏎ key or *click* with the *mouse* outside the expression to perform the *factorization*.

III. *Click* the

button for the

factor

keyword in *operator palette no. 8* (*keyword palette*) and enter the factorization variable in the free *right-hand placeholder*.
Finally, press the ⏎ key or *click* with the *mouse* outside the expression to perform the *factorization*.

Factorization of *polynomials* (see Section 16.2) with real coefficients is understood to be the *writing* as *product* of

*   *linear factors* (for the real zeros)

*   *quadratic polynomials* (for the complex zeros),

i.e., (for $a_n = 1$)

$$\sum_{k=0}^{n} a_k \cdot x^k \; =$$

$$(x - x_1)^{k_1} \cdot (x - x_2)^{k_2} \cdot \ldots \cdot (x - x_r)^{k_r} \cdot (x^2 + b_1 \cdot x + c_1) \cdot \ldots \cdot (x^2 + b_s \cdot x + c_s)$$

where

$x_1, \ldots, x_r$

are the *real zeros* of *multiplicity*

$k_1, k_2, ..., k_r$.

Such a *factorization* obeys the fundamental theorem of *algebra*. Because the *factorization* of *polynomials* is closely related to the determination of zeros (see Section 16.2), it is not surprising that MATHCAD sometimes cannot solve this problem (see Examples 13.14c and d). This is caused by there being no finite algorithm to determine the zeros for polynomials greater than the fourth degree. However, MATHCAD can in many cases also determine integer zeros for polynomials having a higher degree.

♦

Example 13.14:

a) We use the *method I.* of MATHCAD:

$$a^3 - 3 \cdot a^2 \cdot b + 3 \cdot a \cdot b^2 - b^3 \quad \textit{by factoring, yields} \quad (a - b)^3$$

b) We factorize a third degree polynomial by using all three methods:

* Use of *method I.*:

$$x^3 + x^2 + x + 1 \quad \textit{by factoring, yields} \quad (x + 1) \cdot \left(x^2 + 1\right)$$

* Use of *method II.*:

$$x^3 + x^2 + x + 1 \text{ factor } \rightarrow (x + 1) \cdot \left(x^2 + 1\right)$$

* Use of *method III.*:

$$x^3 + x^2 + x + 1 \text{ factor}, x \rightarrow (x + 1) \cdot \left(x^2 + 1\right)$$

c) In contrast to b, MATHCAD *cannot factorize* the following *polynomial*

$$x^3 + x^2 - 2 \cdot x - 1$$

This is because the three real zeros of the polynomial are non-integer.

d) Although the following fifth degree polynomial with one real zero can be factorized using *method I.*, this is performed incompletely, because MATHCAD cannot determine the real zero:

$$x^5 + x + 1 \quad \textit{by factoring, yields} \quad \left(x^2 + x + 1\right) \cdot \left(x^3 - x^2 + 1\right)$$

Methods II. and III. do not return any factorization.

e) *Factorization* is also used to *decompose* a *natural number* N into its *primary factors*:

12345   *by factoring, yields* $3 \cdot 5 \cdot 823$

♦

13.7 Reduce to a Common Denominator

Because the procedure here is the same as for the simplification of expressions, we refer the reader to Section 13.2. This procedure can be used to reduce fractional rational expressions to a common denominator.

Example 13.15:

$$\frac{1}{x-1} + \frac{1}{x+1} \quad simplifies\ to \quad 2 \cdot \frac{x}{((x-1) \cdot (x+1))}$$

♦

13.8 Substitution

Substitution refers to the *replacement* of certain *subexpressions* (constants or variables) in an expression (target expression) by another expression. MATHCAD permits such *substitutions* to be *performed* in one of the following *ways*:

I. *Mark* the *expression* to be *substituted* with *editing lines* and then *copy* it in the usual WINDOWS manner into the *clipboard*, e.g., by clicking the *copy button*

in MATHCAD's standard toolbar.

Then *mark* with *editing lines* one of the *subexpressions* (constants or variables) to be replaced in the *target expression* and *activate* the *menu sequence*

Symbolics ⇒ Variable ⇒ Substitute

II. The **substitute** *keyword* for the *substitution* is activated by clicking the

button in the *operator palette no.8*; the *target expression* is entered in the *left-hand placeholder* and the *substitution* entered in the *right-hand placeholder* using the *equal operator* (see Example 13.16).

Because of its simpler handling, Method II. with the

substitute

keyword is preferable. This avoids cumbersome storing using the clipboard.

♦

Let us consider the detailed processing for substitution in MATHCAD using the following example.

Example 13.16:

a) We wish to *substitute* in the *function*

$$x^3 + x^2 + x + 1$$

for the *variable* x the *expression*

$$\frac{a+b}{c+d}$$

$$x^3 + x^2 + x + 1 \text{ substitute, } x = \frac{a+b}{c+d} \;\rightarrow\; \frac{(a+b)^3}{(c+d)^3} + \frac{(a+b)^2}{(c+d)^2} + \frac{(a+b)}{(c+d)} + 1$$

b) You can also use substitution if you wish to replace a variable in an expression with a number, for example, if you wish to use the value 3 for x in the *function*

$$x^3 + x^2 + x + 1$$

i.e., calculate the function value at x=3:

$$x^3 + x^2 + x + 1 \text{ substitute, } x = 3 \;\rightarrow\; 40$$

◆

There are further uses for *substitution* in addition to those shown in the examples, such as for the solution of systems of equations.

◆

13.9 Transformation of Trigonometric Expressions

MATHCAD also permits the *transformation/manipulation* of *expressions* with *trigonometric functions:*
The same method is used here as for expanding and multiplying, where, in particular, the

expand

keyword is appropriate. Activate expand by *clicking* the

expand

button in *operator palette no. 8 (keyword palette).*

Then enter the *expression* A in the *left-hand placeholder* and use the

button from the same operator palette (see Example 13.17) to enter an *attribute* (*modifier*) in the *form*

assume=trig

in the *right-hand placeholder.* This attribute indicates to MATHCAD that trigonometric expressions are being used. However, problems can also be solved without specifying the attribute (see Examples 13.17a and c).

Finally, press the ⏎ key or *click* with the *mouse* outside the expression to perform the *transformation.*

☞

However, MATHCAD cannot transform every trigonometric expression. It is quite possible that the transformation succeeds or fails for similar function expressions, as shown in Example 13.17d. It would be desirable that a future development of MATHCAD solves this problem.

♦

Example 13.17:

a) sin(x+y) yields

$$\sin(x + y) \text{ expand}, \text{assume} \equiv \text{trig} \;\rightarrow\; \sin(x) \cdot \cos(y) + \cos(x) \cdot \sin(y)$$

The following scheme returns the same result:

$$\sin(x + y) \text{ expand}, x, y \;\rightarrow\; \sin(x) \cdot \cos(y) + \cos(x) \cdot \sin(y)$$

b) $\sin(x)^2$ yields

$$\sin(x)^2 \text{ expand}, \text{assume} \equiv \text{trig} \;\rightarrow\; \sin(x)^2$$

i.e., MATHCAD returns *no result.*

c) sin 2x yields

$$\sin(2 \cdot x) \text{ expand}, \text{assume} \equiv \text{trig} \;\rightarrow\; 2 \cdot \sin(x) \cdot \cos(x)$$

The following scheme returns the same result:

$$\sin(2 \cdot x) \text{ expand}, x \;\rightarrow\; 2 \cdot \sin(x) \cdot \cos(x)$$

d) Whereas version 7 solves the problem

$$\tan\left(\frac{x}{2}\right) \text{ expand, assume} \blacksquare \text{trig} \rightarrow \frac{(1 - \cos(x))}{\sin(x)}$$

MATHCAD cannot solve the similar problem

$$\sin\left(\frac{x}{2}\right) \text{ expand, assume} \blacksquare \text{trig} \rightarrow \sin\left(\frac{1}{2} \cdot x\right)$$

MATHCAD Version 8 cannot solve either problem.
◆

14 Sums and Products

Because the *calculation* of finite *sums* and *products* of the form

$$\sum_{k=1}^{n} a_k = a_1 + a_2 + \ldots + a_n \quad \text{or} \quad \prod_{k=1}^{n} a_k = a_1 \cdot a_2 \cdot \ldots \cdot a_n$$

where the elements

$$a_k \qquad (k = 1 , 2 , \ldots , n)$$

are real or complex numbers require only a finite number of computational steps (finite algorithm), they do not present any difficulties to MATHCAD. However, the calculation can take a very long time for large values of n. Perform the following steps to use MATHCAD to calculate finite *sums* and *products*:

- *First click* the *sum button*

  or the *product button*

  from the *operator palette no. 5*

  and *complete* the associated *placeholders* using the usual mathematical notation in the *symbols*

$$\sum_{\blacksquare = \blacksquare}^{\blacksquare} \blacksquare \qquad \text{or} \qquad \prod_{\blacksquare = \blacksquare}^{\blacksquare} \blacksquare$$

  that appear in the worksheet.
- Then *mark* the complete *expression* with *editing lines.*
- Use one of the *following ways* to *initiate* the *calculation:*
  - * *Exact calculation* using one of the following actions:
    - − *Input* of the *menu sequence*

Symbolics ⇒ Evaluate ⇒ Symbolically

– *Input* of the *menu sequence*
   Symbolics ⇒ Simplify

– *Input* of the *symbolic equal sign* → and then press the ⏎-key.

* *Numerical calculation* using

– *Input* of the *menu sequence*
   Symbolics ⇒ Evaluate ⇒ Floating Point...

– *Input* of the *numerical equal sign* =

The

    or

buttons from the *operator palette no. 5* can also be used for the *numerical calculation* of *sums* and *products*:

- After clicking these operators, the following *symbols* appear in the worksheet:

$$\sum_{\blacksquare}^{\blacksquare} \blacksquare \qquad \text{bzw.} \qquad \prod_{\blacksquare}^{\blacksquare} \blacksquare$$

  in which the *placeholder* after the symbol has the same meaning as above.

- Because only the index designation is entered in the *placeholder* below the symbol, a *range variable* must also be used to define the range for the index. A sum or product can also be formed using non-integer indexes (i.e., arbitrary range variables).

These sums and products are also known as *range sums* or *range products*. They are calculated only numerically by entering the numerical equal sign =. Example 14.1 illustrates some applications.

♦

A multiple click on the appropriate operators *nests* the *sums* and *products*, as we can see in Example 14.1d for the calculation of a *double sum*.

♦

Whereas specific numerical values must be assigned to all quantities (also n) for a *numerical calculation* of finite *sums* and *products*, these also permit general calculations to be calculated exactly (see Example 14.1c).

♦

Example 14.1:

a) If you enter the appropriate values in the placeholders in the

$$\sum_{\blacksquare=\blacksquare}^{\blacksquare} \blacksquare$$

symbol that appears after clicking the *sum button* in *operator palette no. 5*, you can *calculate* the *sum*

$$\sum_{k=1}^{30} \frac{1}{2^k}$$

*   *exactly* by entering the symbolic equal sign → and then pressing the ⏎-key:

$$\sum_{k=1}^{30} \frac{1}{2^k} \rightarrow \frac{1073741823}{1073741824}$$

*   *numerically* by entering the numerical equal sign = :

$$\sum_{k=1}^{30} \frac{1}{2^k} = 0.999999999068677$$

You achieve the same numerical result using *range sums* in the following way:

$$k := 1..30$$

$$\sum_{k} \frac{1}{2^k} = 0.999999999068677$$

b) If you enter the appropriate values in the *placeholders* in the

$$\prod_{\blacksquare=\blacksquare}^{\blacksquare} \blacksquare$$

symbol that appears after clicking the *product button* in the *operator palette no. 5*, you can calculate the *product*

$$\prod_{k=1}^{10} \left(\frac{5}{6}\right)^k$$

* *exactly* by entering the symbolic equal sign → and then pressing the ⏎-key:

$$\prod_{k=1}^{10} \left(\frac{5}{6}\right)^k \ \rightarrow \ \frac{27755575615628913510590791702270507 8125}{628519521356600533556105353315002621 7291776}$$

* *numerically* by entering the numerical equal sign = :

$$\prod_{k=1}^{10} \left(\frac{5}{6}\right)^k \ = 4.416024430827729 \cdot 10^{-5}$$

You can achieve the same numerical result using *range product* in the following manner:

$k := 1..10$

$$\prod_{k} \left(\frac{5}{6}\right)^k \ = 4.416024430827729 \cdot 10^{-5}$$

c) *Exact calculation* can be used to *calculate general sums* and *products*, as shown by the following examples:

c1)

$$\sum_{k=0}^{n} \frac{1}{a^k} \ \rightarrow \ -\left(\frac{1}{a}\right)^{(n+1)} \cdot \frac{a}{(-1+a)} + \frac{a}{(-1+a)}$$

c2)

$$\left[\prod_{k=2}^{n} 1 - \frac{1}{k}\right] \ \rightarrow \ \frac{\Gamma(n)}{\Gamma(n+1)}$$

Γ here represents the *Gamma function* that returns the value $\Gamma(n+1)$ = n ! for an integer n.

d) We calculate the *double sum*

$$\sum_{i=1}^{2} \sum_{k=1}^{3} (i+k)$$

by *nesting* the *sum operator.*

* *exactly* using the symbolic equal sign:

$$\sum_{i = 1}^{2} \left[\sum_{k = 1}^{3} i + k \right] \rightarrow 21$$

* *numerically* using the numerical equal sign:

$$\sum_{i = 1}^{2} \left[\sum_{k = 1}^{3} i + k \right] = 21 \quad \blacksquare$$

e) MATHCAD also permits   *range sums* or *products* to be realized over *any range variables:*

* We calculate the *sum* of the *squares* of the *numbers* 0.1, 0.2, ... , 1 :

$$x := 0.1, 0.2 .. 1$$

$$\sum_{x} x^2 - 3.85$$

* We calculate the *product* of the *numbers* 0.1, 0.2, ... , 1 :

$$x := 0.1, 0.2 .. 1$$

$$\prod_{x} x = 3.629 \cdot 10^{-4}$$

$\blacklozenge$

15 Vectors and Matrices

Vectors and *matrices* play a fundamental role in business, engineering and science (see [3], [4]). Consequently, MATHCAD contains comprehensive facilities for their representation and calculation. We discuss these capabilities in this chapter.

In general, we will consider *matrices* of the *type (m,n)*, i.e., matrices with *m rows* and *n columns:*

$$\begin{pmatrix} a_{11} & a_{12} & \cdots & a_{1n} \\ a_{21} & a_{22} & \cdots & a_{2n} \\ \vdots & \vdots & \vdots & \vdots \\ a_{m1} & a_{m2} & \cdots & a_{mn} \end{pmatrix}$$

These are also known as *m×n matrices*.

Vectors can be considered to be *special cases* of *matrices*, because

* *row vectors*

  are 1×n matrices, i.e.

  $$(x_1, x_2, ..., x_n)$$

* *column vectors*

  are n×1 matrices, i.e.

  $$\begin{pmatrix} x_1 \\ x_2 \\ \vdots \\ x_n \end{pmatrix}$$

Consequently, the following facts specified for matrices naturally also apply for vectors, provided the vector type permits this. This affects input, addition, multiplication and transposition.

♦

Note that MATHCAD accepts only *column vectors* for the calculation with the *computational operations* provided for *vectors*.

♦

MATHCAD uses also the collective term *array* for both vectors and matrices.
♦

Note, although MATHCAD always starts with 0 (default setting) for the *indexing* of the *components* or *elements* for *calculations* with *vectors* and *matrices*, the indexing in mathematics usually begins with the *starting index* 1. Consequently, MATHCAD provides the capability to specify a different starting value for the indexing (usually 1) to *replace* the *default value* 0 for the *built-in variable* **ORIGIN**; this value is entered in the

Array Origin

field for the

Built-In Variables

that is displayed in the

Math Options

dialogue box
initiated with the

Math ⇒ Options...

In order that the indexing applies for the complete worksheet, you should then activate the *menu sequence*

Math ⇒ Calculate Worksheet

The *assignment*

ORIGIN := 1

assigns 1 locally to the *starting index*.
♦

15.1 Input

Vectors and *matrices* must first be *entered* into the *worksheet* before you can use them in calculations. MATHCAD provides two methods to perform this input:

* *read* from a *data medium* (hard disk, diskette or CD-ROM)

* *input* from the *keyboard*

We have already discussed in Section 9.1 the *input* (*reading*) of vectors and matrices from data media.
We discuss in the following section the capabilities MATHCAD offers for the input using the keyboard.

15.1.1 Input of Vectors

MATHCAD provides the following two *capabilities* for the *input* of (column) *vectors* into the *worksheet* using the keyboard:

I. After activating the *menu sequence*

Insert ⇒ Matrix...

or
clicking the *matrix button*

in *operator palette no. 4 (vector and matrix palette)*
displays the

Insert Matrix

dialogue box in which the *number of components* is to be entered in

Rows:

and a 1 in

Columns:

because vectors in MATHCAD are considered to be *column vectors*. Then, after *clicking* the *button*

Insert

a *vector* of the *form* (e.g., for four components):

is inserted in the worksheet at the location determined by the cursor. The specific components of the vector are entered in its placeholders using the keyboard. A mouse click or the

⟨⇆⟩-key

can be used to switch between the placeholders.
This dialogue box is also used to *insert* or *delete components* in an existing vector.

II. After entering the *index* as *range variable*, e.g., for n components

i := 1 .. n

vectors can be *created* by *clicking* the

button from *operator palette no. 1* or *4*. The designation of the vector
(e.g., **x**) and the index (e.g., i) are entered in the displayed placeholder

$\blacksquare_\blacksquare$

The *assignment operators* := or ≡ (see Section 10.3) then can be used to
assign values to the *components*

x_i

of the *vector* **x**. This can be done with

* a function f(i), provided that the i th component of the vector is a
  function f of the index i (see Example 15.1c), i.e.,

  $x_i := f(i)$

* the input of a comma after every number following the assignment
  operator when numbers are to be assigned to the components from
  the keyboard (input as *number table*); this produces an *input table*
  (see Example 15.1b).

It is desirable to use the *assignment operators* := or ≡ to assign *vector sym-
bols / vector names* (typically lowercase letters **a**, **b** , ... , **x** , **y** , **z**) to *vectors*
(see Example 15.1a), which can be used in further calculations. The := and
≡ operators perform a *local* assignment or *global* assignment respectively
(see Section 10.3).

◆

Example 15.1c shows the *three representation forms* for *vectors* that MATH-
CAD permits after the input of the numerical equal sign =:

• as *scrolling output table* (with row and column number) after the input
  of the vector name when the vector has more than 9 components

• as *output table* after the input of the vector name with index

• as *column vector* by suppression of the row and column number, by
  activating the *menu sequence*

Format ⇒ Result...

and setting

Matrix

in the **Matrix display style** *field* of the displayed

Result Format

dialogue box.

◆

Example 15.1:

The following examples all assume **ORIGIN:**=1, i.e., the indexing starts with
1.

a) *Method I.* is most often used to *create vectors*. A vector with the name **x** entered using this method assumes the following form in the worksheet:

$$x := \begin{bmatrix} 1 \\ 2 \\ 3 \\ 4 \end{bmatrix}$$

b) We *create* a *vector* using *Method II.* with a *number table* that contains 6 values.

We can use here the *input tables* provided by MATHCAD:

$i := 1..6$

$x_i :=$

3
4
6
8
2
5

$$x = \begin{bmatrix} 3 \\ 4 \\ 6 \\ 8 \\ 2 \\ 5 \end{bmatrix}$$

The *input table* is obtained by *entering* a *comma* after every number following the assignment operator. The *control variables* i defined as 1 to 6 can be used with *input tables* to create additional vectors, which can also have fewer than 6 components:

$y_i :=$

2
4
6

$$y = \begin{bmatrix} 2 \\ 4 \\ 6 \end{bmatrix}$$

$z_i :=$

4
5
6
3

$$z = \begin{bmatrix} 4 \\ 5 \\ 6 \\ 3 \end{bmatrix}$$

c) We *create* a *vector* **x** whose components can be represented as a function of the index i:

$i := 1..10$

$x_i := i^2$

and then show the three different *representation forms* :

* *Representation* as *scrolling output table* (with row and column number):

$$
X =
\begin{array}{c|c}
 & 1 \\
\hline
1 & 1 \\
\hline
2 & 4 \\
\hline
3 & 9 \\
\hline
4 & 16 \\
\hline
5 & 25 \\
\hline
6 & 36 \\
\hline
7 & 49 \\
\hline
8 & 64 \\
\hline
9 & 81 \\
\hline
10 & 100 \\
\end{array}
\quad \blacksquare
$$

The new MATHCAD Version 8 available to the author did not succeed with the specified methods in setting the value 1 as the starting value for the indexing; 0 was always used as the initial value.

+ *Representation* as *output table:*

$$x_i$$

1
4
9
16
25
36
49
64
81
100

This representation is achieved by entering the *numerical equal sign* = after x_i.

* *Representation* as *column vector* after **Matrix** has been set in the **Matrix display style** *field* of the **Result Format** *dialogue box*.

$$x = \begin{bmatrix} 1 \\ 4 \\ 9 \\ 16 \\ 25 \\ 36 \\ 49 \\ 64 \\ 81 \\ 100 \end{bmatrix} \blacksquare$$

♦

15.1.2 Input of Matrices

We consider *mxn matrices*, i.e., matrices with *m rows* and *n columns* having the form

$$\begin{pmatrix} a_{11} & a_{12} & \cdots & a_{1n} \\ a_{21} & a_{22} & \cdots & a_{2n} \\ \vdots & \vdots & \vdots & \vdots \\ a_{m1} & a_{m2} & \cdots & a_{mn} \end{pmatrix}$$

MATHCAD provides the following three *methods* for the *input* of *matrices* in the worksheet using the keyboard:

I.   *Direct input* for matrices with not more than 10 rows and columns:
     After activating the *menu sequence*

     Insert ⇒ Matrix...

     or
     clicking the *matrix button*

     in *operator palette no. 4 (vector and matrix palette)*
     causes the

     Insert Matrix

     dialogue box to appear, in which the *number* of *rows* and *columns* for the matrix should be entered in

     Rows:

     and

Columns:

respectively. *Clicking* the

Insert

button creates a *matrix* **A** of the *form* (e.g., for 5 rows and 6 columns):

$$\mathbf{A} := \begin{pmatrix} \bullet & \bullet & \bullet & \bullet & \bullet & \bullet \\ \bullet & \bullet & \bullet & \bullet & \bullet & \bullet \\ \bullet & \bullet & \bullet & \bullet & \bullet & \bullet \\ \bullet & \bullet & \bullet & \bullet & \bullet & \bullet \\ \bullet & \bullet & \bullet & \bullet & \bullet & \bullet \end{pmatrix}$$

in the worksheet at the cursor position. You enter the specific *elements*

a_{ik} (the designation $A_{i,k}$ is used in MATHCAD)

in its *placeholders* from the keyboard. A mouse click or the

-key

can be used to switch between the placeholders.

This dialogue box is also used to *insert* or *delete* rows/columns in a given matrix.

II. *Input* with an *assignment* using range variables i and k of the form (nested loops)

i := 1 .. m   k := 1 .. n

$A_{i,k} := f(i, k)$

where the general *elements*

$A_{i,k}$

of the *matrix* **A** are entered using the

button from *operator palette no. 1* or *4*.

This creates a m×n matrix **A** for which the function f(i,k) specifies the *rules* used to *define* the *elements* of the *matrix* **A** that can depend on the row index i and column index k (see Example 15.2b).

This method can also be used to enter *matrices* that have more than 10 rows and columns.

However, the *display* is made as *scrolling output table* unless you have set *display* as *matrix* by setting

Matrix

in the

Matrix display style

field in the *dialogue box* that appears when you select the *menu sequence*

Format ⇒ Result...

III. *Input* using the *menu sequence*

Insert ⇒ Component ⇒ Input Table

that displays a *table* with a placeholder in the worksheet. Enter the matrix designation in the placeholder. The number of rows and columns for the table can be set by moving while keeping the mouse button pressed or with the cursor buttons. Finally, the *elements* of the *matrix* are *entered* in the table.

This procedure can also be used to enter *matrices* that have more than 10 rows and columns.

Reading from disk or diskette (see Section 9.1) is another method of entering matrices that have more than 10 rows and columns.

♦

It is desirable to use the *assignment operators* := or ≡ to assign a *matrix symbol* or *matrix name* (normally an uppercase letter **A**, **B**, ...) to a matrix that can be used later for calculations (see Example 15.2). Whereas the := operator performs a *local* assignment, the ≡ operator performs a *global* assignment (see Section 10.3).

♦

Example 15.2:

The following examples all assume

ORIGIN:=1

i.e., the indexing starts at 1.

a) Although *method I.* is most often used to *create matrices*, it can be used only for matrices not having more than 10 rows and columns.
   A matrix **A** created using *method I.* has the following general form in the worksheet:

$$A := \begin{bmatrix} 1 & 3 & 5 \\ 7 & 9 & 2 \end{bmatrix}$$

b) We use *method II.* to create a matrix **A** with 14 rows and 12 columns:

   i := 1 .. 14   k := 1 .. 12

   $\mathbf{A}_{i,k} := i - k$

   Representation as *scrolling output table*

	2	3	4	5	6	7	8	9	10	11	12
1	-1	-2	-3	-4	-5	-6	-7	-8	-9	-10	-11
2	0	-1	-2	-3	-4	-5	-6	-7	-8	-9	-10
3	1	0	-1	-2	-3	-4	-5	-6	-7	-8	-9
4	2	1	0	-1	-2	-3	-4	-5	-6	-7	-8
5	3	2	1	0	-1	-2	-3	-4	-5	-6	-7
6	4	3	2	1	0	-1	-2	-3	-4	-5	-6
7	5	4	3	2	1	0	-1	-2	-3	-4	-5
8	6	5	4	3	2	1	0	-1	-2	-3	-4
9	7	6	5	4	3	2	1	0	-1	-2	-3
10	8	7	6	5	4	3	2	1	0	-1	-2
11	9	8	7	6	5	4	3	2	1	0	-1
12	10	9	8	7	6	5	4	3	2	1	0
13	11	10	9	8	7	6	5	4	3	2	1
14	12	11	10	9	8	7	6	5	4	3	2

A = (above table)   ∎

The new MATHCAD Version 8 available to the author did not succeed with the specified methods in setting the value 1 as the starting value for the indexing, 0 was always used as the initial value.

Display as matrix:

$$
A = \begin{bmatrix}
0 & -1 & -2 & -3 & -4 & 5 & -6 & -7 & -8 & -9 & -10 & -11 \\
1 & 0 & -1 & -2 & -3 & -4 & -5 & -6 & -7 & -8 & -9 & -10 \\
2 & 1 & 0 & -1 & -2 & -3 & -4 & -5 & -6 & 7 & -8 & -9 \\
3 & 2 & 1 & 0 & -1 & -2 & -3 & -4 & -5 & -6 & -7 & -8 \\
4 & 3 & 2 & 1 & 0 & -1 & -2 & -3 & -4 & -5 & -6 & -7 \\
5 & 4 & 3 & 2 & 1 & 0 & -1 & -2 & -3 & -4 & -5 & -6 \\
6 & 5 & 4 & 3 & 2 & 1 & 0 & -1 & -2 & -3 & -4 & -5 \\
7 & 6 & 5 & 4 & 3 & 2 & 1 & 0 & -1 & -2 & -3 & -4 \\
8 & 7 & 6 & 5 & 4 & 3 & 2 & 1 & 0 & -1 & -2 & -3 \\
9 & 8 & 7 & 6 & 5 & 4 & 3 & 2 & 1 & 0 & -1 & -2 \\
10 & 9 & 8 & 7 & 6 & 5 & 4 & 3 & 2 & 1 & 0 & -1 \\
11 & 10 & 9 & 8 & 7 & 6 & 5 & 4 & 3 & 2 & 1 & 0 \\
12 & 11 & 10 & 9 & 8 & 7 & 6 & 5 & 4 & 3 & 2 & 1 \\
13 & 12 & 11 & 10 & 9 & 8 & 7 & 6 & 5 & 4 & 3 & 2
\end{bmatrix} \quad \blacksquare
$$

c) We use *method III.* to create a *matrix* **A** with 4 rows and 3 columns using the menu sequence

Insert ⇒ Component ⇒ Input Table

We enter the elements of the matrix in the displayed table and the name of the matrix in the upper left-hand placeholder:

We obtain the following *matrix:*

$$
A = \begin{bmatrix} 1 & 2 & 3 \\ 4 & 5 & 6 \\ 7 & 8 & 9 \\ 10 & 11 & 12 \end{bmatrix} \quad \blacksquare
$$

◆

15.2 Vector and Matrix Functions

MATHCAD contains *vector* and *matrix functions* that can be used to perform certain calculations on *vectors* **v** and *matrices* **A** and **B**. The most important functions:

*   **augment (A , B)**

    create a *new matrix* in which the *columns* of **A** and **B** are written next to
    each other (the function can be used only when **A** and **B** both have the
    same number of rows).

*   **cols (A)**

    calculate the *number* of *columns* for *matrix* **A**.

*   **diag (v)**

    create a *diagonal matrix,* whose *diagonals* are formed by the *vector* **v**.

*   **identity** (n)

    create an *identity matrix* (*unit matrix*) having n rows and n columns.

*   **last (v)**

    determine the *index* of the *last element* of *vector* **v**.

*   **length (v)**

    calculate the *number* of *components* of *vector* **v**.

* **max (A)**

  calculate the *maximum* of the *elements* of *matrix* **A**.

* **min (A)**

  calculate the *minimum* of the *elements* of *matrix* **A**.

* **rank (A)**

  calculate the *rank* of *matrix* **A**.

* **rows (A)**

  calculate the *number* of *rows* of *matrix* **A**.

* **stack (A , B)**

  form a *new matrix* in which the *rows* of **A** and **B** are under each other (the function can be used only when **A** and **B** both have the same number of columns).

* **submatrix (A , i , j , k , l)**

  form a *submatrix* of **A** that contains the rows i to j and the columns k to l, where i ≤ j and k ≤ l.

* **tr (A)**

  calculate the *trace* of *matrix* **A**.

Before these functions can be used, the appropriate matrices/vectors must have been assigned to **A, B** and **v** or the matrices/vectors specified directly as function arguments instead of **A, B** or **v**.

You can obtain the function *result* by marking the *function expression* with *editing lines* and then initiating the *numerical calculation* using the *numerical equal sign* = or the *menu sequence*

Symbolics ⇒ Evaluate ⇒ Floating Point...

◆

Because the *exact calculation* using the *symbolic equal sign* → or by activating the *menu sequence*

Symbolics ⇒ Evaluate ⇒ Symbolically

does not function for all *vector* and *matrix functions*, it is better to use the numerical calculation.

◆

Example 15.3:

Let us now try the given functions for the following matrices and vectors; the indexing starts at 1, i.e., **ORIGIN** := 1.

$$
\mathbf{A} := \begin{pmatrix} 1 & 2 & 3 \\ 4 & 5 & 6 \\ 7 & 8 & 9 \end{pmatrix} \qquad \mathbf{B} := \begin{pmatrix} 3 & 4 & 7 \\ 1 & 5 & 2 \end{pmatrix} \qquad \mathbf{C} := \begin{pmatrix} 9 & 11 \\ 21 & 34 \\ 7 & 43 \end{pmatrix}
$$

$$D := \begin{pmatrix} 5 & 7 & 11 \\ 21 & 45 & 7 \\ 9 & 3 & 1 \end{pmatrix} \qquad v := \begin{pmatrix} 3 \\ 5 \\ 9 \end{pmatrix} \qquad w := \begin{pmatrix} 1 \\ 4 \\ 8 \\ 3 \end{pmatrix}$$

a) Determine the *number* of *rows*:

   rows (B) = 2   **rows (w)** = 4

b) Determine the *number* of *columns*:

   cols (A) = 3   **cols (v)** = 1

c) Create an 5×5 *identity matrix* (*unit matrix*):

$$\text{identity } (5) \;=\; \begin{vmatrix} 1 & 0 & 0 & 0 & 0 \\ 0 & 1 & 0 & 0 & 0 \\ 0 & 0 & 1 & 0 & 0 \\ 0 & 0 & 0 & 1 & 0 \\ 0 & 0 & 0 & 0 & 1 \end{vmatrix}$$

d) Calculate the *rank*:

   rank (A) = 2   **rank (B)** = 2   **rank (C)** = 2   **rank (D)** = 3

e) Create a *diagonal matrix* whose diagonal is formed from vector **w**:

$$\text{diag } (w) \;=\; \begin{pmatrix} 1 & 0 & 0 & 0 \\ 0 & 4 & 0 & 0 \\ 0 & 0 & 8 & 0 \\ 0 & 0 & 0 & 3 \end{pmatrix}$$

f) Determine the *maximum* and *minimum elements*:

   max (D) = 45   **max (v)** = 9   **min (A)** = 1

g) Calculate the *number* of *components* for **w** and the index of the *last component* of **v**:

   length (w) = 4   **last (v)** = 3

h) *Extend* the given *matrix* **A** with the *vector* **v**:

$$\text{augment } (A, v) \;=\; \begin{pmatrix} 1 & 2 & 3 & 3 \\ 4 & 5 & 6 & 5 \\ 7 & 8 & 9 & 9 \end{pmatrix}$$

i) Form the *union* of the two *matrices* **A** and **D** to produce a new matrix:

$$\text{stack}(\mathbf{A}, \mathbf{D}) = \begin{pmatrix} 1 & 2 & 3 \\ 4 & 5 & 6 \\ 7 & 8 & 9 \\ 5 & 7 & 11 \\ 21 & 45 & 7 \\ 9 & 3 & 1 \end{pmatrix}$$

j) Create a *submatrix* of matrix **A** :

$$\text{submatrix}\left(\begin{pmatrix} 1 & 2 & 3 \\ 4 & 5 & 6 \\ 7 & 8 & 9 \end{pmatrix}, 2, 3, 1, 2 \right) = \begin{pmatrix} 4 & 5 \\ 7 & 8 \end{pmatrix}$$

or

$$\text{submatrix}(\mathbf{A}, 2, 3, 1, 2) = \begin{pmatrix} 4 & 5 \\ 7 & 8 \end{pmatrix}$$

♦

15.3 Computational Operations

Before we discuss the usual computational operations with matrices as used with MATHCAD, we must first consider how you can extract elements, columns or rows from a matrix defined in the worksheet.

If you wish to use *elements* of a *matrix* **A** defined in the worksheet, these must have the same designation as the matrix, the two indices must be separated with a comma. These elements are created using the

button from *operator palette no. 1* or *4* (the indices can also be enclosed within parentheses). Thus,

$$A_{i, k} \qquad \text{or} \qquad A_{(i, k)}$$

must be entered for the element from the i-th row and the k-th column of matrix **A**.

This differs from the mathematical notation where the elements of a matrix **A** are designated as a_{ik}.

♦

You can *extract columns* (column vectors) from a given *matrix* **A** by clicking the

button in *operator palette no. 4.* The following symbol appears in the worksheet

$\blacksquare^{<\blacksquare>}$

The name of the matrix **A** is entered in its large lower placeholder and the number of the required column entered in its small upper placeholder.

You must then mark the expression with editing lines, enter the symbolic or numerical equal sign and press the ⏎ key.

If you wish to extract *rows* (rows vectors) from a *matrix* **A**, you can also use the previously described method provided you have *transposed* the *matrix beforehand* (see Section 15.3.2).

◆

MATHCAD automatically recognizes an attempt to use elements/columns that lie outside the defined indices and issues an appropriate *error message.*

◆

The following example illustrates the procedure for the above described operations.

Example 15.4:

Let us consider the following *matrix:*

$$\mathbf{A} := \begin{pmatrix} 3 & 12 & 6 & 9 & 0 \\ 1 & 23 & 35 & 2 & 7 \\ 9 & 2 & 43 & 3 & 8 \\ 3 & 5 & 71 & 27 & 49 \end{pmatrix}$$

a) With the *starting value* 0 for the *indexing,* i.e.,

ORIGIN := 0

the extraction

• *of elements provides the following capabilities:*

    * *with numerical equal sign*

$$\mathbf{A}_{1,2} = 35 \qquad \mathbf{A}_{0,0} = 3 \qquad \mathbf{A}_{3,4} = 49$$

$$\mathbf{A}_{(1,2)} = 35 \qquad \mathbf{A}_{(0,0)} = 3 \qquad \mathbf{A}_{(3,4)} = 49$$

    * *with symbolic equal sign*

$$A_{1,2} \to 35 \qquad A_{0,0} \to 3 \qquad A_{3,4} \to 49$$

$$A_{(1,2)} \to 35 \qquad A_{(0,0)} \to 3 \qquad A_{(3,4)} \to 49$$

- *of columns provides the following capabilities:*
  * *with numerical equal sign*

$$A^{<0>} = \begin{pmatrix} 3 \\ 1 \\ 9 \\ 3 \end{pmatrix} \qquad A^{<2>} = \begin{pmatrix} 6 \\ 35 \\ 43 \\ 71 \end{pmatrix}$$

  * *with symbolic equal sign*

$$A^{<0>} \to \begin{pmatrix} 3 \\ 1 \\ 9 \\ 3 \end{pmatrix} \qquad A^{<2>} \to \begin{pmatrix} 6 \\ 35 \\ 43 \\ 71 \end{pmatrix}$$

b) With the *starting value* 1 for the *indexing*, i.e.,

ORIGIN := 1

the extraction

- *of elements provides the following capabilities:*
  * *with numerical equal sign*

$$A_{2,3} = 35 \qquad A_{1,1} = 3 \qquad A_{4,5} = 49$$

$$A_{(2,3)} = 35 \qquad A_{(1,1)} = 3 \qquad A_{(4,5)} = 49$$

  * *with symbolic equal sign*

$$A_{2,3} \to 35 \qquad A_{1,1} \to 3 \qquad A_{4,5} \to 49$$

$$A_{(2,3)} \to 35 \qquad A_{(1,1)} \to 3 \qquad A_{(4,5)} \to 49$$

- *of columns provides the following capabilities:*
  * *with numerical equal sign*

$$A^{<1>} = \begin{pmatrix} 3 \\ 1 \\ 9 \\ 3 \end{pmatrix} \qquad A^{<3>} = \begin{pmatrix} 6 \\ 35 \\ 43 \\ 71 \end{pmatrix}$$

* *with symbolic equal sign*

$$\mathbf{A}^{<1>} \rightarrow \begin{pmatrix} 3 \\ 1 \\ 9 \\ 3 \end{pmatrix} \qquad \mathbf{A}^{<3>} \rightarrow \begin{pmatrix} 6 \\ 35 \\ 43 \\ 71 \end{pmatrix}$$

◆

In the following sections we discuss the procedure used in MATHCAD to perform *computational operations* with *matrices*.

Note for *computational operations* that

* *addition* (subtraction) $\mathbf{A} \pm \mathbf{B}$

    is possible only when the $\mathbf{A}$ and $\mathbf{B}$ *matrices* have the *same type (m,n)*

* *multiplication* $\mathbf{A} \cdot \mathbf{B}$

    is possible only when the $\mathbf{A}$ and $\mathbf{B}$ *matrices* are *chained*, i.e., $\mathbf{A}$ must have the same number of columns as the number of rows in $\mathbf{B}$

* *formation* of *inverses*

    is possible only for *non-singular square matrices*.

    ◆

For practical reasons, assign letters $\mathbf{A}$, $\mathbf{B}$, $\mathbf{C}$, ... as matrix symbols or names to the matrices before performing computational operations and then use these letters for the associated operations.

You can obviously assign a new matrix name to the result of such a computational operation.

◆

MATHCAD issues an error message if the type of the matrices does not agree with the operation to be performed.

◆

Although MATHCAD can also perform the addition, multiplication and transposition for larger matrices without difficulty, because the computational effort and memory requirements grow rapidly with size, you quickly meet difficulties *calculating determinants* and *inverses* for an n-row square matrix for large n.

◆

15.3.1   Addition and Multiplication

MATHCAD performs the *addition* and *multiplication* of two *matrices* **A** and **B** in the worksheet as follows:

- Enter

  A + B

  or

  A * B

  in the worksheet and mark the expression with editing lines.
- The *input* of the *symbolic equal sign* → or *numerical equal sign* = with the subsequent pressing of the ⏎-key returns the *exact* or *numerical result* respectively.

  The *menu sequence* for the *exact* or *numerical calculation*

  Symbolics ⇒ Evaluate ⇒...

  returns a *result* only when the matrix was *entered directly*.

MATHCAD also permits the

*   *multiplication* of a *matrix* **A** with a *scalar* (number) t, i.e.,

  t · **A**   or   **A** · t

*   *addition* of a *matrix* **A** with a *scalar* (number) t, i.e.,

  t + **A**   or   **A** + t

During the *multiplication*, every element of **A** is multiplied by the *scalar* t; during the *addition*, the scalar t is added to every element of **A**. These two operations are performed in a similar manner as the operations between two matrices; you enter the appropriate scalar instead of one of the matrices.

◆

Example 15.5:

a) Let us consider the addition and multiplication of matrices that have numbers as elements:

$$A := \begin{pmatrix} 1 & 2 & 3 \\ 4 & 5 & 6 \end{pmatrix} \quad B := \begin{pmatrix} 5 & 8 & 5 \\ 8 & 9 & 2 \end{pmatrix} \quad C := \begin{pmatrix} 1 & 2 \\ 3 & 4 \\ 7 & 8 \end{pmatrix}$$

$$A + B = \begin{pmatrix} 6 & 10 & 8 \\ 12 & 14 & 8 \end{pmatrix} \qquad A \cdot C = \begin{pmatrix} 28 & 34 \\ 61 & 76 \end{pmatrix}$$

If we use the *symbolic equal sign* instead of the *numerical equal sign*, we obtain for integers the same form of result:

$$A + B \rightarrow \begin{pmatrix} 6 & 10 & 8 \\ 12 & 14 & 8 \end{pmatrix} \qquad A \cdot C \rightarrow \begin{pmatrix} 28 & 34 \\ 61 & 76 \end{pmatrix}$$

b) If the matrices do not just contain integers but also fractions as elements, you can recognize the *difference* between the use of the *numerical equal sign* and the *symbolic equal sign:*

$$A := \begin{pmatrix} \dfrac{1}{2} & \dfrac{1}{3} \\ \dfrac{1}{4} & \dfrac{1}{5} \end{pmatrix} \qquad B := \begin{pmatrix} 1 & \dfrac{2}{7} \\ 3 & 2 \end{pmatrix}$$

$$A + B = \begin{pmatrix} 1.5 & 0.619 \\ 3.25 & 2.2 \end{pmatrix} \qquad A + B \rightarrow \begin{pmatrix} \dfrac{3}{2} & \dfrac{13}{21} \\ \dfrac{13}{4} & \dfrac{11}{5} \end{pmatrix}$$

$$A \cdot B = \begin{pmatrix} 1.5 & 0.81 \\ 0.85 & 0.471 \end{pmatrix} \qquad A \cdot B \rightarrow \begin{pmatrix} \dfrac{3}{2} & \dfrac{17}{21} \\ \dfrac{17}{20} & \dfrac{33}{70} \end{pmatrix}$$

c) Let us consider the *addition* and *multiplication* of a *matrix* A *with a scalar*:

$$A := \begin{pmatrix} 1 & 2 & 3 \\ 4 & 5 & 6 \end{pmatrix}$$

$$1 + A = \begin{pmatrix} 2 & 3 & 4 \\ 5 & 6 & 7 \end{pmatrix} \qquad A + 1 = \begin{pmatrix} 2 & 3 & 4 \\ 5 & 6 & 7 \end{pmatrix}$$

$$1 + A \rightarrow \begin{pmatrix} 2 & 3 & 4 \\ 5 & 6 & 7 \end{pmatrix} \qquad A + 1 \rightarrow \begin{pmatrix} 2 & 3 & 4 \\ 5 & 6 & 7 \end{pmatrix}$$

$$6 \cdot A = \begin{pmatrix} 6 & 12 & 18 \\ 24 & 30 & 36 \end{pmatrix} \qquad A \cdot 6 = \begin{pmatrix} 6 & 12 & 18 \\ 24 & 30 & 36 \end{pmatrix}$$

$$6 \cdot A \rightarrow \begin{pmatrix} 6 & 12 & 18 \\ 24 & 30 & 36 \end{pmatrix} \qquad A \cdot 6 \rightarrow \begin{pmatrix} 6 & 12 & 18 \\ 24 & 30 & 36 \end{pmatrix}$$

Instead of a number value for the scalar, the symbolic equal sign also permits the use of a variable (e.g., t):

$$t + A \rightarrow \begin{pmatrix} 1+t & 2+t & 3+t \\ 4+t & 5+t & 6+t \end{pmatrix} \qquad A + t \rightarrow \begin{pmatrix} 1+t & 2+t & 3+t \\ 4+t & 5+t & 6+t \end{pmatrix}$$

$$t + \begin{pmatrix} 1 & 2 & 3 \\ 4 & 5 & 6 \end{pmatrix} \rightarrow \begin{pmatrix} 1+t & 2+t & 3+t \\ 4+t & 5+t & 6+t \end{pmatrix}$$

$$\begin{pmatrix} 1 & 2 & 3 \\ 4 & 5 & 6 \end{pmatrix} + t \rightarrow \begin{pmatrix} 1+t & 2+t & 3+t \\ 4+t & 5+t & 6+t \end{pmatrix}$$

$$t \cdot \mathbf{A} \rightarrow \begin{pmatrix} t & 2 \cdot t & 3 \cdot t \\ 4 \cdot t & 5 \cdot t & 6 \cdot t \end{pmatrix} \qquad \mathbf{A} \cdot t \rightarrow \begin{pmatrix} t & 2 \cdot t & 3 \cdot t \\ 4 \cdot t & 5 \cdot t & 6 \cdot t \end{pmatrix}$$

$$t \cdot \begin{pmatrix} 1 & 2 & 3 \\ 4 & 5 & 6 \end{pmatrix} \rightarrow \begin{pmatrix} t & 2 \cdot t & 3 \cdot t \\ 4 \cdot t & 5 \cdot t & 6 \cdot t \end{pmatrix}$$

$$\begin{pmatrix} 1 & 2 & 3 \\ 4 & 5 & 6 \end{pmatrix} \cdot t \rightarrow \begin{pmatrix} t & 2 \cdot t & 3 \cdot t \\ 4 \cdot t & 5 \cdot t & 6 \cdot t \end{pmatrix}$$

d) If you assign the result of an addition or multiplication of matrices to a new matrix name, you can also perform these operations using its definition:

$$\mathbf{A} := \begin{pmatrix} 1 & 2 \\ 3 & 4 \end{pmatrix} \qquad \mathbf{B} := \begin{pmatrix} 5 & 6 \\ 7 & 8 \end{pmatrix}$$

$$i := 1..2 \qquad k := 1..2$$

$$\mathbf{C}_{i,k} := \mathbf{A}_{i,k} + \mathbf{B}_{i,k} \qquad \mathbf{D}_{i,k} := \sum_{j=1}^{2} \mathbf{A}_{i,j} \cdot \mathbf{B}_{j,k}$$

$$\mathbf{C} = \begin{pmatrix} 6 & 8 \\ 10 & 12 \end{pmatrix} \qquad \mathbf{D} = \begin{pmatrix} 19 & 22 \\ 43 & 50 \end{pmatrix}$$

◆

15.3.2 Transposition

Transposition is a further operation for matrices, i.e., the *exchange* of *rows* and *columns*.

MATHCAD permits transposition of a matrix **A** in one of the following *ways*:

I. *Mark* the directly entered *matrix* with *editing lines* and use the *menu sequence*

Symbolics ⇒ Matrix ⇒ Transpose

to return the exact result.

II. *Use* the

button from *operator palette no. 4* to display the

$$\blacksquare^{T}$$

symbol at the required position in the worksheet and enter **A** in the symbol's *placeholder* (where the appropriate matrix has been previously assigned to **A**). The matrix can also be directly entered here in place of **A**

Then *mark* the *complete* expression with *editing lines.*

* *Enter* the *symbolic equal sign* → or *numerical equal sign* = and then press the ⏎-key to return the *exact* or *numerical result* respectively.

* The *menu sequence* for the *exact* or *numerical calculation*

Symbolics ⇒ Evaluate ⇒...

returns a *result* only when the matrix was entered directly.

15.3.3 Inverse

The *calculation* of the *inverse*

$$\mathbf{A}^{-1}$$

of a given *matrix*

A

is possible only for square (n-row) matrices where

det **A** ≠ 0

must also be satisfied (*non-singular matrix*).

The *algorithms* provided for the *calculation* of *inverses* become very time consuming for large values of n. However, MATHCAD is very helpful for the calculation, provided n is not too large.

♦

MATHCAD permits the calculation of an inverse in one of the following *ways:*

I. *Mark* the directly entered *matrix* **A** with *editing lines* and use the *menu sequence*

Symbolics ⇒ Matrix ⇒ Invert

to return the *exact result.*

II. Enter

$\mathbf{A}^{-1}$

and *mark* this *expression* with *editing lines*. Enter the *symbolic equal sign* → and then press the ⏎-key to perform the *exact calculation* (the appropriate matrix must have been previously assigned to **A** or the matrix was entered directly instead of **A**).

III. Enter

$\mathbf{A}^{-1}$

and *mark* this *expression* with *editing lines*. The *menu sequence* for *exact calculation*

Symbolics ⇒ Evaluate ⇒ Symbolically

returns a *result* only when the matrix **A** was entered directly.

IV. Enter

$\mathbf{A}^{-1}$

and *mark* this *expression* with *editing lines*. The input of the *numerical equal sign* = *calculates numerically* the inverse (the appropriate matrix must have been previously assigned to **A** or the matrix was entered directly instead of **A**).

V. Enter

$\mathbf{A}^{-1}$

and *mark* this *expression* with *editing lines*. The *menu sequence* for *numerical calculation*

Symbolics ⇒ Evaluate ⇒ Floating Point...

returns a *result* only when the matrix **A** was entered directly.

☞

MATHCAD issues an error message if the *matrix* to be inverted is *singular* or *non-square*.

♦

☞

To *check* the calculated inverse, we suggest that you calculate the products $\mathbf{A}\cdot\mathbf{A}^{-1}$ or $\mathbf{A}^{-1}\cdot\mathbf{A}$ that must return the unit matrix **E**.

♦

☞

The *calculation* of *inverses* is a special case of the calculation of integer powers $\mathbf{A}^n$ of a matrix **A** that MATHCAD also supports (see Example 15.6b).

♦

Example 15.6:

This example uses the following *non-singular matrix*

$$A := \begin{bmatrix} 1 & 2 \\ 5 & 4 \end{bmatrix}$$

a) Calculate the inverse using *methods II.* and *IV.*

$$A^{-1} \rightarrow \begin{bmatrix} \dfrac{-2}{3} & \dfrac{1}{3} \\[2mm] \dfrac{5}{6} & \dfrac{-1}{6} \end{bmatrix} \qquad A^{-1} = \begin{bmatrix} -0.667 & 0.333 \\ 0.833 & -0.167 \end{bmatrix} \; \blacksquare$$

Note, because of rounding errors, only an approximate unit matrix is produced for the numerical calculation when you perform a *check*:

$$\begin{bmatrix} 1 & 2 \\ 5 & 4 \end{bmatrix} \cdot \begin{bmatrix} \dfrac{-2}{3} & \dfrac{1}{3} \\[2mm] \dfrac{5}{6} & \dfrac{-1}{6} \end{bmatrix} \rightarrow \begin{bmatrix} 1 & 0 \\ 0 & 1 \end{bmatrix}$$

$$\begin{bmatrix} 1 & 2 \\ 5 & 4 \end{bmatrix} \cdot \begin{bmatrix} -0.667 & 0.333 \\ 0.833 & -0.167 \end{bmatrix} = \begin{bmatrix} 0.999 & -1 \cdot 10^{-3} \\ -3 \cdot 10^{-3} & 0.997 \end{bmatrix} \; \blacksquare$$

b) *Matrices* can be *raised* to a *power* without difficulty:

$$A^2 \rightarrow \begin{bmatrix} 11 & 10 \\ 25 & 26 \end{bmatrix} \qquad A^3 \rightarrow \begin{bmatrix} 61 & 62 \\ 155 & 154 \end{bmatrix}$$

$\blacklozenge$

15.3.4   Scalar, Vector and Scalar Triple Product

The following *products* can be *computed* for any (column) *vectors:*

$$\mathbf{a} = \begin{pmatrix} a_1 \\ \vdots \\ a_n \end{pmatrix} \qquad \mathbf{b} = \begin{pmatrix} b_1 \\ \vdots \\ b_n \end{pmatrix} \qquad \mathbf{c} = \begin{pmatrix} c_1 \\ \vdots \\ c_n \end{pmatrix}$$

- *Scalar product*

$$\mathbf{a} \circ \mathbf{b} = \sum_{i=1}^{n} a_i b_i$$

- *Vector product* (for n=3)

$$\mathbf{a} \times \mathbf{b} = \begin{vmatrix} \mathbf{i} & \mathbf{j} & \mathbf{k} \\ a_1 & a_2 & a_3 \\ b_1 & b_2 & b_3 \end{vmatrix} = (a_2 b_3 - a_3 b_2, a_3 b_1 - a_1 b_3, a_1 b_2 - a_2 b_1)$$

$$= (a_2b_3 - a_3b_2)\mathbf{i} + (a_3b_1 - a_1b_3)\mathbf{j} + (a_1b_2 - a_2b_1)\mathbf{k}$$

- *Scalar triple product* (for n=3)

$$(\mathbf{a} \times \mathbf{b}) \circ \mathbf{c} = \begin{vmatrix} a_1 & a_2 & a_3 \\ b_1 & b_2 & b_3 \\ c_1 & c_2 & c_3 \end{vmatrix}$$

If the vectors **a** and **b** have been previously entered in the worksheet as *column vectors*, MATHCAD calculates *scalar* and *vector products* as follows:

- *Calculation* of the *scalar product:*
  Direct input of

  a * b

  or using the

  button from *operator palette no. 4.* You enter here the vectors **a** and **b** in the two displayed *placeholders*

  ▪ · ▪

  i.e., a · b

- *Calculation* of the *vector product:*
  Direct input of

  a × b

  The multiplication sign × is realized with the Ctrl-⑧-key combination or using the

  button from *operator palette no. 4.* You enter here the vectors **a** and **b** in the two displayed *placeholders*

  ▪ × ▪

  i.e., **a × b**

Mark the entered scalar or vector product with editing lines, enter the *symbolic equal sign* → or the *numerical equal sign* = and then press the ⏎-key to return the *result.* You can achieve the same result by entering the *menu sequence*

Symbolics ⇒ Evaluate ⇒...

to perform the *exact* or *numerical calculation.*

Because the *scalar triple product* can be obtained using the calculation of the given determinant or by calculating the vector product **a × b** with the subsequent scalar multiplication with the vector **c**, the scalar triple product of three *vectors* **a, b** and **c** does not require any special technique.

♦

Example 15.7:

Calculate for the *vectors*

$$\mathbf{a} = \begin{pmatrix} 1 \\ 3 \\ 5 \end{pmatrix} \qquad \mathbf{b} = \begin{pmatrix} 1 \\ 3 \\ 7 \end{pmatrix} \qquad \mathbf{c} = \begin{pmatrix} 9 \\ 6 \\ 8 \end{pmatrix}$$

a) the *scalar product* **a** ∘ **b**

b) the *vector product* **a** × **b**

c) the *scalar triple product* (**a** × **b**) ∘ **c**

These calculations can be realized in the MATHCAD worksheet in the following form:

$$\mathbf{a} := \begin{pmatrix} 1 \\ 3 \\ 5 \end{pmatrix} \qquad \mathbf{b} := \begin{pmatrix} 1 \\ 3 \\ 7 \end{pmatrix} \qquad \mathbf{c} := \begin{pmatrix} 9 \\ 6 \\ 8 \end{pmatrix}$$

scalar product   **a** · **b** = 45

$$\text{vector product} \qquad \mathbf{a} \times \mathbf{b} = \begin{pmatrix} 6 \\ -2 \\ 0 \end{pmatrix}$$

scalar triple product   (**a** × **b**) · **c** = 42

♦

15.4 Determinants

Let us calculate the *determinant*

$$\det \mathbf{A} = \begin{vmatrix} a_{11} & a_{12} & \cdots & a_{1n} \\ a_{21} & a_{22} & \cdots & a_{2n} \\ \vdots & \vdots & \vdots & \vdots \\ a_{n1} & a_{n2} & \cdots & a_{nn} \end{vmatrix}$$

that is defined for the square *matrix* **A**

$$
\mathbf{A} \; = \; \begin{pmatrix} a_{11} & a_{12} & \cdots & a_{1n} \\ a_{21} & a_{22} & \cdots & a_{2n} \\ \vdots & \vdots & \vdots & \vdots \\ a_{n1} & a_{n2} & \cdots & a_{nn} \end{pmatrix}
$$

The algorithms available for the *calculation* of *determinants* (e.g., transformation to triangular form, use of the Laplace law of expansion) become very time-consuming for increasing values of n. However, MATHCAD is a great help for the calculation provided n is not too large.

♦

MATHCAD uses the *magnitude* (*absolute value*) *button*

from *operator palette no. 1* or *4* to form the *determinant* of a *matrix* **A**: If the *matrix* **A**

* is already in the worksheet, enter just the *designation* **A** of the matrix within the magnitude bars, i.e.,

  |**A**|

* is not yet in the worksheet, click the *matrix button*

  from *operator palette no. 4* to write the matrix symbol inside the magnitude bars; the *elements* are then entered in the free placeholders.

  ♦

MATHCAD can *calculate* the *determinant* of a *matrix* **A** using the following methods:

I. *Mark* the directly entered *matrix* **A** with *editing lines* and activate the *menu sequence*

   Symbolics ⇒ Matrix ⇒ Determinant

   to return the *exact calculation*.

II. *Enter*

   |**A**|

   and *mark* with *editing lines*, enter the *symbolic equal sign* → and then *press* the ⏎-key to *calculate* the *determinant exactly*; the appropriate matrix must have been previously assigned to **A** or the matrix entered directly.

III. After *marking* |**A**| with *editing lines*, the *menu sequence* for *exact calculation*

Symbolics ⇒ Evaluate ⇒ Symbolically

returns a *result* only when the matrix **A** has been entered directly.

IV. After entering

|**A**|

and *marking* with *editing lines*, enter the *numerical equal sign* = to *calculate* the *determinant numerically*, the appropriate matrix must have been previously assigned to **A** or the matrix entered directly.

V. After *marking* |**A**| with *editing lines*, the *menu sequence* for *numerical calculation*

Symbolics ⇒ Evaluate ⇒ Floating Point...

returns a *result* only when the matrix **A** has been entered directly.

MATHCAD issues an *error message* if you inadvertently try to calculate the *determinant* of a *non-square matrix*.

We now demonstrate in the following example the use of the provided methods to calculate determinants.

Example 15.8:

a) The *exact calculation* of the *determinant* of a matrix using the **Symbolics** menu, i.e., with *method I.*, is successful only for a directly entered matrix:

$$\begin{pmatrix} 1 & 2 & 3 \\ 4 & 6 & 5 \\ 7 & 1 & 3 \end{pmatrix} \qquad \textit{has determinant} \qquad -55$$

b) The *determinant* of the matrix from example a is now calculated exactly using the symbolic equal sign → (*method II.*).
   This functions using both:

$$A := \begin{pmatrix} 1 & 2 & 3 \\ 4 & 6 & 5 \\ 7 & 1 & 3 \end{pmatrix} \qquad |A| \rightarrow -55$$

and also with the direct use of the entered determinant:

$$\left| \begin{pmatrix} 1 & 2 & 3 \\ 4 & 6 & 5 \\ 7 & 1 & 3 \end{pmatrix} \right| \rightarrow -55$$

c) We now perform the *numerical calculation* of the *determinant* for the matrix from example a (method IV.). This functions both with the input of the *numerical equal sign* =:

$$A := \begin{pmatrix} 1 & 2 & 3 \\ 4 & 6 & 5 \\ 7 & 1 & 3 \end{pmatrix} \qquad |A| = -55$$

and also with the direct use of the entered determinant:

$$\left| \begin{pmatrix} 1 & 2 & 3 \\ 4 & 6 & 5 \\ 7 & 1 & 3 \end{pmatrix} \right| = -55$$

♦

15.5   Eigenvalues and Eigenvectors

The *calculation* of *eigenvalues* λ and the associated *eigenvectors* is another important task involving square matrices **A**.
The *eigenvalues* for a matrix **A** are those real or complex numbers
$$\lambda_i$$
that have non-trivial (i.e., non-zero) solution vectors
$$\mathbf{x}^i$$
for the system of linear homogeneous equations
$$(\mathbf{A} - \lambda_i \cdot \mathbf{E}) \cdot \mathbf{x}^i = 0$$
These solutions are known as *eigenvectors*.
Therefore, the eigenvalues
$$\lambda_i$$
are determined as solutions of the *characteristic polynomial*
$$\det(\mathbf{A} - \lambda \cdot \mathbf{E}) = 0$$
and MATHCAD must solve for every eigenvalue the given system of linear homogeneous equations.
MATHCAD provides the following *functions* to *calculate eigenvalues* and *eigenvectors* for a matrix **A** :

I.   **eigenvals (A)**

   to *calculate* the *eigenvalues*

II. **eigenvec (A, λ)**

   to *calculate* the *eigenvector* associated with the *eigenvalue* λ

III. **eigenvecs (A)**

to *calculate all eigenvectors.*

Before using these functions, the required matrix must have been assigned to **A** or the matrix directly entered instead of **A**.

Entering the *numerical equal sign* = after the function returns the *numerical result.*

Although *eigenvalues* can be *calculated exactly* if you enter the *symbolic equal sign* → after the function **eigenvals**, because the exact calculation often fails for the following reasons, this calculation method is not recommended.

♦

Note for all calculations with MATHCAD

* *The eigenvalues are the zeros of the characteristic polynomial of degree n (for an n-row matrix A). This leads to the problems discussed in Section 16.2 for n ≥ 5.*

* *Eigenvectors are determined only to one factor, which MATHCAD normalizes to the length 1.*

♦

Let us now consider in the following example the calculation of eigenvalues and eigenvectors.

Example 15.9:

a) Calculate the eigenvalues and associated eigenvectors for the following *matrix*

$$\mathbf{A} := \begin{pmatrix} 2 & -5 \\ 1 & -4 \end{pmatrix}$$

$$\text{eigenvals}(\mathbf{A}) = \begin{pmatrix} 1 \\ -3 \end{pmatrix} \qquad \text{eigenvecs}(\mathbf{A}) = \begin{pmatrix} 0.981 & 0.707 \\ 0.196 & 0.707 \end{pmatrix}$$

$$\text{eigenvec}(\mathbf{A},1) = \begin{pmatrix} -0.981 \\ -0.196 \end{pmatrix} \qquad \text{eigenvec}(\mathbf{A},-3) = \begin{pmatrix} 0.707 \\ 0.707 \end{pmatrix}$$

b) For the *matrix* **A** with the triple eigenvalue 2

$$\mathbf{A} := \begin{pmatrix} 3 & 4 & 3 \\ -1 & 0 & -1 \\ 1 & 2 & 3 \end{pmatrix}$$

MATHCAD returns:

$$\text{eigenvals}(\mathbf{A}) = \begin{pmatrix} 2 \\ 2 \\ 2 \end{pmatrix} \qquad \text{eigenvecs}(\mathbf{A}) = \begin{pmatrix} 0.577 & -0.577 & 0.577 \\ -0.577 & 0.577 & -0.577 \\ 0.577 & -0.577 & 0.577 \end{pmatrix}$$

i.e., the single eigenvector (without normalization)

$$\begin{pmatrix} 1 \\ -1 \\ 1 \end{pmatrix}$$

whereas MATHCAD calculates an *incorrect eigenvector* with the **eigenvec** function

$$\text{eigenvec}(\mathbf{A}, 2) = \begin{pmatrix} 1 \\ 0 \\ 0 \end{pmatrix}$$

c) For the *symmetric matrix,*

$$\mathbf{A} := \begin{pmatrix} 1 & \sqrt{2} & -\sqrt{6} \\ \sqrt{2} & 2 & \sqrt{3} \\ -\sqrt{6} & \sqrt{3} & 0 \end{pmatrix}$$

MATHCAD returns:

$$\text{eigenvals}(\mathbf{A}) = \begin{pmatrix} 3 \\ 3 \\ -3 \end{pmatrix} \qquad \text{eigenvecs}(\mathbf{A}) = \begin{pmatrix} -0.265 & 0.772 & 0.577 \\ 0.725 & 0.554 & -0.408 \\ 0.635 & -0.31 & 0.707 \end{pmatrix}$$

$$\text{eigenvec}(\mathbf{A}, -3) = \begin{pmatrix} 0.577 \\ -0.408 \\ 0.707 \end{pmatrix} \qquad \text{eigenvec}(\mathbf{A}, 3) = \begin{pmatrix} -0.373 \\ 0.639 \\ 0.673 \end{pmatrix}$$

Here the **eigenvec** function calculates only a single eigenvector for the double eigenvalue 3, whereas **eigenvecs** calculates two eigenvectors.

d) Consider the calculation of eigenvalues when complex values occur:

$$\mathbf{A} := \begin{pmatrix} -1 & -8 \\ 2 & -1 \end{pmatrix}$$

$$\text{eigenvals}(\mathbf{A}) \; = \; \begin{pmatrix} -1 + 4\text{i} \\ -1 - 4\text{i} \end{pmatrix}$$

The associated eigenvectors obviously also contain complex values:

$$\text{eigenvecs}(\mathbf{A}) := \begin{pmatrix} 0.894 & 0.894 \\ -0.447\text{i} & 0.447\text{i} \end{pmatrix}$$

♦

The examples for eigenvalues and eigenvectors show that it is desirable to use the

eigenvecs (A)

function to calculate all eigenvectors, and so obtain all existing linear independent eigenvectors for multiple eigenvalues.

Also note that different results are possible for the eigenvectors associated with an eigenvalue, because these are not determined in the length and can also be formed by linear combinations. This frequently causes differences between the calculation of all eigenvectors and the calculation of individual eigenvectors, as shown in Example 15.9c. Consequently, you should check the calculated eigenvectors to detect any MATHCAD error (Example 15.9b).

♦

16 Equations and Inequalities

Equations and *inequalities* play a fundamental role not only in *engineering* and *science* but also in *economics* (see [3], [4]).

Whereas mainly *linear equations/inequalities* occur in economics, *nonlinear equations/inequalities* are found more often in engineering and science.

Normally not one but several equations/inequalities with several unknowns occur in applications, which are known as *systems*.

Usually *indexed variables* are used in systems of equations for the unknowns. MATHCAD is the only system that provides this representation; it even provides two forms for indexed variables (see Section 8.2).

Whereas finite solution algorithms exist for systems of linear equations, this is not normally the case for systems of nonlinear equations. Consequently, MATHCAD provides numerical methods to determine approximate solutions for nonlinear equations; these methods are discussed in Section 16.4.

♦

16.1 Systems of Linear Equations and Analytical Geometry

Finite algorithms to determine the exact solution exist for systems of linear equations. The *Gaussian algorithm* is the best known of these. Consequently, the *exact solution* does not cause any difficulty to MATHCAD when the *number* of *equations* and variables is not too large (maximum 100 equations). The *exact solution* has the major advantage that the *rounding error problem* which plays a large role in numerical methods does not occur.

A general *system of linear equations* that consists of *m linear equations* with *n unknowns/variables*

$x_1, \ldots, x_n$

has the *form*

$$a_{11} \cdot x_1 + a_{12} \cdot x_2 + \dots + a_{1n} \cdot x_n = b_1$$
$$a_{21} \cdot x_1 + a_{22} \cdot x_2 + \dots + a_{2n} \cdot x_n = b_2$$
$$\vdots \qquad\quad \vdots \qquad\qquad\quad \vdots$$
$$a_{m1} \cdot x_1 + a_{m2} \cdot x_2 + \dots + a_{mn} \cdot x_n = b_m$$

and written in *matrix notation*

A · x = b

where

$$A = \begin{pmatrix} a_{11} & a_{12} & \cdots & a_{1n} \\ a_{21} & a_{22} & \cdots & a_{2n} \\ \vdots & \vdots & \cdots & \vdots \\ a_{m1} & a_{m2} & \cdots & a_{mn} \end{pmatrix} \qquad x = \begin{pmatrix} x_1 \\ x_2 \\ \vdots \\ x_n \end{pmatrix} \qquad b = \begin{pmatrix} b_1 \\ b_2 \\ \vdots \\ b_m \end{pmatrix}$$

and

* **A** is called the *coefficient matrix*
* **x** is called the *vector* of the *unknowns* (*solution vector*)
* **b** is called the *right-hand side vector*

Depending on the *coefficient matrix* **A** and the right-hand side vector **b** the *solution theory* for *systems of linear equations* gives conditions when

* *a unique solution*
* *several solutions*
* *no solution*

exist.

We will use the following example to consider these three possible cases for the solution of systems of linear equations.

Example 16.1:

a) The *system of equations*

$$3 \cdot x_1 + 2 \cdot x_2 = 14$$
$$4 \cdot x_1 - 5 \cdot x_2 = -12$$

has a *unique solution*

$$x_1 = 2, \ x_2 = 4$$

b) The *system of equations*

$$x_1 + 3 \cdot x_2 = 3$$
$$3 \cdot x_1 + 9 \cdot x_2 = 9$$

has an *infinite number* of *solutions* of the form

$x_1 = 3 - 3 \cdot \lambda$, $x_2 = \lambda$,(λ arbitrary real number),

because the second equation is a multiple of the first.

c) The *system of equations*

$$x_1 + 3 \cdot x_2 = 3$$
$$3 \cdot x_1 + 9 \cdot x_2 = 4$$

has *no solution*, because the equations contradict each other.

♦

MATHCAD provides the following capabilities for the *exact solution* of *systems* of *linear equations*:

I. For *systems* with a *nonsingular square coefficient matrix* **A**, for which exactly one solution exists, a *solution possibility* exists in the calculation of the *inverse matrix* $\mathbf{A}^{-1}$.
The *solution vector* **x** then results as the product of $\mathbf{A}^{-1}$ and **b**, i.e.,

$$\mathbf{x} = \mathbf{A}^{-1} \cdot \mathbf{b}$$

II. The *menu sequence*

Symbolics ⇒ Variable ⇒ Solve

can be used to *solve* an *equation* for one variable (which must have been marked with editing lines). This permits the *stepwise solution* of a *system* of *equations* by solving one equation for one variable and then using the *menu sequence*

Symbolics ⇒ Variable ⇒ Substitute

or the

substitute

keyword to set the result using the clipboard in the other equations (see Section 13.8).
This method (known as *elimination method*) is long-winded for systems with several unknowns. Therefore this method cannot be recommended for this reason and because it does not have any advantages over the following Method III.

III. An *effective solution method* from MATHCAD is performed in the following steps:

* Enter the

given

command in the worksheet. Note that this must be done in *math mode*.

* Then enter the *system of equations* to be solved below it. Use the *button* for the *equal operator*

from *operator palette no. 2* or the *key combination*

Ctrl-⊞

to enter the *equal sign* in the equations.

* Then enter the *function*

  find (...)

  below the system of equations (also while in *math mode*); the variables (separated by commas) for which the solution is to be found must appear in the argument. The region between **given** and **find** is called *solve block*.

* Enter the *symbolic equal sign* → and press the ⏎-key to return the *exact result*, provided that the system of equations can be solved. You can also use the *assignment*

  x := find (...) →

  to assign the *computed result* to a *solution . vector* **x** (see Examples 16.2b and d).

IV. Another of MATHCAD's *effective solution methods* uses the **solve** *keyword* and is processed similarly to Method III.:

* Click the

  button in the *operator palette no. 8* from the *math toolbar*, in order to *activate* the **solve** *keyword*.

* Then write the *equation system* to be solved in the left-hand placeholder of the

  ■ solve, ▮ →

  symbol and the *unknown variables* in the right-hand placeholder, where each of these are entered as components of a vector (see Example 16.2b).

  The *equal sign* in the individual equations must be entered using the *button* for the *equal operator*

  ▮=▮

  from *operator palette no. 2* or the *key combination*

  Ctrl-⊞

  You can also use the

  x := ■ solve, ■ →

  assignment to assign the *computed result* to a *solution vector* **x** (see Example 16.2b).

*   Enter the *symbolic equal sign* → and press the ⏎-key to return the *exact result* provided that the system of equations can be solved (see Example 16.2b).

☞

The *variables* for the equations in the worksheet can be entered as

*   either *indexed*

$x_1, x_2, x_3, \ldots$

*   or *nonindexed*

x1, x2, x3, …

The two forms described in Section 8.2 (array index or literal index) can be used when indexed variables are used.

◆

Consider the solution methods using the following example.

Example 16.2:

a) Use *method I.* to solve the *system* of *linear equations*

$$x_1 + 3 \cdot x_2 + 3 \cdot x_3 = 2$$
$$x_1 + 3 \cdot x_2 + 4 \cdot x_3 = 1$$
$$x_1 + 4 \cdot x_2 + 3 \cdot x_3 = 4$$

with *nonsingular, square coefficient matrix*

*   First calculate the *inverse coefficient matrix* (see Section 15.3.3):

$$\begin{bmatrix} 1 & 3 & 3 \\ 1 & 3 & 4 \\ 1 & 4 & 3 \end{bmatrix}^{-1} \rightarrow \begin{bmatrix} 7 & -3 & -3 \\ -1 & 0 & 1 \\ -1 & 1 & 0 \end{bmatrix}$$

*   and then the *multiplication* of the calculated *inverses* with the *vector* from the *right-hand side* of the system (see Section 15.3.1):

$$\begin{pmatrix} 7 & -3 & -3 \\ -1 & 0 & 1 \\ -1 & 1 & 0 \end{pmatrix} \cdot \begin{pmatrix} 2 \\ 1 \\ 4 \end{pmatrix} \rightarrow \begin{pmatrix} -1 \\ 2 \\ -1 \end{pmatrix}$$

yields the following *solution:*

$$x_1 = -1, \quad x_2 = 2, \quad x_3 = -1$$

b) Solve the system from example a using

●   *Method III.:*

Enter the following *solve block* in the worksheet:

given

$$x1 + 3 \cdot x2 + 3 \cdot x3 = 2$$

$$x1 + 3 \cdot x2 + 4 \cdot x3 = 1$$

$$x1 + 4 \cdot x2 + 3 \cdot x3 = 4$$

$$\textbf{find } (x1, x2, x3) \ \rightarrow \ \begin{pmatrix} -1 \\ 2 \\ -1 \end{pmatrix}$$

or by *assignment* to a *solution vector* **x**

$$\mathbf{x} := \textbf{find } (x1, x2, x3) \ \rightarrow \ \begin{pmatrix} -1 \\ 2 \\ -1 \end{pmatrix}$$

- *Method IV.:*

  The **solve** *keyword* is used by clicking the appropriate button in operator palette no. 8 and completing the two placeholders as follows; the three equations and three unknown variables are each entered as vectors with three components:

$$\begin{bmatrix} x1 + 3 \cdot x2 + 3 \cdot x3 \equiv 2 \\ x1 + 3 \cdot x2 + 4 \cdot x3 \equiv 1 \\ x1 + 4 \cdot x2 + 3 \cdot x3 \equiv 4 \end{bmatrix} \text{ solve, } \begin{bmatrix} x1 \\ x2 \\ x3 \end{bmatrix} \ \rightarrow (-1 \quad 2 \quad -1)$$

  The *assignment* to a *solution vector* **x** is performed as follows:

$$x := \begin{bmatrix} x1 + 3 \cdot x2 + 3 \cdot x3 \equiv 2 \\ x1 + 3 \cdot x2 + 4 \cdot x3 \equiv 1 \\ x1 + 4 \cdot x2 + 3 \cdot x3 \equiv 4 \end{bmatrix} \text{ solve, } \begin{bmatrix} x1 \\ x2 \\ x3 \end{bmatrix} \ \rightarrow (-1 \quad 2 \quad -1)$$

  The symbolic equal sign can be used to display the *solution vector* **x**:

  $$x \rightarrow (-1 \quad 2 \quad -1)$$

c) Consider the *system* of *linear equations*

$$x_1 + 2 \cdot x_2 + x_3 = 1$$
$$2 \cdot x_1 + x_2 + 3 \cdot x_3 = 0$$

for which the solution cannot be determined uniquely (one variable can be freely chosen) and use *Method III.* here.

Enter the following *solve block* in the worksheet:

given

$$x1 + 2 \cdot x2 + x3 = 1$$

$2 \cdot x1 + x2 + 3 \cdot x3 = 0$

$$\textbf{find} \, (x1, x2, x3) \; \rightarrow \; \begin{pmatrix} \dfrac{-5}{3} \cdot x3 - \dfrac{1}{3} \\[2ex] \dfrac{1}{3} \cdot x3 + \dfrac{2}{3} \\[2ex] x3 \end{pmatrix}$$

Because we have only two equations for the three unknowns x1, x2 and x3, one of the variables can be chosen arbitrary. MATHCAD has selected x3.

d) In the following example we use *Method III.* with *indexed variables* to solve a system of linear equations (see Section 8.2):

d1)Use *indexed variables* as components of a vector **x**, i.e., use the *array index:*

given

$x_1 + x_2 = 1$

$x_1 - x_2 = 0$

$$\textbf{find} \, (x_1, x_2) \; \rightarrow \; \begin{pmatrix} \dfrac{1}{2} \\[1.5ex] \dfrac{1}{2} \end{pmatrix}$$

or *assign* to the *solution vector* **x**

$$\textbf{x} \; := \; \textbf{find} \, (x_1, x_2) \; \rightarrow \; \begin{pmatrix} \dfrac{1}{2} \\[1.5ex] \dfrac{1}{2} \end{pmatrix}$$

We obtain

$$x_1 = \frac{1}{2}, \; x_2 = \frac{1}{2}$$

only after this assignment to the solution vector **x**.

d2)Use *indexed variables* with *literal index:*

given

$x_1 + x_2 = 1$

$$x_1 - x_2 = 0$$

$$\mathbf{find}(x_1, x_2) \;\rightarrow\; \begin{pmatrix} \dfrac{1}{2} \\ \dfrac{1}{2} \end{pmatrix}$$

or through an *assignment* to a *solution vector* **x**

$$\mathbf{x} := \mathbf{find}(x_1, x_2) \;\rightarrow\; \begin{pmatrix} \dfrac{1}{2} \\ \dfrac{1}{2} \end{pmatrix}$$

e) In the following example we use indexed variables (array index), however, the *system of equations has no solution*. MATHCAD correctly recognizes this and issues an *error message.*

given

$$x_1 + x_2 = 1$$

$$2 \cdot x_1 + 2 \cdot x_2 = 0$$

$$\boxed{\mathrm{find}\left(x_1, x_2\right) \;\rightarrow\;}$$

No answer found.: Can't find a solution to this system of
equations. Try a different guess value o

◆

A *system* of *linear equations* can be solved without difficulty using MATH-
CAD provided the number of equations and variables does not exceed 100.
We recommend that you always use *Method III.* or *IV.* These methods are
the simplest to use and are suitable for all types of systems of equations.
Because problems occurred with the Version 8 available to the author in
conjunction with the use of the array index for the indexing of the variables, this effect required the literal index.

◆

Many *problems* from *analytical geometry* can be solved without difficulty
using MATHCAD:

- *Determine* the *intersection points* for
  * straight lines
  * straight lines and planes
  * planes.

  This results in the *solution* of a *system* of *linear equations.*

- *Calculate* the *distance* between
  * point – straight line
  * point – plane
  * straight line – straight line
  * straight line – plane

  This is performed using the *Hessian standard form* or with the given *computational formulae*.

- *Principal axis transformations* for conic equations.
  This can be performed using functions for eigenvalue calculation for matrices.

The following example illustrates the use of MATHCAD to solve some of these problems from analytical geometry.

Example 16.3:

a) *Determine* the *intersection point* of the two straight lines in the plane:

  G1: y = 2·x+3   and   G2: y = –x–1

  The following *system of linear equations* (2 equations with 2 unknowns) must be solved:

  given

  $$y = 2 \cdot x + 3$$

  $$y = -x - 1$$

  $$\mathbf{find}\,(x, y) \;\rightarrow\; \begin{pmatrix} \dfrac{-4}{3} \\ \dfrac{1}{3} \end{pmatrix}$$

  The calculated *intersection point* can be *shown graphically* by drawing the two straight lines (see Section 18.1):

  $$x := -2, -1.999 \ .. \ 2$$

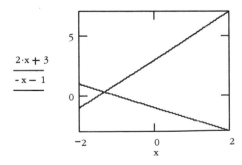

b) *Determine* the *distance* between the *point*

P=(3, 3, 4)

and the *plane*

$x + y + z - 1 = 0$

using the *Hessian standard form for a plane:*

$$\frac{x + y + z - 1}{\sqrt{3}} = 0$$

The following function can be used to calculate the distance between a point (x,y,z) from this plane

$$\text{Distance}(x, y, z) := \frac{x + y + z - 1}{\sqrt{3}}$$

The distance of the point (3,3,4) yields:

$\text{Distance}(3, 3, 4) \rightarrow 3 \cdot \sqrt{3} = 5.196$

c) *Investigate* the *positional relationship* of the *two straight lines* in the space:

$$G1: \begin{array}{l} x + y + z = 1 \\ 2x + y - z = 0 \end{array} \qquad G2: \begin{array}{l} x - y - z = 3 \\ x - 2y + 3z = 2 \end{array}$$

This results in the solution of the following *system* of *linear equations* (4 equations with 3 unknowns):

given

$x + y + z = 1$

$2 \cdot x + y - z = 0$

$x - y - z = 3$

$x - 2 \cdot y + 3 \cdot z = 2$

$\text{find}(x, y, z) \rightarrow$ ▪

| did not find solution |

Because MATHCAD has *not found* any *solution,* the two *straight lines* G1 and G2 are *skewed,* i.e., they have *no point of intersection.*

d) Determine the *point of intersection* between the *straight line* G given by the two points

P1 (1 , 2 , 3) and P2 (2 , 4 , 5)

and the *plane* E given by the three points

P3 (2 , 3 , 1) , P4 (3 , 0 , 2) and P5 (4 , 5 , 6)

Unfortunately, the following elegant solution method using vector algebra cannot be realized, because MATHCAD cannot solve the resulting equations in vector form:

Assign radius vectors to the given points:

$$a := \begin{pmatrix} 1 \\ 2 \\ 3 \end{pmatrix} \quad b := \begin{pmatrix} 2 \\ 4 \\ 5 \end{pmatrix} \quad c := \begin{pmatrix} 2 \\ 3 \\ 1 \end{pmatrix} \quad d := \begin{pmatrix} 3 \\ 0 \\ 2 \end{pmatrix} \quad e := \begin{pmatrix} 4 \\ 5 \\ 6 \end{pmatrix}$$

Thus, the straight line G (parameter t) and the plane E (parameters u and v) have the following parametric forms:

$G \blacksquare a + t \cdot (b - a)$

$E \blacksquare c + u \cdot (d - c) + v \cdot (e - c)$

The point of intersection between the two results from the solution of the vector equation with the unknowns
t, u and v :

given

$c + u \cdot (d - c) + v \cdot (e - c) \blacksquare a + t \cdot (b - a)$

$find(t, u, v) \rightarrow \blacksquare$

| invalid arguments |

Because MATHCAD cannot solve this vector equation, you must write the components. This can be solved as follows

$$(c + u \cdot (d - c) + v \cdot (e - c) - a - t \cdot (b - a)) \rightarrow \begin{pmatrix} 1 + u + 2 \cdot v - t \\ 1 - 3 \cdot u + 2 \cdot v - 2 \cdot t \\ -2 + u + 5 \cdot v - 2 \cdot t \end{pmatrix}$$

given

$1 + u + 2 \cdot v - t \blacksquare 0$

$1 - 3 \cdot u + 2 \cdot v - 2 \cdot t \blacksquare 0$

$-2 + u + 5 \cdot v - 2 \cdot t \blacksquare 0$

$$\text{find}(t,u,v) \rightarrow \begin{pmatrix} \dfrac{36}{7} \\[2ex] \dfrac{-9}{7} \\[2ex] \dfrac{19}{7} \end{pmatrix}$$

Thus, for example, the point of intersection can be determined as follows:

$$a + \dfrac{36}{7} \cdot (b-a) \rightarrow \begin{pmatrix} \dfrac{43}{7} \\[2ex] \dfrac{86}{7} \\[2ex] \dfrac{93}{7} \end{pmatrix}$$

16.2 Polynomials

Polynomial functions (polynomials) of the *degree n* $P_n(x)$ with real coefficients, that are also known as *integral rational function,* can be written in the form

$$P_n(x) = \sum_{k=0}^{n} a_k \cdot x^k = a_n \cdot x^n + a_{n-1} \cdot x^{n-1} + \dots + a_1 \cdot x + a_0 , \quad (a_n \neq 0).$$

An important *problem* for *polynomials* consists in the *calculation* of the real and complex zeros, i.e., of the *solutions* x_i of the associated *polynomial equation*

$$P_n(x) = 0$$

It can be *proved* that a *polynomial* of *degree n* has n *zeros*, which can be *real* or *complex* and *multiple.*

Solution formulae for the *determination* of *zeros* exist only for *polynomials* with a *degree* less than or equal to *four* (i.e., to n=4). The best known is that for *quadratic equations* (n=2)

$$x^2 + a_1 \cdot x + a_0 = 0$$

with the two *solutions*

$$x_{1,2} = -\frac{a_1}{2} \pm \sqrt{\frac{a_1^2}{4} - a_0}$$

The *solution formulae* for *polynomials* becomes much more complicated for n=3 and 4. Because general polynomials with a degree greater than four cannot be solved by radicals, there are no formulae for the zero determination for n≥5.

Consequently, we cannot expect MATHCAD to be always able to find exact solutions for n≥5.

♦

Factorization of polynomials is understood to be in the form of *product* of

* *linear factors* (for the real zeros)

* *quadratic factors* (for the complex zeros),

i.e., (for a_n =1)

$$\sum_{k=0}^{n} a_k \cdot x^k =$$

$$(x - x_1)^{k_1} \cdot (x - x_2)^{k_2} \cdot \ldots \cdot (x - x_r)^{k_r} \cdot (x^2 + b_1 \cdot x + c_1) \cdot \ldots \cdot (x^2 + b_s \cdot x + c_s)$$

where

$$x_1, \ldots, x_r$$

are the *real zeros* of *multiplicity*

$$k_1, k_2, \ldots, k_r$$

The *fundamental theorem* of *algebra* ensures the existence of such a *factorization* and is closely associated with the determination of the *zeros*.

Consequently, it is not surprising that MATHCAD cannot always perform the factorization (see Examples 13.14c and d).

However, MATHCAD can often factorize polynomials of higher degree that have integer zeros (see Section 13.6).

♦

MATHCAD can *calculate* the zeros for a polynomial in the following ways: Application of

* *factorization* to determine the *real zeros* (see Section 13.6)

* *special functions* of MATHCAD for *polynomials* (see Section 16.4)

* *commands/functions* of MATHCAD for general *nonlinear equations* (see Section 16.3), because polynomial equations form a special case.

 ♦

Example 16.4 illustrates the problems involved with the determination of zeros for polynomials. We *use* the following *four methods* here:

I. *Factorization* (see Section 13.6)

Use the *menu sequence*

Symbolics ⇒ Factor

or the

factor

keyword.

II. *Activate the menu sequence*

Symbolics ⇒ Variable ⇒ Solve

to determine zeros for equations with one unknown (see Section 16.3)

III. Use of the **solve** keyword (see Section 16.1):

*   Click the

button in *operator palette no. 8* to activate the

solve

keyword

*   Then enter the *polynomial equation* in the left-hand *placeholder* and the *unknown variable* x in the right-hand *placeholder* in the displayed

■ solve, ▮ →

symbol.
The *button* for the *equal operator*

from *operator palette no. 2* or the *key combination*

Ctrl-⊕

must be used to enter the *equal sign* in the polynomial equation.

*   Then press the ⏎-key to return the *exact result* provided MATH-CAD is successful.

IV. Use of the *solve block* with **given** and **find** to solve general equations.

Example 16.4:

a) The *polynomial*

$$x^5 - 5 \cdot x^4 - 5 \cdot x^3 + 25 \cdot x^2 + 4 \cdot x - 20$$

has the *zeros* -1, -2, 1, 2, and 5. MATHCAD calculates these using

* *Method I.:*

  Factoring using the *menu sequence*

  Symbolics ⇒ Factor

  returns

  $$x^5 - 5 \cdot x^4 - 5 \cdot x^3 + 25 \cdot x^2 + 4 \cdot x - 20 \qquad by\ factoring,\ yields$$

  $$(x - 5) \cdot (x - 1) \cdot (x - 2) \cdot (x + 2) \cdot (x + 1)$$

* *Method II.:*

  using the *menu sequence*

  Symbolics ⇒ Variable ⇒ Solve

  returns

  $$x^5 - 5 \cdot x^4 - 5 \cdot x^3 + 25 \cdot x^2 + 4 \cdot x - 20 \quad has\ solution(s) \quad \begin{pmatrix} 5 \\ 1 \\ 2 \\ -2 \\ -1 \end{pmatrix}$$

* *Method III.:*

  Use of the **solve** *keyword* returns

  $$x^5 - 5 \cdot x^4 - 5 \cdot x^3 + 25 \cdot x^2 + 4 \cdot x - 20 \blacksquare 0 \ solve, x \ \rightarrow \begin{bmatrix} 5 \\ 1 \\ 2 \\ -2 \\ -1 \end{bmatrix}$$

* *Method IV.:*

  Use of the *solve block* with **given** and **find**:

  given

  $$x^5 - 5 \cdot x^4 - 5 \cdot x^3 + 25 \cdot x^2 + 4 \cdot x - 20 \blacksquare 0$$

  $$find(x) \rightarrow (5 \quad 1 \quad 2 \quad -2 \quad -1)$$

b) For the *polynomial*

$$x^9 + x + 1$$

MATHCAD determines one real and eight complex *zeros*. As the following graph shows, the only *real zero* lies between -1 and -0.5:

$$x := -1, -0.999 .. 0$$

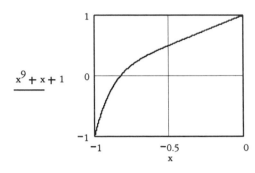

$$x^9 + x + 1 \quad \textit{has solution(s)}$$

$$\begin{bmatrix}
-.82430056322968701271 \\
-.78815677587643473001 \quad -.50285258764546612087 \cdot 1i \\
-.78815677587643473001 \quad +.50285258764546612087 \cdot 1i \\
-.28138444440632605813 \quad -.98259597890434460256 \cdot 1i \\
-.28138444440632605813 \quad +.98259597890434460256 \cdot 1i \\
.47229060916692797336 \quad -.95393081242028581325 \cdot 1i \\
.47229060916692797336 \quad +.95393081242028581325 \cdot 1i \\
1.0094008927306763211 \quad -.39206252357421550811 \cdot 1i \\
1.0094008927306763211 \quad +.39206252357421550811 \cdot 1i
\end{bmatrix}$$

c) Consider the *polynomial*

$$x^7 - x^6 + x^2 - 1$$

whose *function curve* has the following form

$$x := -1.2, -1.1999 .. 1.2$$

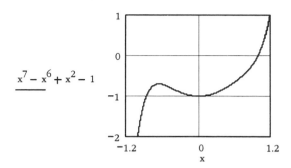

You can see that the only real *zero* is at 1. We now attempt to determine the zeros using:

* *Factorization*

$$x^7 - x^6 + x^2 - 1 \quad by\ factoring,\ yields \quad (x-1) \cdot (x^6 + x + 1)$$

The other *complex zeros* are not determined.

* The *menu sequence*

  Symbolics ⇒ Variable ⇒ Solve

  or the **solve** *keyword*

$$x^7 - x^6 + x^2 - 1 \equiv 0 \quad has\ solution(s) \quad \text{"untranslatable"}$$

$$x^7 - x^6 + x^2 - 1 \equiv 0 \ solve, x \ \rightarrow \ \text{"untranslatable"}$$

In contrast to the factorization, this method does not calculate any zeros.

d) Although the *menu sequence*

Symbolics ⇒ Variable ⇒ Solve

causes MATHCAD to find the following four *complex zeros*:

$$x^4 + 1 \quad has\ solution(s) \quad \begin{pmatrix} \frac{1}{2} \cdot \sqrt{2} + \frac{1}{2} \cdot i \cdot \sqrt{2} \\ \frac{-1}{2} \cdot \sqrt{2} + \frac{1}{2} \cdot i \cdot \sqrt{2} \\ \frac{1}{2} \cdot \sqrt{2} - \frac{1}{2} \cdot i \cdot \sqrt{2} \\ \frac{-1}{2} \cdot \sqrt{2} - \frac{1}{2} \cdot i \cdot \sqrt{2} \end{pmatrix}$$

for the *polynomial*

$$x^4 + 1$$

it *cannot* specify the resulting *factorization*:

$$\left(x^2 - x \cdot \sqrt{2} + 1\right) \cdot \left(x^2 + x \cdot \sqrt{2} + 1\right)$$

but just returns the unchanged polynomial:

$x^4 + 1$ *by factoring, yields* $x^4 + 1$

e) For the *polynomial*

$$x^4 - 1$$

MATHCAD calculates the *zeros* using the *menu sequence*
Symbolics $\Rightarrow$ **Variable** $\Rightarrow$ **Solve** or the **solve** *keyword*

$x^4 - 1$ *has solution(s)* $\begin{pmatrix} 1 \\ -1 \\ i \\ -i \end{pmatrix}$

and performs the *factorization*:

$x^4 - 1$ *by factoring, yields* $(x - 1) \cdot (x + 1) \cdot (x^2 + 1)$

f) The *polynomial equation*

$$x^5 - x + 1 = 0$$

is solved; the following graph shows the real zero:

$x := -1.5, -1.499 .. 1$

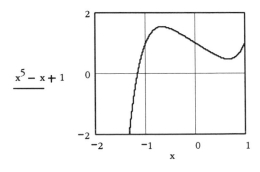

MATHCAD returns the *solutions:*

$x^5 - x + 1$ *has solution(s)*

$$
\begin{bmatrix}
-1.1673039782614186843 \\
-.18123244446987538390 \quad - \ 1.0839541013177106684 \ \cdot 1i \\
-.18123244446987538390 \quad + \ 1.0839541013177106684 \ \cdot 1i \\
.76488443360058472603 \quad - \ .35247154603172624932 \ \cdot 1i \\
.76488443360058472603 \quad + \ .35247154603172624932 \ \cdot 1i
\end{bmatrix}
$$

♦

Even when MATHCAD calculates the zeros of a polynomial, the result must be considered critically.

I suggest that you plot the function curve for the polynomial function and then use it to determine the approximate values for the real zeros, which you then can compare with the calculated values. I also suggest that you make a test, i.e., set the calculated zeros in the polynomial.

♦

If the *exact determination* of *zeros fails*, the same messages appear as issued for the solution of nonlinear equations (see Section 16.3). In this case you can resort to *numerical analysis* (which we discuss in Section 16.4).

♦

16.3 Nonlinear Equations

Let us first consider the problems associated with the solution of one *nonlinear equation* having the form

f(x) = 0

with one unknown (variable) x that can produce real and/or complex solutions. This problem is equivalent to determining the *zeros* for the *function*

f(x)

With the exception of the *special cases* of *linear equations* and *polynomial equations* (up to degree 4) that we considered in Sections 16.1 and 16.2, no finite algorithms exist that determine all solutions for an arbitrary nonlinear equation.

♦

MATHCAD provides the following three capabilities (see Example 16.5a) for the *exact solution* of a nonlinear *equation* having the *form*

f(x) = 0

I. Use of the **Symbolics** *menu*:

* *First* use the *button* for the *equal operator*

from *operator palette no. 2* to place the *equation* in the *worksheet*, i.e., f(x)=0.

It suffices just to enter the function f(x)

* *Then mark* with editing lines a *variable* x in the function expression f(x).

* *Finally*, enter the *menu sequence*

Symbolics ⇒ Variable ⇒ Solve

to initiate the calculation.

II. Use of the **solve** *keyword* (see Section 6.1):

* Click the

button in *operator palette no. 8* to activate the

solve

keyword

* Then enter the *polynomial equation* in the left-hand *placeholder* and the *unknown variable* x in the right-hand *placeholder* in the displayed

 →

symbol (see Example 16.5a).

The *button* for the *equal operator*

from *operator palette no. 2* or the *key combination*

Ctrl-⊞

must be used to enter the *equal sign* in the equation.

* Then press the ⏎-key to return the *exact result* provided MATH-CAD is successful (see Example 16.5a).

III. Use of the *solve block* with **given** and **find** to solve general equations (see Example 16.5a).

Let us use the following Example 16.5 to show MATHCAD's effectiveness in solving a nonlinear equation.

Example 16.5:

a) The *graphic display* of the *function* x − sin(x)

$$x := -1, -0.999 .. 1$$

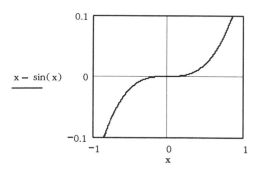

shows the real *zero* x = 0 that MATHCAD also calculates using the given methods:

* *Method I.:*

  Use the *menu sequence*

  Symbolics → Variable → Solve

  to return

  $$x - \sin(x) \quad \text{has solution(s)} \quad 0$$

* *Method II.:*

  Use the **solve** *keyword* to return

  $$x - \sin(x) \equiv 0 \text{ solve, } x \rightarrow 0$$

* *Method III.:*

  Use of the *solve block* with **given** and **find** to return

  given

  $$x - \sin(x) \equiv 0$$

  $$\text{find}(x) \rightarrow 0$$

b) Consider the *equation*

$$\cos x - e^{-x} = 0$$

MATHCAD yields only the *solution* 0 with the given methods:

* *Method I.:*

  $$\cos(x) - e^{-x} \quad \text{has solution(s)} \quad 0$$

* *Method II.:*

  $$\cos(x) - e^{-x} \equiv 0 \text{ solve, } x \rightarrow 0$$

*   *Method III.:*

    given

    $\cos(x) - e^{-x} = 0$

    find$(x) \rightarrow 0$

although additional (infinitely many) real solutions exist, as we can see from the graph:

   $x := -2, -1.999 .. 30$

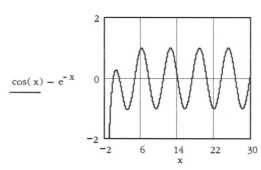

c) Consider the *equation*

   $3 \cdot \cosh x - \sinh x - 9 = 0$

   Substituting the definitions for sinh x and cosh x and applying the *transformation*

   $y = e^{x}$

   that can be rewritten as the *quadratic equation*

   $y^2 - 9 \cdot y + 2 = 0$

   MATHCAD returns the *two solutions*

   $3 \cdot \cosh(x) - \sinh(x) - 9$ *has solution(s)* $\begin{bmatrix} \ln\left(\frac{9}{2} + \frac{1}{2} \cdot \sqrt{73}\right) \\ \ln\left(\frac{9}{2} - \frac{1}{2} \cdot \sqrt{73}\right) \end{bmatrix}$

d) Consider the *equation*

   $1 + x - \sqrt{1 + x} = 0$

   MATHCAD determines the two real *solutions:*

$$1 + x - \sqrt{1 + x} \qquad \textit{has solution(s)} \qquad \begin{bmatrix} -1 \\ 0 \end{bmatrix}$$

that the *graphical display* also supplies:

$$x := -2, -1.999 .. 2$$

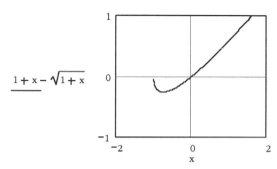

e)   MATHCAD does not supply *any solution* for the *equation*

$$e^x + \ln x = 0$$

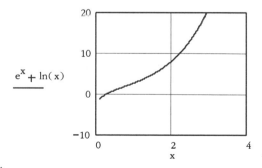

even though a real solution exists as the following display shows:

$$x := 0.1, 0.1001 .. 3$$

The last examples show that MATHCAD, like other computer algebra systems, has *difficulties* with the *exact solution* of *nonlinear equations*.
Because finite solution algorithms exist only for the specified special cases, this is not surprising.

Even when a result is obtained, this can be incorrect or incomplete. Consequently, you should always perform an additional check by resubstitution of the obtained solutions or make a graphical check by plotting the function f(x) (see Section 18.1).

♦

In most practical problems, one equation with just one unknown does not occur, but rather several equations with several unknowns.

No generally usable finite algorithm exists for the exact solution of these *systems* of *nonlinear equations* (m equations with n unknowns) having the *form*

$$u_1(x_1, x_2,..., x_n) = 0$$
$$u_2(x_1, x_2,..., x_n) = 0$$
$$\vdots$$
$$u_m(x_1, x_2,..., x_n) = 0$$

We have already discussed in Section 16.1 systems of linear equations as a special case.

♦

Let us now consider some *examples* of *systems of nonlinear equations*.

Example 16.6:

a) System of two *algebraic equations* with *two unknowns* x , y

$$x^2 - y^2 = 3$$
$$x^4 + y^4 = 17$$

b) System of three *algebraic equations* with *three unknowns* x , y , z

$$x + y + z = 3$$
$$2 \cdot x^2 - y^2 - z^2 = 0$$
$$x^4 + 2 \cdot y^4 - z^4 = 2$$

c) System of two *transcendental equations* with *two unknowns* x , y

$$\sin x + e^y = 1$$
$$2 \cdot \cos x + \ln(y + 1) = 2$$

d) System of two *algebraic* and *transcendental equations* with *two unknowns* x,y

$$x^4 + y^4 + 3 \cdot x + y = 0$$
$$\cos x + e^y = 2$$

♦

The procedure used in MATHCAD for the *exact solution* of general *systems of nonlinear equations* is analogue to the solution of systems of linear equations. The *methods II.*, *III.* and *IV.* used in Section 16.1 can also be used here; it suffices to replace the linear equations by the given nonlinear equations.

Consequently, we do not repeat the description here, but rather start with the exact solution of the systems of equations specified in Example 16.6. We only use the general *methods III.* and *IV.*, because *method II.* does not offer any advantages and is more complicated.

Example 16.7:

a) Solve the system from Example 16.6a using

- the *solve block* with **given** and **find**:

    given

    $$x^2 - y^2 = 3 \qquad\qquad x^4 + y^4 = 17$$

    $$\text{find}(x,y) \;\rightarrow\; \begin{pmatrix} 2 & -2 & 2 & -2 & 1i & -1i \\ 1 & 1 & -1 & -1 & 2\cdot 1i & -2\cdot 1i \end{pmatrix}$$

- the **solve** *keyword*:

    $$\begin{bmatrix} x^2 - y^2 \blacksquare 3 \\ x^4 + y^4 \blacksquare 17 \end{bmatrix} \text{solve}, \begin{bmatrix} x \\ y \end{bmatrix} \;\rightarrow\; \begin{bmatrix} 2 & 1 \\ -2 & 1 \\ 2 & -1 \\ -2 & -1 \\ 1i & 2\cdot 1i \\ -1i & -2\cdot 1i \end{bmatrix}$$

    Although this system of polynomial equations is solved, MATHCAD omits the two conjugated complex solutions.

b) MATHCAD finds only one real *solution* $x = 1$, $y = 1$, $z = 1$ for the solution of the system with three equations from Example 16.6b

    $$x + y + z \blacksquare 3$$

    $$2\cdot x^2 - y^2 - z^2 \blacksquare 0$$

    $$x^4 + 2\cdot y^4 - z^4 \blacksquare 2$$

c) MATHCAD does not find the real solution $x = 0$, $y = 0$ but only complex solutions for the *system* of *transcendental equations* from Example 16.6c

$$\sin(x) + e^y \blacksquare 1$$

$$2 \cdot \cos(x) + \ln(y + 1) \blacksquare 2$$

However only the solution (0,0) is easy for the given system of equations but this can be obtained by trial. Because MATHCAD cannot use this heuristic method for the solution, its failure is obvious.

d) MATHCAD finds just the solution $x = 0$, $y = 0$ for the system from Example 16.6d from an algebraic and a transcendental equation:

- Using **given** and **find**:

  given

  $$x^4 + y^4 + 3 \cdot x + y \blacksquare 0$$

  $$\cos(x) + e^y \blacksquare 2$$

  find $(x,y) \rightarrow \begin{pmatrix} 0 \\ 0 \end{pmatrix}$

- Using the **solve** *keyword*:

  $$\left[\begin{array}{c} x^4 + y^4 + 3 \cdot x + y \blacksquare 0 \\ \cos(x) + e^y \blacksquare 2 \end{array} \right] \text{solve}, \left[\begin{array}{c} x \\ y \end{array} \right] \rightarrow (0 \quad 0)$$

  ◆

The calculated examples show that MATHCAD can have difficulties when transcendental equations occur. However, it can be seen that MATHCAD Version 8 is an improvement over the previous versions with regard to the equation solution. However, the effect shown in Example 16.7a can occur, in which conjugate complex solutions are omitted.

◆

If MATHCAD does not *find any exact solution* for an equation, one of the following responses can occur:

* A *message* appears indicating that the *result* is MAPLE-specific and whether it is to be stored in the clipboard.
  The clipboard contents can then be copied into the worksheet in the usual WINDOWS manner.

The display in MAPLE syntax can sometimes be used to deduce the solutions. However, frequently the equation to be solved is merely redis-
played.

* A *message* appears indicating that *no solution* was obtained.
* The *computation does not complete*. Press the (Esc)-key to *terminate*.
    ◆

16.4 Numerical Solution Methods

The *numerical methods* provide the only possibility in many practical
problems to obtain solutions for nonlinear equations. Although finite solution algorithms exist for linear equations, MATHCAD provides for the solution of a linear system (see Section 16.1)

A · x = b

with *nonsingular square coefficient matrix* **A**

the *numerical function*

lsolve (A , b)

that returns a *solution vector* when you enter the *numerical equal sign* =.

Example 16.8:

Use the **lsolve** *numerical function* on the *linear system of equations*

$$x_1 + 3 \cdot x_2 + 3 \cdot x_3 = 2$$
$$x_1 + 3 \cdot x_2 + 4 \cdot x_3 = 1$$
$$x_1 + 4 \cdot x_2 + 3 \cdot x_3 = 4$$

from Example 16.2a:

$$\mathbf{lsolve}\left(\begin{pmatrix} 1 & 3 & 3 \\ 1 & 3 & 4 \\ 1 & 4 & 3 \end{pmatrix}, \begin{pmatrix} 2 \\ 1 \\ 4 \end{pmatrix} \right) = \begin{pmatrix} -1 \\ 2 \\ -1 \end{pmatrix}$$

The computation becomes more obvious when you assign the symbols **A**
and **b** to the coefficient matrix and the vector of the right-hand sides:

$$\mathbf{A} := \begin{pmatrix} 1 & 3 & 3 \\ 1 & 3 & 4 \\ 1 & 4 & 3 \end{pmatrix} \qquad \mathbf{b} := \begin{pmatrix} 2 \\ 1 \\ 4 \end{pmatrix}$$

$$\mathbf{lsolve}\,(\mathbf{A}\,,\mathbf{b}) \;=\; \begin{pmatrix} -1 \\ 2 \\ -1 \end{pmatrix}$$

♦

As result of the problems discussed in Sections 16.2 and 16.3, in most cases you are limited to *numerical methods* (in particular, *iteration methods*) that return approximate values for the solutions of *nonlinear equations*:

* You require an *estimate value* for a *solution* as the *starting value* for this method. If the method converges, it improves this starting value. If the equation has several solutions, the convergence of this method will not provide all the solutions.

* It is well known from *numerical analysis* that *numerical methods* (e.g., *regula falsi, Newton method*) do not necessarily converge, i.e., do not return a result, even when the starting value is near a solution.

* The *selection* of suitable *starting values* is also a problem.
  It can be simplified for one unknown (i.e., for the solution of a equation of the form u(x) = 0) when you display the function u(x) and then obtain the approximate values for the zeros.

  ♦

MATHCAD provides the following capabilities for the *numerical solution* of *nonlinear equations:*

* *Numerical solution* of *one equation* with *one unknown* of the *form* u(x) = 0:

  * Provided it converges, the *numerical function*

    root (u(x), x) =

    returns a real or complex approximate solution when

    x := x_a

    was used beforehand to assign a real or complex *starting value*

    x_a

    to the variable x.
    MATHCAD uses here the *Regula falsi* as *numerical method* (iteration method). Although the Regula falsi requires two starting values, MATHCAD requests just one.

  * The *numerical function*

    polyroots (**a**) =

    can be used instead of **root** when the function u(x) is a *polynomial function*, i.e.,

$$u(x) = a_n \cdot x^n + a_{n-1} \cdot x^{n-1} + \ldots + a_1 \cdot x + a_0$$

The vector **a** in the argument contains the *coefficients* of the *polynomial function* u(x) in the following sequence

$$\mathbf{a} := \begin{pmatrix} a_0 \\ a_1 \\ \ldots \\ a_n \end{pmatrix}$$

This numerical function has the *advantage* compared with **root** that it does not require any *starting values*.

- *Numerical solution* for *systems of equations* :
  With two exceptions, the procedure is performed analogue to the exact solution using **given** and **find**:

  * *Assignment* of *starting values* before **given**

  ✝ The *numerical equal sign* = is entered instead of the symbolic equal sign › after **find**.

This results in the following procedure:

Assign the starting values to all variables

x_1 , x_2 , ... , x_n

given

Enter the equations
Use the *button* for the *equal operator*

from *operator palette no. 2* to enter the *equal sign*

find (x_1 , x_2 , ... , x_n) =

MATHCAD uses the Levenberg-Marquardt method for the numerical solution. This method allows both equations and inequalities (see Example 16.10).

♦

If a *system of equations* does not have any *solutions*, the *numerical function*

minerr (x_1 , x_2 , ... , x_n)

can be used instead of

find (x_1 , x_2 , ... , x_n)

minerr minimizes the sum of squares from the left-hand side of equations for the given system, i.e.,

$$\sum_{i=1}^{m} u_i^2(x_1, x_2,..., x_n) \rightarrow \underset{x_1, x_2,..., x_n}{\text{Minimum}}$$

and so determines a *generalized solution* (see Example 16.9f).

Provided they converge, the given numerical functions return a calculated solution vector when the *numerical equal sign* = is entered.
Instead of using this numerical equal sign, you can also *assign* the calculated solution to a *solution vector*, e.g.,

z := root (u(x) , x)

or

z := polyroots (**a**)

or

z := find (x_1 , x_2 , ... , x_n)

or

z := minerr (x_1 , x_2 , ... , x_n)

♦

Note the *specification* of *real* or *complex starting values* for those numerical methods which require starting values normally return real or complex approximations (see Example 16.9e).

♦

We now use a number of examples to illustrate the operation of MATH-CAD's numerical functions.

Example 16.9:

a) Because MATHCAD calculates only one exact solution 1, we attempt to calculate numerically the *zeros* of the *function*

$$u(x) = 2 \cdot x + |x - 1| - |x + 1|$$

by assigning various starting values.
However, as the following graph shows, all x-values in the interval [-1,1] are zeros for this function:

x := - 2 , - 1.999 .. 2

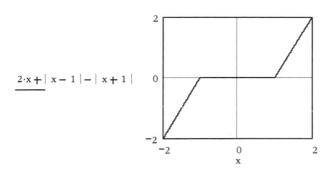

$2 \cdot x + | x - 1 | - | x + 1 |$

- Begin with the *starting value* 0:
  MATHCAD obtains with

  * **root**:

    x := 0

    $\text{root}(2 \cdot x + | x - 1 | - | x + 1 |, x) = 1 \cdot 10^{-3}$

  * **given** and **find**:

    x := 0

    given

    $2 \cdot x + | x - 1 | - | x + 1 | = 0$

    find(x) = 0

- Zeros are also calculated for nonzero *starting values*, e.g., with **root**:

  x := - 2

  $\text{root}(2 \cdot x + | x - 1 | - | x + 1 |, x) = -1$

  x := 2

  $\text{root}(2 \cdot x + | x - 1 | - | x + 1 |, x) = 1$

b) MATHCAD does not provide an exact solution for the *polynomial equation* (see Example 16.4c)

$$x^7 - x^6 + x^2 - 1 = 0$$

The **polyroots** *numerical function* returns six complex solutions and one real solution 1:

$$a := \begin{bmatrix} -1 \\ 0 \\ 1 \\ 0 \\ 0 \\ 0 \\ -1 \\ 1 \end{bmatrix} \qquad \text{polyroots}(a) = \begin{bmatrix} -0.791 - 0.301i \\ -0.791 + 0.301i \\ -0.155 + 1.038i \\ -0.155 - 1.038i \\ 0.945 + 0.612i \\ 0.945 - 0.612i \\ 1 \end{bmatrix}$$

c) Take the *equation*

$$e^x + \ln x = 0$$

in Example 16.5e, MATHCAD does not calculate any *solution*.
The **root** *numerical function*, used with various starting values, returns:

$$x := 1$$

$$\text{root}(e^x + \ln(x), x) = 0.2698744$$

or

$$x := 0.1$$

$$\text{root}(e^x + \ln(x), x) = 0.2698328$$

d) In Example 16.7c, MATHCAD does not return any exact solutions, not even (0,0).

Using the *solve block* with **given** and **find** returns for the

* *starting values* (2,5) :

$$x := 2 \qquad\qquad y := 5$$

given

$$\sin(x) + e^y \equiv 1$$

$$2 \cdot \cos(x) + \ln(y + 1) \equiv 2$$

$$\text{find}(x, y) = \begin{pmatrix} 0 \\ 0 \end{pmatrix}$$

i.e., the solution (0,0)

* *starting values* (0,1) :

x := 0   y := 1

$$\text{find}(x,y) = \begin{pmatrix} -0.626 \\ 0.461 \end{pmatrix}$$

i.e., an additional real solution x = −0.626 , y = 0.461.

e) Consider the *numerical function* **root** when real or complex starting values are given using quadratic equations as example:

e1) For the *quadratic equation*

$$x^2 - 2x + 2 = 0$$

that has the *complex solutions* 1−i and 1+i, MATHCAD returns the numerical solution:

* for real *starting values:*

  no solution

* for *complex starting values*, e.g.,

  x := i

  $\text{root}(x^2 - 2 \cdot x + 2, x) = 1 + i$

  and

  x := −i

  $\text{root}(x^2 - 2 \cdot x + 2, x) = 1 - i$

  a complex solution in each case.

e2) Take the *quadratic equation*

$$x^2 - 3x + 2 = 0$$

that has the *real solutions* 1 and 2, MATHCAD returns the numerical solution:

* for real *starting values*, e.g.,

  x := 5

  $\text{root}(x^2 - 3 \cdot x + 2, x) = 2$

  and

  x := −1

  $\text{root}(x^2 - 3 \cdot x + 2, x) = 1$

* for *complex starting values*, e.g.,

  x := −i

  $\text{root}(x^2 - 3 \cdot x + 2, x) = 1$

These examples show that it is better to use **root** for real or complex starting values when looking for real or complex solutions. The quadratic polynomials used here serve only demonstrative purposes. Thus, you should first use **polyroots** to try to find the *numerical solution* of *polynomial equations*, because it does not require any starting values. This means the following for our two examples

$$\text{polyroots}\left(\begin{pmatrix} 2 \\ -2 \\ 1 \end{pmatrix}\right) = \begin{pmatrix} 1 - 1i \\ 1 + 1i \end{pmatrix} \qquad \text{polyroots}\left(\begin{pmatrix} 2 \\ -3 \\ 1 \end{pmatrix}\right) = \begin{pmatrix} 1 \\ 2 \end{pmatrix}$$

f) Consider an overdeterminated *system of equations* (without solution) and determine a *generalized solution* using the numerical function **minerr**:

x := 2   y := 1

given

$$x^2 - y^2 = 3$$
$$x^4 + y^4 = 17$$
$$x + y = 5$$

$$\textbf{minerr}\,(x\,,\,y) = \begin{pmatrix} 1.976 \\ 1.17 \end{pmatrix}$$

The obtained *generalized solution*

x = 1.976   y = 1.17

obviously cannot satisfy the three equations, but rather returns a *minimum* of the *sum* of *squares*

$$(x^2 - y^2 - 3)^2 + (x^4 + y^4 - 17)^2 + (x + y - 5)^2$$

♦

To summarize, MATHCAD is *not always successful* for the *numerical solution* of nonlinear equations. This is to be expected, because we know from *numerical analysis* that the used methods *do not necessarily converge*, even when the starting values are near a solution.

The examples show that it is better to solve a system of equations numerically using various starting values.

♦

16.5   Inequalities

Consider an *inequality* of the form

$$u\,(x) \le 0$$

and *systems* of *inequalities* of the form

$$u_1(x_1\,,\,x_2\,,...,\,x_n)\;\le\;0$$
$$u_2(x_1\,,\,x_2\,,...,\,x_n)\;\le\;0$$
$$\vdots$$
$$u_m\,(x_1\,,\,x_2\,,...,\,x_n)\;\le\;0$$

The *exact* and *numerical solutions* for these *inequalities* are obtained in MATHCAD analogue to the solution of *equations*. You just replace the equations with the appropriate inequalities, where the *inequalities* are entered in the worksheet using the *inequality buttons*

   or

from *operator palette no. 2*.

If *no exact solution* of an inequality *was found*, MATHCAD reacts in the same manner as for equations.

Because *no finite solution algorithm* also does not exist for the solution of general nonlinear inequalities, this situation occurs frequently.

In this case you can use MATHCAD's numerical methods for the approximate solution of equations (see Section 16.4). You can also plot the function u(x) (see Section 18.1) to obtain information about the solutions.

♦

We now use a number of examples to illustrate the problems involved with the solution of inequalities.

Example 16.10:

The following *inequalities* in a to e are solved using the *menu sequence*

Symbolics ⇒ Variable ⇒ Solve or the **solve** *keyword*.

a) The inequality

$$x^2 - 9\cdot x + 2 \ge 0 \qquad \textit{has solution(s)} \qquad \begin{pmatrix} x \le \dfrac{9}{2} - \dfrac{1}{2}\cdot\sqrt{73} \\[2mm] \dfrac{9}{2} + \dfrac{1}{2}\cdot\sqrt{73} \le x \end{pmatrix}$$

$$x^2 - 9 \cdot x + 2 \ge 0 \text{ solve}, x \quad \rightarrow \quad \left[\begin{array}{c} x \le \dfrac{9}{2} - \dfrac{1}{2} \cdot \sqrt{73} \\ \dfrac{9}{2} + \dfrac{1}{2} \cdot \sqrt{73} \le x \end{array} \right]$$

is solved without difficulty using both methods. The *graphical display* confirms the obtained result:

$$x := -1, -0.999 \ .. \ 10$$

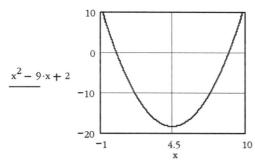

b) Both methods solve the following *inequality.*

$$|x - 1| + |x + 1| \ge 3 \qquad \textit{has solution(s)} \qquad \left[\begin{array}{c} x \le \dfrac{-3}{2} \\ \dfrac{3}{2} \le x \end{array} \right]$$

$$|x - 1| + |x + 1| \ge 3 \text{ solve}, x \quad \rightarrow \quad \left[\begin{array}{c} \dfrac{3}{2} \le x \\ x \le \dfrac{-3}{2} \end{array} \right]$$

The *graphical display* confirms the result:

$$x := -3, -2.999 .. 3$$

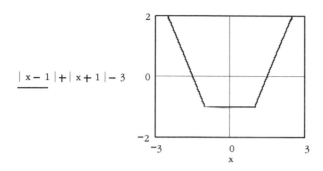

$$|x - 1| + |x + 1| - 3 \qquad 0$$

c) Both methods solve the following *inequality:*

$$|x - 1| + |x + 1| \leq 3 \qquad \textit{has solution(s)} \qquad \left(\frac{-3}{2} \leq x\right) \cdot \left(x \leq \frac{3}{2}\right)$$

$$|x - 1| + |x + 1| \leq 3 \text{ solve}, x \quad \rightarrow \quad \left(\frac{-3}{2} \leq x\right) \cdot \left(x \leq \frac{3}{2}\right)$$

d) If we make a minor change to the *inequality* from c

$$|x - 1| - |x + 1| \leq 3$$

so that it is satisfied for all values of x, MATHCAD does not find *any solution:*

$$|x - 1| - |x + 1| \leq 3 \qquad \textit{has solution(s)} \qquad \text{"untranslatable"}$$

$$|x - 1| - |x + 1| \leq 3 \text{ solve}, x \quad \rightarrow \quad \text{"untranslatable"}$$

e) The simple *inequality*

$$x^3 + \sin x \leq 0$$

that is satisfied for all $x \leq 0$, as you can see from the following graph:

$$x := -3, -2.999 .. 3$$

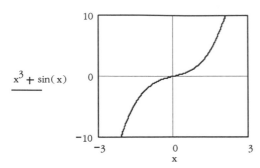

is solved by MATHCAD with both methods:

$$x^3 + \sin(x) \leq 0 \quad \textit{has solution(s)} \quad x \leq 0$$

$$x^3 + \sin(x) \leq 0 \text{ solve}, x \quad \rightarrow x \leq 0$$

f) If you only require real and positive solutions for the system of equations from Example 16.7a, you must add the two inequalities x≥0 and y≥0 so that a *system of inequalities* is to be solved.

- We first *try* for the *exact solution* using

  * the *solve block* with **given** and **find**

    given

    $$x^2 - y^2 = 3$$

    $$x^4 + y^4 = 17$$

    $$x \geq 0 \quad y \geq 0$$

    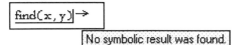

    No symbolic result was found.

  * the **solve** *keyword*:

$$\begin{bmatrix} x^2 - y^2 = 3 \\ x^4 + y^4 = 17 \\ x \geq 0 \\ y \geq 0 \end{bmatrix} \text{ solve}, \begin{bmatrix} x \\ y \end{bmatrix} \rightarrow (\, 2 \quad 1 \,)$$

A MATHCAD problem occurs here; the two methods for the exact
solution provide different results: one method calculates the solution
whereas the other method fails.

* Let us now *determine numerically* a *solution* for the *starting values*
  (3,4):

$$x := 3 \qquad y := 4$$

given

$$x^2 - y^2 = 3$$

$$x^4 + y^4 = 17$$

$$x \geq 0 \qquad y \geq 0$$

$$\textbf{find}\,(x,y) \;=\; \binom{2}{1}$$

The solution (2,1) is found here.

♦

☞

As shown in Example 16.10d, MATHCAD sometimes fails for inequalities
that contain magnitudes. Example 16.10f also shows MATHCAD's failure
when the **given** and **found** methods are used.

Because the zeros of the functions are required for the solution of inequalities, the same problems occur for inequalities as for equations.

♦

17 Functions

Functions play a *fundamental role* in all *applications* (see [3,4]). The wide range of functions integrated in MATHCAD simplify many of your tasks. We differentiate here between *general* and *mathematical functions.*

We have already discussed many of these functions and will also meet new functions in the following chapters. We discuss in Section 17.1 some important *general functions* and provide in Section 17.2 an overview of *elementary* and *special mathematical functions.*

All *functions integrated* in MATHCAD (*built-in* or *predefined functions*) shown in the

Insert Function

dialogue box can be opened in two ways

* using the *menu sequence*

    Insert ⇒ Function ...

* by clicking the

    button in the *standard toolbar.*

♦

You can enter the *designations* of the *functions integrated* in MATHCAD (built-in or predefined functions) at the position in the worksheet marked by the cursor in two different ways:

I. *Direct input* using the *keyboard.*

II. Insert with mouse click in the

    Insert Function

dialogue box. Because you see a short *explanation* next to the *function syntax,* we recommend the use of Method II.

♦

17.1 General Functions

MATHCAD contains many general functions, some of which we have already discussed, such as the *input/output functions* (*file access functions*)
described in Chapter 9.
We discuss in the following sections some rounding, sorting and string
functions that are useful when working with MATHCAD.

17.1.1 Rounding Functions

The *rounding functions* include the following functions that round up or
down a number or calculate remainders, where x and y are floating point
numbers:

- **floor** (x)

  calculate the *largest integer* $\leq x$

- **ceil** (x)

  calculate the *smallest integer* $\geq x$

- **mod** (x , y)

  calculate the remainder produced by the *division* x : y; the result has
  the same sign as x.

- **round** (x , n)

  rounds the number x to n decimal places.

- **trunc** (x)

  returns the integer part of a number x by removing the fraction part.

These *rounding functions* are used as follows:

* *First enter* the *function* with its arguments in the worksheet; the *argu-
  ments* can be variables, constants or numbers.
* *Then mark* the *function* with *editing lines.*
* The *input* of the *numerical equal sign* = returns the *result.*

The following example shows the effect of the rounding functions.

Example 17.1:

x := 5.87   y := −11.57

floor (x) = 5   **floor** (y) = −12

ceil (x) = 6   **ceil** (y) = −11

mod (x , y) = 5.87   **mod** (5 , 2) = 1

mod (6 , 3) = 0   **mod** (3 , 6) = 3

round (x , 1) = 5.9   **round** (y , 1) = – 11.6

trunc (x) = 5   **trunc** (y) = – 11
♦

17.1.2 Sorting Functions

MATHCAD contains a number of *sorting functions* that can be used to sort *vector* components or *matrix elements*. The most important of these functions:

- **sort (x)**

  sort the *components* of a *vector* **x** into *ascending order* of its number values.

- **csort (A , n)**

  sort the *rows* of a *matrix* **A** so that the *elements* in the *n-th column* are in *ascending order.*

- **rsort (A , n)**

  sort the *columns* of a *matrix* **A** so that the *elements* in the *n-th row* are in *ascending order.*

- **reverse (x)**

  arrange the *components* of a *vector* **x** in *reverse order,* i.e., the last element becomes the first element.

- **reverse (A)**

  arrange the rows of a matrix **A** in *reverse order.*

- **reverse (sort (x))**

  These nested *functions sort* the *components* of a *vector* **x** in *descending order.*

These *functions* are used as follows:

* *First enter* the *function* with its arguments in the worksheet.
* Then *mark* the *function* with *editing lines.*
* *Enter* the *numerical equal sign* = to return the *result.*

Let us consider the given sort functions in the following example.

Example 17.2:

Note that *vectors* are always entered in *column form* in the examples.
All the shown vectors and matrices use 1 as the starting value for indexing, i.e.,

ORIGIN := 1

a) The components of the following three vectors

$$\mathbf{x} := \begin{pmatrix} 1 \\ 3 \\ 5 \\ 9 \\ 2 \\ 3 \\ 6 \\ 4 \\ 2 \\ 7 \\ 8 \\ 4 \end{pmatrix} \qquad \mathbf{y} := \begin{pmatrix} -1 \\ -3 \\ -9 \\ -11 \\ 4 \\ 5 \\ 23 \\ 21 \\ 17 \\ -21 \\ -31 \\ 4 \\ 13 \\ -19 \\ 26 \\ -15 \\ 25 \end{pmatrix} \qquad \mathbf{z} := \begin{pmatrix} 1.23 \\ -7.65 \\ 2.34 \\ 6 \\ 87 \\ -56 \\ 34 \\ 7.31 \\ 23 \\ 1 \\ 7 \\ -9 \\ 2.94 \end{pmatrix}$$

are

*   sorted according to size using the **sort** *function* (ascending order):

sort(x) =			sort(y) =			sort(z) =		
		1			1			1
	1	1		1	-31		1	-56
	2	2		2	-26		2	-9
	3	2		3	-21		3	-7.65
	4	3		4	-19		4	1
	5	3		5	-15		5	1.23
	6	4		6	-11		6	2.34
	7	4		7	-9		7	2.94
	8	5		8	-3		8	6
	9	6		9	-1		9	7
	10	7		10	4		10	7.31
	11	8		11	4		11	23
	12	9		12	5		12	34
				13	13		13	87
				14	17			
				15	21			
				16	23			

* shown in reverse order using the **reverse** *function*:

reverse(x) =			reverse(y) =			reverse(z) =		
		1			1			1
	1	4		1	25		1	2.94
	2	8		2	-15		2	-9
	3	7		3	-26		3	7
	4	2		4	-19		4	1
	5	4		5	13		5	23
	6	6		6	4		6	7.31
	7	3		7	-31		7	34
	8	2		8	-21		8	-56
	9	9		9	17		9	87
	10	5		10	21		10	6
	11	3		11	23		11	2.34
	12	1		12	5		12	-7.65
				13	4		13	1.23
				14	-11			
				15	-9			
				16	-3			

* sorted into descending order using the nested *functions*
reverse (sort ( )):

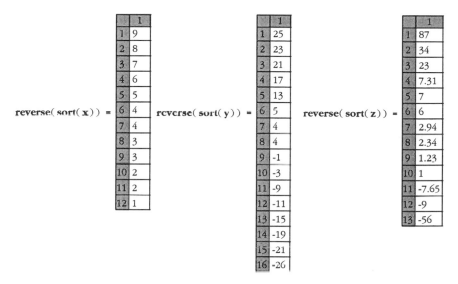

You can again observe the effect that not all components of the result vectors are displayed directly on the screen for the *representation* with *indices*. A *scrolling output table* is specified in this display form. This can be seen with the vector **y** that has 17 components. Only 16 *components* are shown directly in the scrolling output table for the *display* of the *result* with specification of the *indices*.

If you wish to *suppress* the *indices* for the *results*, i.e., display the *results* in *matrix form*, you can achieve this by *activating* the

Format ⇒ Result...

menu sequence and then setting

Matrix

in the **Matrix display style** *field* in the displayed **Result Format** *dialogue box* (see Section 15.1).

The *representation* in *matrix form* is preferable when you wish to display all elements, as we demonstrate in the following:

$$\mathbf{sort(x)} = \begin{pmatrix} 1 \\ 2 \\ 2 \\ 3 \\ 3 \\ 4 \\ 4 \\ 5 \\ 6 \\ 7 \\ 8 \\ 9 \end{pmatrix}$$

$$\mathbf{sort(y)} = \begin{pmatrix} -31 \\ -26 \\ -21 \\ -19 \\ -15 \\ -11 \\ -9 \\ -3 \\ -1 \\ 4 \\ 4 \\ 5 \\ 13 \\ 17 \\ 21 \\ 23 \\ 25 \end{pmatrix}$$

$$\mathbf{sort(z)} = \begin{pmatrix} -56 \\ -9 \\ -7.65 \\ 1 \\ 1.23 \\ 2.34 \\ 2.94 \\ 6 \\ 7 \\ 7.31 \\ 23 \\ 34 \\ 87 \end{pmatrix}$$

$$\mathbf{reverse(x)} = \begin{pmatrix} 4 \\ 8 \\ 7 \\ 2 \\ 4 \\ 6 \\ 3 \\ 2 \\ 9 \\ 5 \\ 3 \\ 1 \end{pmatrix}$$

$$\mathbf{reverse(y)} = \begin{pmatrix} 25 \\ -15 \\ -26 \\ -19 \\ 13 \\ 4 \\ -31 \\ -21 \\ 17 \\ 21 \\ 23 \\ 5 \\ 4 \\ -11 \\ -9 \\ -3 \\ -1 \end{pmatrix}$$

$$\mathbf{reverse(z)} = \begin{pmatrix} 2.94 \\ -9 \\ 7 \\ 1 \\ 23 \\ 7.31 \\ 34 \\ -56 \\ 87 \\ 6 \\ 2.34 \\ -7.65 \\ 1.23 \end{pmatrix}$$

$$\text{reverse}(\text{sort}(\mathbf{x})) = \begin{pmatrix} 9 \\ 8 \\ 7 \\ 6 \\ 5 \\ 4 \\ 4 \\ 3 \\ 3 \\ 2 \\ 2 \\ 1 \end{pmatrix} \quad \text{reverse}(\text{sort}(\mathbf{y})) = \begin{pmatrix} 25 \\ 23 \\ 21 \\ 17 \\ 13 \\ 5 \\ 4 \\ 4 \\ -1 \\ -3 \\ -9 \\ -11 \\ -15 \\ -19 \\ -21 \\ -26 \\ -31 \end{pmatrix} \quad \text{reverse}(\text{sort}(\mathbf{z})) = \begin{pmatrix} 87 \\ 34 \\ 23 \\ 7.31 \\ 7 \\ 6 \\ 2.94 \\ 2.34 \\ 1.23 \\ 1 \\ -7.65 \\ -9 \\ -56 \end{pmatrix}$$

b) We obtain the following results if we use the provided *sorting functions* for a given *matrix* **A**:

$$\mathbf{A} := \begin{pmatrix} 4 & 2 & 6 \\ 1 & 5 & 3 \\ 7 & 9 & 8 \end{pmatrix} \qquad\qquad \text{reverse}(\mathbf{A}) = \begin{pmatrix} 7 & 9 & 8 \\ 1 & 5 & 3 \\ 4 & 2 & 6 \end{pmatrix}$$

$$\text{csort}(\mathbf{A}, 3) = \begin{pmatrix} 1 & 5 & 3 \\ 4 & 2 & 6 \\ 7 & 9 & 8 \end{pmatrix} \qquad \text{rsort}(\mathbf{A}, 2) = \begin{pmatrix} 4 & 6 & 2 \\ 1 & 3 & 5 \\ 7 & 9 & 8 \end{pmatrix}$$

♦

17.1.3 String Functions

MATHCAD has a number of functions (*string functions*) that can be used to perform operations on *strings* (see Section 8.4); concatenation is an example of such a function. Because we do not use these functions in the book, we refer the reader to the MATHCAD help that provides a detailed explanation of these functions.

17.2 Mathematical Functions

MATHCAD has a number of *mathematical functions* from which we discuss *real functions* for *real variables* in this section. These functions are normally divided into two groups:

* *Elementary functions* (see Section 17.2.1)

  These include power, logarithmic and exponential functions, trigonometric and hyperbolic functions and their inverse functions, all of which are available in MATHCAD.

* *Special functions* (see Section 17.2.2)

  MATHCAD provides the Bessel functions.

Because the *elementary functions* occur in many applications, they adopt the dominating role.

MATHCAD provides *additional functions*, such as

* *functions* of *complex variables*
* *statistics functions*
* *matrix functions*

that also belong to the *mathematical functions* and which we discuss in the appropriate chapters.

♦

We assume in our discussions in this book that the user has basic knowledge of *real functions*. Consequently, we do not provide an exact mathematical definition of these functions. The following *simplified definition* suffices for our applications:

A rule f that assigns exactly one real number z from a set B (range) to each n-tuple of real numbers

$$x_1, x_2, \dots, x_n$$

from a given set A (domain of definition), is called a real function f of n real variables.

We use the *following designations* for *real functions:*

* *functions of one real variable* x

  $y = f(x)$

* *functions of two real variables* x , y

  $z = f(x,y)$

* *functions of n real variables* $x_1, x_2, \dots, x_n$

  $z = f(x_1, x_2, \dots, x_n)$

where we use the non-precise mathematical notation and designate the function value rather than f as function. MATHCAD also uses this notation.

MATHCAD provides three methods of writing variables for the *representation* of *functions* having *n variables*

$$z = f(x_1, x_2, \ldots, x_n)$$

* *variables without indexes* in the form

  x1, x2, ... , xn

  i.e.,   z = f (x1, x2, ... , xn)

* *indexed variables* with *array index*

  $$x_1, x_2, \ldots, x_n$$

  i.e., $z = f(x_1, x_2, \ldots, x_n)$

* *indexed variables* with *literal index*

  $$x_1, x_2, \ldots, x_n$$

  i.e., $z = f(x_1, x_2, \ldots, x_n)$

Section 8.2 provided a detailed discussion of these representation forms for variables.

♦

MATHCAD provides comprehensive tools to *investigate real functions*; we discuss these tools in the course of the book.
These include

* *graphic representations* (Chapter 18)
* *determination of zeros* (Chapter 16)
* *calculation of limits* (Section 19.4)
* *differentiation* (Section 19.1)
* *discussion of curves* (Section 19.5).

♦

17.2.1 Elementary Functions

The *elementary functions* include the following functions of a real variable x :

* *power functions* and their *inverses* (*root functions*)
* *exponential functions* and their *inverses* (*logarithmic functions*)
* *trigonometric functions* and their *inverses*
* *hyperbolic functions* and their *inverses*.

Because you can obtain the *syntax* of these *functions* from the

Insert Function

dialogue box as already described at the start of the chapter, we do not need to specify the syntax.

However, please note that MATHCAD requires that the arguments of these functions are always included within parentheses.

♦

Once a function with its argument is contained in the worksheet and has been marked with editing lines, the *input* of the

* *symbolic equal sign* → with the subsequent pressing of the ⏎-key

  returns the *exact function value*

* *numerical equal sign* =

  returns the *numerical function value.*

♦

Please note the following *peculiarities* when working with *elementary functions* in MATHCAD:

- *Representation*
    * The *e-function*

      e^x

      can be entered as either e^x or exp (x)

    * Only the form a^x is permitted as input for *general power functions* to any base a>0, i.e.,

      a^x

    * No function names exist for the calculation of *roots* and the *magnitudes (absolute value)*, rather they are created by clicking the *root button*

          (*square root*)   (*n-th root*)

      or the *magnitude (absolute value) button*

      from *operator palette no.1.*

- *Arguments*

    The *arguments* for *trigonometric functions* are specified in *radians* (*rad* unit). If you wish to enter such arguments in *degrees*, the argument must be multiplied with *deg* (Example 17.3a).

The function value of their *inverse functions* is output in radians. If you wish to convert the result to degrees, enclose the expression with editing lines and enter *deg* in the displayed *units placeholder* (Example 17.3b).

Chapter 11 provides a more detailed discussion of working with units.

♦

Example 17.3:

a) Calculate the sin function for 90 degrees, first for *radians* and then for *degrees:*

$$\sin\left(\frac{\pi}{2}\right) = 1 \qquad \sin(90 \cdot \deg) = 1$$

b) The function values for arc functions are output in radians:

$$\text{asin}(1) = 1.571 \quad \blacksquare \quad \text{acos}(0) = 1.571 \quad \blacksquare$$

If you require the result in *degrees*, enter *deg* in the *units placeholder* (black box):

$$\text{asin}(1) = 90 \cdot \deg \qquad \text{acos}(0) = 90 \cdot \deg$$

c) The *conversion from radians* into degrees and vice versa:

$$\frac{\pi}{2} \cdot \text{rad} = 90 \cdot \deg$$

$$180 \cdot \deg = 3.142 \cdot \text{rad}$$

♦

17.2.2 Special Functions

MATHCAD contains as *special functions* the *first* and *second type Bessel functions* required for the solution of differential equations. Because you can obtain the syntax of these functions from the MATHCAD help when you enter *Bessel* as a search expression we do not need to specify the syntax.

17.2.3 Definition of Functions

Although MATHCAD provides many functions, it is often necessary for effective working to *define additional functions*. We have already received a first impression of this in Section 10.3.

Let us consider *two characteristic cases* for which a *function definition* is desirable:

I. if in the course of a session, you make frequent use of *formulae* or *expressions* that are not integrated in MATHCAD;

II. if you receive *expressions* as the *result* of a *calculation* (e.g., differentiation or integration of a function) that are required in subsequent calculations.

Such *function definitions* have the *advantage* that it suffices to use the selected *function designation* in the subsequent calculations rather than entering the complete expression each time.

♦

Function definitions are realized using the *assignment operators*

*   :=   (local)

*   ≡   (global)

by entering the colon or using the *operator palette no. 2*; local and global definitions have the same properties as for variables (see Sections 8.2 and 10.3). Note that MATHCAD distinguishes between uppercase and lowercase for the selected *function designations*.

♦

Ensure that your *defined functions* do not have the names of *built-in* (*predefined*) *functions*, such as **sin, cos, ln, floor** (*reserved names*), because these would then no longer be available. Also note that MATHCAD does not differentiate between function names and variable names. If, for example, you define a *function* v(x) and then a *variable* v, the function v(x) is now no longer available (see Example 17.4f).

♦

We now summarize both *methods* for *function definitions* in MATHCAD:

I. A *given expression* :

$$A(x_1, x_2,..., x_n)$$

is *assigned* to the *function* f locally or globally using

$$f(x_1, x_2,..., x_n) := A(x_1, x_2,..., x_n)$$

or

$$f(x_1, x_2,..., x_n) \equiv A(x_1, x_2,..., x_n)$$

If the *function* to be defined consists of *several analytical expressions*, such as

$$f(x_1,x_2,...,x_n) = \begin{cases} A_1(x_1,x_2,...,x_n) & \text{if } (x_1,x_2,...,x_n) \in D_1 \\ A_2(x_1,x_2,...,x_n) & \text{if } (x_1,x_2,...,x_n) \in D_2 \end{cases}$$

or

$$f(x_1, x_2, ..., x_n) = \begin{cases} A_1(x_1, x_2, ..., x_n) & \text{if } (x_1, x_2, ..., x_n) \in D_1 \\ A_2(x_1, x_2, ..., x_n) & \text{if } (x_1, x_2, ..., x_n) \in D_2 \\ A_3(x_1, x_2, ..., x_n) & \text{if } (x_1, x_2, ..., x_n) \in D_3 \end{cases}$$

you can define it using the **if** statement (see Examples 17.4b and c).

II. There are two methods of assigning a calculated *expression* to a *function*:

   * You can use the *method I.* described above. However, you do not need to re-input the *function expression*, but merely *mark* it with *editing lines* and *copy* it in the usual Windows manner, e.g., by *clicking* the *copy symbol* from the standard toolbar to copy it into the *clipboard*.

      The *insertion* in the assignment at the required position is also performed in the usual Windows manner by *clicking* the *insert symbol* in the standard toolbar.

   * You can realize this *function assignment easier* than the method just described if you perform calculations using the *symbolic equal sign.* Here the calculation of an *expression 1* by entering the *symbolic equal sign* yields the *expression_2*, i.e.,

     Expression_1 → *Expression_2* .

     The function assignment

     f(x, y, ...) := *Expression_1* → *Expression_2*

     assigns the result *Expression_2* to the function f(x, y, ...). Example 17.4e provides a detailed illustration of the procedure involved.
     ♦

As with the built-in functions in MATHCAD, you can enter the symbolic or numerical equal sign to calculate *exactly* or *numerically* the *function values of defined functions* (see Example 17.4a), i.e., using

$$f(x_1, x_2, ..., x_n) \rightarrow \qquad \text{or} \qquad f(x_1, x_2, ..., x_n) =$$
♦

Example 17.4:

a) Let us now consider the various ways of *defining* a *function* with *three variables:*

   a1) Definition using *non-indexed variables* such as u, v and w:

     $f(u, v, w) := u \cdot v^2 \cdot \sin(w)$

The exact or numerical *calculation* of *function values* is performed as follows:

$$f(1,2,3) \rightarrow 4 \cdot \sin(3) \qquad \text{or} \qquad f(1,2,3) = 0.564$$

a2) If you wish to use *indexed variables* (with *array index*) for the definition, this cannot be done in the following form:

$$f(x_1, x_2, x_3) := x_1 \cdot (x_2)^2 \cdot \sin(x_3)$$

$$\boxed{\text{error in list}}$$

Because the variables for the array index are interpreted as being components of a vector **x**, the following method must be used:

Function definition:

$$f(x) := x_1 \cdot (x_2)^2 \cdot \sin(x_3)$$

Exact and numerical calculation of function values:

$$f\left(\begin{pmatrix} 1 \\ 2 \\ 3 \end{pmatrix}\right) \rightarrow 4 \cdot \sin(3) \qquad \text{or} \qquad f\left(\begin{pmatrix} 1 \\ 2 \\ 3 \end{pmatrix}\right) = 0.564$$

The *indexed variables* can be formed using the

button from *operator palette no. 1* or *4*.

a3) If you use *variables* with *literal index* (see Section 8.2), proceed as follows:

Function definition:

$$f(x_1, x_2, x_3) := x_1 \cdot (x_2)^2 \cdot \sin(x_3)$$

Exact and numerical calculation of function values:

$$f(1,2,3) \rightarrow 4 \cdot \sin(3) \qquad f(1,2,3) = 0.564$$

The *literal index* can be formed by entering a period after the variable designation x.

b) If a *function* is formed from *two analytical expressions*, such as

$$f(x) = \begin{cases} x+1 & \text{if } x \leq 0 \\ \dfrac{x^2}{2} + 3 & \text{if } x > 0 \end{cases}$$

that has a *discontinuity* at the origin, the function can be defined in MATHCAD using one of the following ways:

*   with the **if** statement

$$f(x) := \mathbf{if}\left(x \le 0 , x+1 , \frac{x^2}{2} + 3\right)$$

*   with the **if** operator from the *operator palette no. 6*

$$f(x) := \begin{Vmatrix} x \mid 1 & \text{if } x \le 0 \\[2mm] \frac{x^2}{2} + 3 & \text{otherwise} \end{Vmatrix}$$

The *graphic representation* yields

$$x := -2, -1.9999 .. 2$$

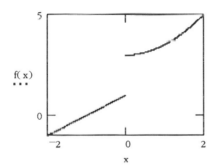

c) If a function is defined by three analytical expressions, such as

$$f(x) = \begin{cases} 0 & \text{if } x \le 0 \\[2mm] 1 & \text{if } 0 < x \le 1 \\[2mm] x^2 & \text{if } 1 < x \end{cases}$$

that has a *discontinuity* at the origin, it can be *defined* and *shown graphically* in MATHCAD

*   by *nesting* the **if** statement

$$f(x) := \mathbf{if}\,(\,x \le 0, 0, \mathbf{if}\,(x \le 1, 1, x^2)\,)$$

*   through the use of the **if** operator from the *operator palette no. 6*

$$f(x) := \begin{cases} 0 & \text{if } x \leq 0 \\ x^2 & \text{if } x > 1 \\ 1 & \text{otherwise} \end{cases}$$

The graphical display yields:

$$x := -2, -1.99 .. 3$$

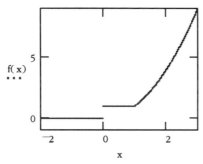

d) In the *compound interest calculation*, the resulting
   end capital K_n

   after n *years* of interest (compound interest) at an
   interest rate p

   starting with the
   initial capital K_0

   can be calculated using the formula

   $$K_n = K_0 \left(1 + \frac{p}{100}\right)^n$$

   If you make frequent use of this formula, it is better to define a *function*

   $$K_n(K_0, p, n) = K_0 \left(1 + \frac{p}{100}\right)^n$$

   with three independent variables

   K_0, p, n

   that returns the *end capital*

   K_n

   as function value.
   You can *define* this *function* in MATHCAD in one of the following ways:

   *   *Without* using *indices* as

$$KN(K0, p, n) := K0 \cdot \left(1 + \frac{p}{100}\right)^n$$

For example, you calculate the end capital after 10 years of interest at 4% starting with the initial capital of $ 25,000 using this defined function:

$$KN(25000, 4, 10) = 3.7006 \cdot 10^4$$

i.e., the capital has increased to $37,006.

* If you wish to use indices for the *initial capital* and *end capital*, it is better to use *literal indices:*

$$K_n(K_0, p, n) := K_0 \cdot \left(1 + \frac{p}{100}\right)^n$$

e) We show in the following the possibilities of the *function definition* for *results* of *calculations* obtained using the *symbolic equal sign* →:

e1)Results of differentiations

$$f(x) := \frac{d}{dx}(e^x + \sin(x)) \rightarrow \exp(x) + \cos(x)$$

If you do not wish to immediately display the result, it suffices to make the function definition

$$f(x) := \frac{d}{dx}(e^x + \sin(x))$$

Enter the symbolic equal sign if you want to display such a defined function later.

$$f(x) \rightarrow \exp(x) + \cos(x)$$

e2)Results of integrations

$$h(x) := \int^{\bullet} x^2 \cdot e^x \, dx \rightarrow x^2 \cdot \exp(x) - 2 \cdot x \cdot \exp(x) + 2 \cdot \exp(x)$$

$$h(x) \rightarrow x^2 \cdot \exp(x) - 2 \cdot x \cdot \exp(x) + 2 \cdot \exp(x)$$

Note this function definition with the symbolic equal sign can be initiated only when the expression to be calculated has been marked with editing lines.

f) We demonstrate in this example the effects when you use the same designation for a function and a variable:

$$v(x) := x^2 + 1$$

$$v(5) = 26$$

$$v := 4$$

$$v(5) =$$

| illegal function name |

You can see that the function $v(x)$ is no longer available once the variable v has been defined.

♦

17.2.4 Approximation of Functions

The *approximation* of *functions* is a comprehensive area of numerical analysis that we cannot cover in detail in this book. However, we will consider a few frequently used methods for functions $f(x)$ of one variable x.

The name already expresses the basic principle behind *approximation*:
A *function* defined *analytically* or by *points* is to be *approximated* by another *function* (*approximate function*) using specific criteria.

♦

Polynomials play a fundamental role for the simpler *functions* normally used as *approximate functions*.
In this book we discuss the following frequently used methods for the *approximation* of *functions*:

- The Taylor series discussed in Section 19.2 provides a method of approximation using polynomials for *analytically* provided *functions*.

- A function defined by n *points* of the plane (pairs of *measured values*)

$$(x_1, y_1), (x_2, y_2), \dots, (x_n, y_n)$$

  requires the construction of an analytical function expression, for which we discuss the following standard methods provided by MATHCAD:

  * *interpolation method*
  * *method of least squares*

Whereas the *method of least squares* in Section 27.2 is considered as part of correlation and regression, we discuss *interpolation* in the following section.

The *interpolation* on the *plane* is based on the *following principle* (*interpolation principle*):

A function f(x) (designated as *interpolation function*) is to be determined so that it contains the given n *points*

$$(x_1, y_1), (x_2, y_2), \dots, (x_n, y_n)$$

i.e., the following relation must apply

$$y_i = f(x_i) \qquad \text{for} \qquad i = 1, 2, \dots, n$$

♦

The individual *interpolation types* differ through the *selection* of the *interpolation function*. Interpolation using polynomials (*polynomial interpolation*) is the best known method.

♦

The *polynomial interpolation* for n given points requires the use of a *polynomial function* of at least the (n−1) th order

$$y = f(x) = a_0 + a_1 \cdot x + \dots + a_{n-1} \cdot x^{n-1}$$

to satisfy the *interpolation principle*. The unknown n *coefficients*

$$a_k \qquad (k = 0, 1, 2, \dots, n-1)$$

are determined making use of the fact that the given n points satisfy the polynomial function. This yields a *linear system* of *equations* with n equations for the unknown n coefficients

$$a_k$$

This provides the first method of determining the interpolation polynomial. *Numerical analysis* provides *more effective methods* to *determine* the *interpolation polynomial.*

♦

The *interpolation functions* provided by MATHCAD require the coordinates of the given n points, which must be read or assigned. In the following section the given points are contained in the column vectors

* **vx** the *x-coordinates* (arranged in increasing sequence)

* **vy** the associated *y-coordinates*

Let us now consider the *properties* of the *interpolation functions:*

* The *interpolation function*

  linterp (vx, vy, x)

  connects the given points with straight lines, i.e., a *linear interpolation* is performed between the points and a *polygonal arc* calculated as approximate function. This function returns the function value of the polygonal arc. MATHCAD *extrapolates* if the x-values lie outside the given values. However, because of the resulting inaccuracy, extrapolation should be avoided wherever possible.

- The *interpolation function*

  interp (vs, vx, vy, x **)**

  performs a *cubic spline interpolation,* i.e., the given points are joined with polynomials of the third degree (cubic polynomials). Thus, a *spline function* of *degree* 3 is created and the **interp** function calculates the associated function value of the calculated spline function for the x-values. MATHCAD provides three possibilities to describe the behaviour of the created spline function at the end points (first and last component of the vector **vx**).

  The *column vector* **vs** required here must have been created previously using one of the following *functions*:

  * **vs := lspline (vx, vy)**

     for a *spline curve* that approaches a straight line at the end points.

  * **vs := pspline (vx, vy)**

     for a *spline curve* that approaches a parabola at the end points.

  * **vs := cspline (vx, vy)**

     for a *spline curve* that can be fully cubic at the endpoints.

We *test* the specified *interpolation functions* in the following example.

Example 17.5:

We approximate the following *five points* in the plane

$(1,2)$, $(2,4)$, $(3,3)$, $(4,6)$, $(5,5)$

that, for example, have been determined by measuring, with the *interpolation functions* from MATHCAD:

a) Construct a *polygonal arc* using *linear interpolation* that connects the given points.

  We use here the *interpolations function*

  linterp

  and initially enter the coordinates of the points:

$$
vx := \begin{pmatrix} 1 \\ 2 \\ 3 \\ 4 \\ 5 \end{pmatrix} \qquad vy := \begin{pmatrix} 2 \\ 4 \\ 3 \\ 6 \\ 5 \end{pmatrix}
$$

The *polygonal arc* is then calculated between x=0 and x=6 with the step size 0.001 and drawn with the given points in a common coordinate system:

$x := 0, 0.001 .. 6 \qquad i := 1 .. 5$

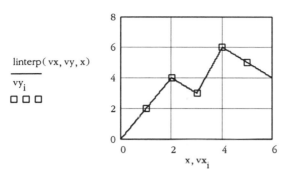

linterp(vx, vy, x)

$\overline{vy_i}$

□ □ □

x, vx$_i$

b) Approximation of the given points using *cubic splines* with the function

   interp

   where we consider the three different methods for the behaviour at the
   end points x=1 and x=5:

   We first enter the coordinates for the points:

$$vx := \begin{pmatrix} 1 \\ 2 \\ 3 \\ 4 \\ 5 \end{pmatrix} \qquad vy := \begin{pmatrix} 2 \\ 4 \\ 3 \\ 6 \\ 5 \end{pmatrix}$$

   b1) The *numerical function* **lspline** yields:

$vs := \textbf{lspline} \, (\, vx \, , \, vy \,)$

$$vs = \begin{bmatrix} 0 \\ 3 \\ 0 \\ 0 \\ -6.964 \\ 9.857 \\ -8.464 \\ 0 \end{bmatrix} \quad \blacksquare$$

$$x := 0, 0.001 .. 6 \qquad i := 1 .. 5$$

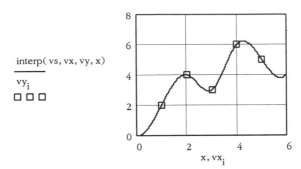

b2)The *numerical function* **pspline** yields:

$$vs := \textbf{pspline} \, (\, vx \, , \, vy \,)$$

$$vs = \begin{bmatrix} 0 \\ 3 \\ 1 \\ -5.4 \\ -5.4 \\ 9 \\ -6.6 \\ -6.6 \end{bmatrix} \blacksquare$$

$$x := 0, 0.001 .. 6 \qquad i := 1 .. 5$$

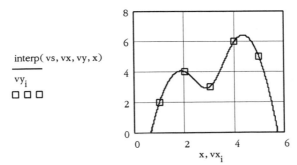

b3) The numerical function **cspline** yields:

$$vs := \textbf{cspline} \, (\, vx \, , \, vy \,)$$

$$vs = \begin{bmatrix} 0 \\ 3 \\ 2 \\ -13.75 \\ -3 \\ 7.75 \\ -4 \\ -15.75 \end{bmatrix} \blacksquare$$

$$x := 0, 0.001 .. 6 \qquad i := 1 .. 5$$

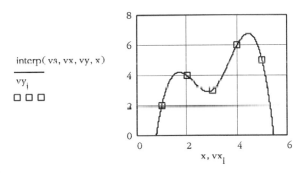

$$interp(vs, vx, vy, x)$$
$$\overline{vy_i}$$
□ □ □

Although MATHCAD does not provide any integrated numerical function for *polynomial interpolation*, you can find one in the **Numerical Recipes** *electronic book* with the name **polint** that we use in the following example.

♦

Example 17.6:

We approximate the points from Example 17.5 using a *interpolation polynomial of the degree four*. We find the required *interpolation function*

polint

in the **Numerical Recipes** *electronic book*.

The *interpolation function* for the *polynomial interpolation*

polint (vx , vy , x)

returns a result vector for the x-values (vector **vx**) and y-values (vector **vy**) of the given points that contains as first component the function value of the interpolation polynomial at x. The second component contains an error estimation.

We display the calculated *interpolation polynomial of the degree four* in the interval [0,6] and for the given points:

$$vx := \begin{pmatrix} 1 \\ 2 \\ 3 \\ 4 \\ 5 \end{pmatrix} \qquad vy := \begin{pmatrix} 2 \\ 4 \\ 3 \\ 6 \\ 5 \end{pmatrix}$$

$x := 0, 0.001 .. 6 \qquad i := 1 .. 5$

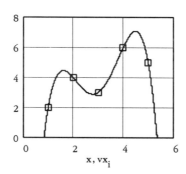

◆

18 Graphics Features

MATHCAD provides comprehensive *graphical capabilities* to draw both 2D and 3D graphs. These include the display of

* *curves* (Section 18.1)
* *surfaces* (Section 18.2)
* *point graphs* (Section 18.3)
* *diagrams* (Section 18.4)
* *animated graphs* (Section 18.5)

that we demonstrate in the following section.

Because it is not possible to explain all the graphical capabilities provided by MATHCAD, we suggest that you experiment by selecting various options, etc. The integrated help also provides further information here.

♦

18.1 Curves

We consider the *graphical representations* of *curves*

* in the *plane* R^2 (*plane curves*)
* in three-dimensional *space* R^3 (*space curves*)

Functions f(x) of one real variable x can be *displayed graphically* in the plane using a Cartesian coordinate system by drawing the *set of points*

$\{ (x,y) \in R^2 \ / \ y = f(x) \ , \ x \in D(f) \}$ (D(f) – *domain of definition*).

The following designations are used for this graphical representation of the function f(x):

* *graph*
* *function curve*
* *curve*.

☞

The graphs just defined cannot represent all possible *curves* of a *plane*, because a *function*

y = f(x)

is defined as a *single valued mapping* (see Section 17.2), so that closed curves, such as circles and ellipses, but also hyperbola, cannot be described:

* A *plane curve* created with the *function*

   y = f(x)

   is designated as a *curve* with an *equation* in *explicit representation*.

* The *implicit representation* provides another possibility. A *plane curve* defined *implicitly* by the *equation*

   F(x,y) = 0

   consists of all points belonging to the set

   $\{ (x,y) \in R^2 \, / \, F(x,y) = 0 \, , \, x \in D(f) \}$
   ◆

Example 18.1:

A *circle* with centre in 0 and radius a > 0 has in *Cartesian coordinates* the *equation*

$x^2 + y^2 = a^2$

i.e.., an equation in *implicit representation*

$F(x,y) = x^2 + y^2 - a^2 = 0$

that does not provide a unique solution for y.
Thus, an *explicit representation* in the form y = f(x) is *not possible* for the circle. Solving for y yields the two semicircles

$y = \sqrt{a^2 - x^2}$ and $y = -\sqrt{a^2 - x^2}$

i.e., two different functions of the form y = f(x).
   ◆
From the previously considered representations for *plane curves* in Cartesian coordinates:

* in *explicit representation*

   y = f(x)

* in *implicit representation*

   F(x,y) = 0

MATHCAD can only draw the explicit one.

The following further representation forms for *plane curves* can be used in MATHCAD:

- *Parametric representations* of the form

    $x = x(t)$, $y = y(t)$ $t \in [a, b]$

    in *Cartesian coordinate systems.*

- *Polar coordinates* of the form

    $r = r(\varphi)$ $\varphi \in [a, b]$

    These belong to the *curvilinear coordinates* that are used in place of *Cartesian coordinates* to simply represent *plane curves*. Where

    *   r

    *   φ

    represent the *radius* and *angle* respectively, i.e., the *radius vector* r is represented as function of the angle φ.

Let us consider several examples for possible curve representations.

Example 18.2:

a) A possible *parametric representation* for the *circle* from Example 18.1 in *polar form:*

  $x = a \cdot \cos t$, $y = a \cdot \sin t$

  where the *parameter* t represents the angle between the radius vector and the positive x–axis and runs from 0 to 2π (360°), i.e.,

  $0 \leq t \leq 2\pi.$

b) The *circle* from Example 18.1 has the simple *representation* in *polar co-ordinates*

  $r = a$

c) The *integral rational function* (*polynomial function*)

  $$y = f(x) = x^5 - 5 \cdot x^4 - 5 \cdot x^3 + 25 \cdot x^2 + 4 \cdot x - 20$$

  for which we calculated the zeros in Example 16.4a is displayed graphically in Figure 18.2.

d) The *fractional rational function*

  $$y = f(x) = \frac{x^4 - x^3 - x - 1}{x^3 - x^2}$$

  has been decomposed in Example 13.11b into partial fractions. Figure 18.3 shows the associated graph.

e) $x(t) = t - \sin t$, $y(t) = 1 - \cos t$ $(-\infty < t < \infty)$

is the *parametric representation* of a *cycloid* that can be represented in implicit form as

$$y = 1 - \cos(x + \sqrt{y \cdot (2 - y)})$$

Figure 18.5 shows the associated graph.

f) The equation of a *lemniscate* has the following representations

* in *implicit form*

$$(x^2 + y^2)^2 - 2 \cdot a \cdot (x^2 - y^2) = 0$$

* as *parametric representation*

$$x = a \cdot \cos t \cdot \sqrt{2 \cdot \cos 2t} \quad , \quad y = a \cdot \sin t \cdot \sqrt{2 \cdot \cos 2t} \qquad (\; 0 \le t \le 2\pi \;)$$

* as *polar coordinates*

$$r = a \cdot \sqrt{2 \cdot \cos 2\varphi} \qquad\qquad (\; 0 \le \varphi \le 2\pi \;)$$

Figure 18.7 shows its graph for a=1.

◆

MATHCAD does not use the *graphic functions* from MAPLE. Rather, a *special system* for *graphical representations* was developed for MATHCAD. We provide a detailed discussion for this system in the following.

◆

For the *graphical representation* of a *given function*

$$y = f(x)$$

the following steps are required in MATHCAD to draw the associated *function curve* in the plane:

* *First* use one the following *activities*
  * activate the *menu sequence*

    Insert ⇒ Graph ⇒ X-Y Plot
  * click the *graphic button*

    in *operator palette no. 3*

  to create a *graphic window* (*graphic frame*) of the form shown in Figure 18.1 at the cursor position in the worksheet.

* *Then* enter the *function designations*

f(x)

in this *graphic window* in the middle *placeholder* of the x-axis

x

and in the middle *placeholder* of the y-axis (indicated by *missing oper-and*) if this function has been defined previously. Otherwise the corresponding function expression must be entered instead of f(x). The remaining (outer) placeholders are used to specify the *dimension* (*axis scaling*). MATHCAD selects the values if you do not enter any values here.

- *Then* use the

  button from *operator palette no. 1* to enter the required *x-domain* (*domain of definition*, e.g., a ≤ x ≤ b) above the *graphic window* in the *form*

  x := a .. b   (*step size* 1)

  or

  x :– a , a + Δx .. b   (*step size* Δx)

  i.e., define x as *range variable* (see Section 8.3).
  If you do not define x as a range variable, MATHCAD creates a so-called *Quick Plot* for the *x-interval* [–10,10].

- *Finally*, if you are in *automatic mode, click* with the *mouse* outside the graphic window or press the ⏎-key, otherwise press the F9-function key, to get MATHCAD to draw the required *function curve*.

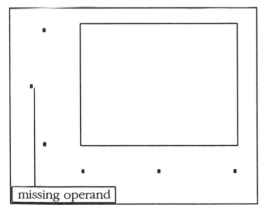

Figure 18.1. Graphic window for curves in Cartesian coordinates

MATHCAD *draws* the *function curve* by calculating the function values f(x) for the values of the x variable defined as range variable and connects the resulting points with straight lines.

If the *graphic* drawn by MATHCAD appears too *coarse*, you can *reduce* the *step size* Δx used in the *definition* of the *range variable* x to increase the number of points used to draw the graph. Figures 18.2, 18.3, 18.4, 18.5 and 18.7 that use 0.01, 0.00003, 0.001 and 0.0001 as step size illustrate this procedure.

A *suitable step size* Δx may also be *required* if the function f(x) has *poles*, in order to exclude them from the function calculation (see Figure 18.3).

◆

If you wish to draw the *curves* for *several functions*

f(x) , g(x) , ...

in the *same coordinate system*, enter these separated by commas in the placeholder for the y-axis in the graphic window.

Figure 18.4 shows the *graphical representation* of *three functions* in a Cartesian coordinate system. MATHCAD uses different line forms or colours to represent the individual functions; you can specify the form of these lines and colours.

◆

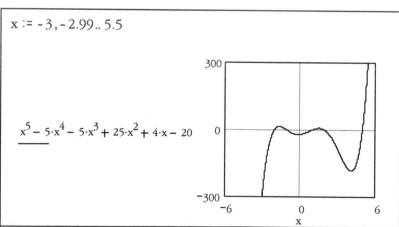

Figure 18.2. Graph of the function from Example 18.2c

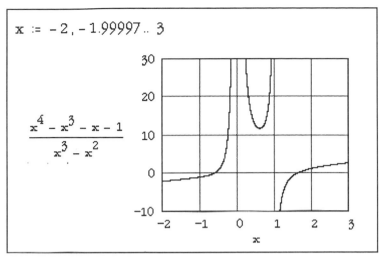

Figure 18.3. Graph of the function from Example 18.2d

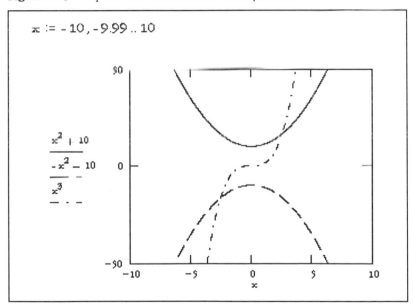

Figure 18.4. Three function curves in a Cartesian coordinate system

As already discussed in Section 8.3, you can use the definition of the variable x as range variable required for the graphical representation of a function f(x) in addition to the *output* of the defined x-*values* and the associated *function values* f(x). You only need to enter x and f(x) in the worksheet and the *numerical equal sign*, as we illustrate in the following Example 18.3.

♦

Example 18.3:

We define the *function*

$$f(x) := -x^2 + 2$$

and its independent variable x in the interval [–2,2] as *range variable* with the *step size* 0.5:

$$x := -2, -1.5 .. 2$$

You can then draw the associated function curve using the discussed method:

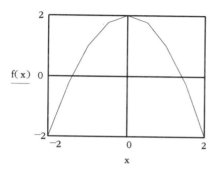

Both the value of the *range variable* x and the associated *function values* f(x) can be *calculated* by entering the numerical equal sign = :

x
– 2
– 1.5
– 1
– 0.5
0
0.5
1
1.5
2

f(x)
– 2
– 0.25
1
1.75
2
1.75
1
– 0.25
– 2

◆

☞

Unlike other computer algebra systems, MATHCAD does not yet have the capability of displaying *plane curves* specified in *implicit form* F(x,y)=0. You must use *parametric representations* or *polar coordinates* in this case (see Examples 18.2e and f.

◆

If a *plane curve* is present as the *parametric representation*

$$x = x(t) , y = y(t) \qquad t \in [a, b]$$

the *graphical representation* in MATHCAD is performed in the following *steps:*

- *First* create a *graphic window* in the worksheet at the cursor position using the method just described (see Figure 18.1).

- *Then* enter the functions x(t) or y(t) in the middle placeholder for the x-axis or y-axis.

- *Then* use the

  button from *operator palette no. 1* to enter the required *t-domain* (*domain of definition*, e.g., a ≤ t ≤ b) in the *form*

  t := a .. b   (*step size* 1)

  or

  t := a , a + Δt .. b   (*step size* Δt)

  above the *graphic window*, i.e., t is defined as *range variable* (see Section 8.3).
  If t has not been defined as range variable, MATHCAD creates a so-called *Quick Plot* for the *t-interval* [−10,10].

- *Finally*, if you are in *automatic mode*, *click* with the *mouse* outside the graphic window or press the ⏎-key, otherwise press the F9-function key, to get MATHCAD to draw the required *function curve*. Figure 18.5 shows an example.

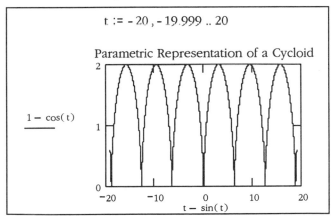

Figure 18.5. Graph of the function from Example 18.2e

If a *plane curve* is specified in *polar coordinates*

r = r(φ)   φ ∈ [a, b]

the *graphical representation* in MATHCAD is performed using the following *steps:*

- *Perform first* one of the following *activities*
  - * activate the *menu sequence*

    Insert ⇒ Graph ⇒ Polar Plot
  - * click on the *graphic button*

    in *operator palette no. 3*

  to create a *graphic window* (*graphic frame*) for *polar coordinates* in the worksheet at the cursor position having the form shown in Figure 18.6.
- *Then* enter

  φ

  in the *lower placeholder* and

  $r(\varphi)$

  in the *left placeholder* (indicated as *missing operand*) of the *graphic window.*

  The remaining (outer) placeholders are used to specify the scale (axis scale). MATHCAD selects these if you do not enter any values.
- *Then, the*

  button from the *operator palette no. 1* must be used to enter above the *graphic window* the required *φ-domain* (*domain of definition*, e.g., a ≤ φ ≤ b) in the *form*

  $\varphi := a \text{ .. } b$ (*step size* 1)

  or

  $\varphi := a , a + \Delta\varphi \text{ .. } b$ (*step size* $\Delta\varphi$)

  i.e., φ defined as *range variable* (see Section 8.3).
  If you do not define φ as range variable, MATHCAD creates a so-called *Quick Plot* for the *φ-interval* [0,2π].
- *Finally,* if you are in *automatic mode, click* with the *mouse* outside the graphic window or press the ⏎-key to obtain the required *function curve,* otherwise press the F9-function key.

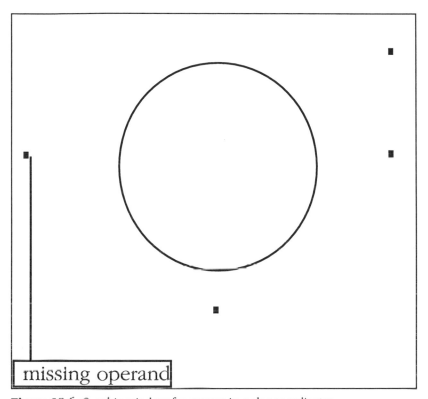

Figure 18.6. Graphic window for curves in polar coordinates

Figure 18.7 shows an example of the *representation* of a *curve* in *polar co-ordinates* in comparison with its *parametric representation*, where we have chosen a *lemniscate*. For reasons of simplicity, we have used the same designation t for both the independent variables.

♦

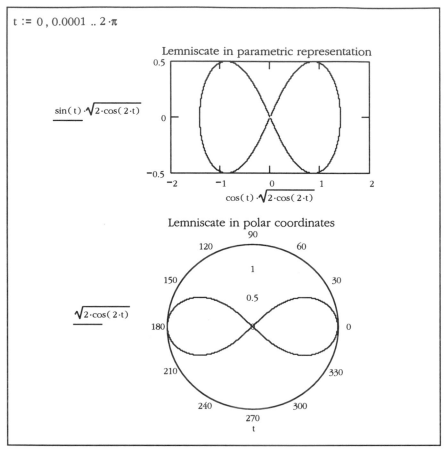

Figure 18.7. Graph of the lemniscate from Example 18.2f

The following steps can be used in MATHCAD to draw *space curves* specified with a *parametric representation* of the *form*

x = x(t) , y = y(t) , z = z(t)   t ∈ [a , b]

- *First perform* one of the following *activities*
    * activate the *menu sequence*

      Insert ⇒ Graph ⇒ 3D Scatter Plot
    * click on the *graphic button*

      in *operator palette no. 3*

to create a *graphic window* (*graphic frame*) in the worksheet at the cursor position having the form as shown in Figure 18.8.

- *Then* enter the *vectors* in the form

  x , y , z   (MATHCAD Version 7)

  (x , y , z)   (MATHCAD Version 8)

  in the *lower placeholder* of the *graphic window* whose components from the coordinates x(t), y(t), z(t) create the solid curve for various parameter values *using* the *graphic window* in the following manner:

  $$N := \qquad\qquad i := 0 .. N \qquad\qquad t_i :=$$

  $$x_i := x(t_i) \qquad y_i := y(t_i) \qquad z_i := z(t_i)$$

  The appropriate values must have been assigned to N and t_i (see Figure 18.8); note that we have selected 0 as starting value for the indexing, i.e.,

  ORIGIN := 0,

- *Finally,* if you are in *automatic mode*, *click* with the *mouse* outside the graphic window or press the ⏎-key to obtain the required *space curve,* otherwise press the F9-function key.

Example 18.4:

We now consider a *circular helix,* as a familiar *space curve.* It has the following *parametric representation:*

$$x(t) = a \cdot \cos t , \; y(t) = a \cdot \sin t , \; z(t) = b \cdot t \qquad (a>0 , b>0)$$

Figure 18.8 shows its curve drawn by MATHCAD for a=1 and b=1. We have chosen here the point form for its representation. The points can be joined by straight lines, if you set this option in the dialogue box displayed when you double-click on the graphic (also see Section 18.3).

♦

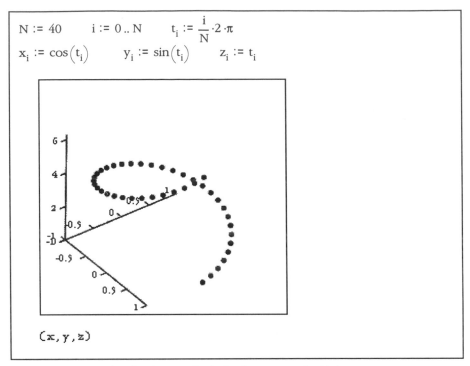

$$N := 40 \qquad i := 0 .. N \qquad t_i := \frac{i}{N} \cdot 2 \cdot \pi$$

$$x_i := \cos(t_i) \qquad y_i := \sin(t_i) \qquad z_i := t_i$$

(x, y, z)

Figure 18.8. Graphic for the circular helix from Example 18.4

The following method can be used to *change* (*manipulate*) the *graphics* created for *Cartesian coordinates* and *polar coordinates*:

- A mouse click initially encloses the graphic with a selection frame (selection rectangle).

- You can then use the following *menu sequences*

    * **Format ⇒ Graph ⇒ X-Y Plot...**

       for *Cartesian coordinates*

    * **Format ⇒ Graph ⇒ Polar Plot...**

       for *polar coordinates*

    * **Format ⇒ Graph ⇒ 3D Plot...**

       for *space curves*

    or double mouse-click on the graphic to *format* the *graphic* in the displayed dialogue box, i.e., scaling of the axes, specification of the colour and form of the curves, assign a title to the graphic, label the coordinate axes, etc.

- In addition, you can use the

    Format ⇒ Graph ⇒ Trace...

menu sequence or click the

button in *operator palette no. 3* to *display* the *coordinates* of a curve point marked with the cursor.

- Furthermore, you can use the

  Format ⇒ Graph ⇒ Zoom...

  menu sequence or click the

button in *operator palette no. 3* to *increase* the size of *parts* of a *graphic* by marking with pressed mouse button the section to be zoomed and then clicking the zoom button in the zoom dialogue box.

You can also *increase* or *decrease* the *size* of the displayed *function curve* by enclosing the curve with a *selection frame* (*selection rectangle*) while keeping the left mouse button pressed and then increase or decrease its size in the usual WINDOWS manner.

Although MATHCAD is very reliable in the graphical representation of function curves, you should critically examine every created graphic to detect any *irregularities* or *errors*. The following two examples illustrate this.

* It is not apparent in the graph that the displayed function is not defined at the zero.

$$x := -2, -1.97 .. 10$$

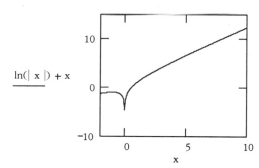

* The next two graphs differ even though we use the same function, however written in different but mathematically correct forms. For some unknown reason, MATHCAD draws fractional powers only for positive x values

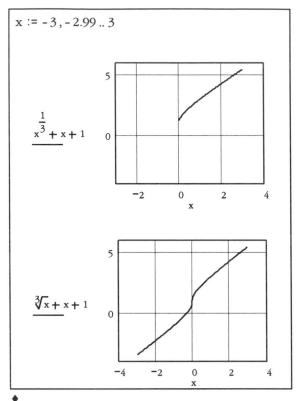

$x := -3, -2.99 .. 3$

$\dfrac{1}{x^{\frac{1}{3}} + x + 1}$

$\dfrac{\sqrt[3]{x} + x + 1}{}$

◆

☞

Section 19.5 in the discussion of curves contains further examples for the graphical representation of plane curves.

You can also use MATHCAD to graphically display curves whose function equation is comprised of several expressions (see Examples 17.4b and c).

◆

18.2 Surfaces

We consider in this section the *graphical representation* of *surfaces* in three-dimensional space (*3D graphs*) whose *function equation* in a Cartesian co-ordinate system can have one of the following forms (compare with the analogy for plane curves):

*   *explicit representation*

    $z = f(x,y)$ with $(x,y) \in D$ (*domain of definition*)

* *implicit representation*

  $F(x,y,z) = 0$ with $(x,y) \in D$ (*domain of definition*)

* *parametric representation*

  $x = x(u,v)$, $y = y(u,v)$, $z = z(u,v)$

  with $a \le u \le b$, $c \le v \le d$

Example 18.5:

a) An *ellipsoid* with a, b and c as semi-axes can be described by the *implicit representation*

$$\frac{x^2}{a^2} + \frac{y^2}{b^2} + \frac{z^2}{c^2} = 1$$

The use of *spherical coordinates* yields the *parametric representation* (with $0 \le u \le 2\pi$, $0 \le v \le \pi$)

$x(u,v) = a \cdot \cos u \cdot \sin v$, $y(u,v) = b \cdot \sin u \cdot \sin v$, $z(u,v) = c \cdot \cos v$

Figure 18.10 shows the graphical representation of an *ellipsoid* using this parametric representation for $a = 1$, $b = 2$ and $c = 3$.

b) Figure 18.11 shows the graphical representation of the *hyperbolic paraboloid* (*saddle surface*)

$z = x^2 - y^2$

over the square D

$-5 \le x \le 5$, $-5 \le y \le 5$

c) Figure 18.12 shows the graphical representation of the surface produced by the *function*

$z = \sin xy$

over the square D

$-2 \le x \le 2$, $-2 \le y \le 2$

d) You can also display surfaces graphically for which the function equation consists of several analytical expressions, such as

$$f(x, y) := \begin{vmatrix} x^2 + y^2 & \text{if} & x^2 + y^2 \le 1 \\ 1 & \text{if} & 1 < x^2 + y^2 \le 4 \\ \sqrt{x^2 + y^2} - 1 & \text{otherwise} \end{vmatrix}$$

We used MATHCAD to define this function in Example 10.3a; Figure 18.13 shows the associated graph.

♦

Proceed as follows in MATHCAD to *display surfaces* that are specified in Cartesian coordinates by a *function*

$z = f(x,y)$ (*explicit representation*)

- *First* create for the surface a *matrix* **M** (e.g., for **ORIGIN** := 0) whose *elements*

  M_{ik}

  are calculated as follows (nested loop):

  $i := 0, \dots , m$ $k := 0, \dots , n$

  $M_{i,k} = f(x_i, y_k)$

  i.e., the *elements* of the *matrix* **M** are formed from the *function values* f(x,y) at the points

  (x_i, y_k)

  that are specified by the user. The values for x and y in the x-y domain D are usually uniformly spaced (see Figures 18.11 and 18.12).

- *After* calculating the *matrix* **M** associated with the surface, use one of the following activities

  * use the *menu sequence*

    Insert ⇒ Graph ⇒ Surface Plot

  * click the

    button in *operator palette no. 3*

  to create a *graphic window* (see Figure 18.9) in which you enter the designation of the calculated matrix **M** in its lower placeholder (see Figures 18.11 and 18.12).

- *Finally, click* outside the graphic window or press the ⏎-key, if you are in *automatic mode*, otherwise press the F9-function key to obtain the required surface.

Figure 18.9. MATHCAD graphic window for surfaces

If the *surface* is available in *parametric representation*

$$x = x(u,v) \; , \; y = y(u,v) \; , \; z = z(u,v)$$

you perform the *graphical representation* as just described. However, instead of the matrix **M**, you must calculate using the *three matrices* **X** , **Y** and **Z** with the elements

$$X_{i,k} = x(u_i, v_k) \; , \; Y_{i,k} = y(u_i, v_k) \; , \; Z_{i,k} = z(u_i, v_k)$$

and then enter these in the placeholders of the graphic window (see Figure 18.10) in the form

X , Y , Z   (MATHCAD Version 7)

(X , Y , Z) (MATHCAD Version 8)
♦

In a similar manner as for curves, the following method can be used to *change* (*manipulate*) the created *3D graphs* (*surfaces*):

*   First click with the mouse on the surface to enclose it within a *selection frame* (*selection rectangle*).

    Then use the following *menu sequence*

    Format ⇒ Graph ⇒ 3D Plot...

or double mouse-click on the surface to *format* the *surface* in the displayed dialogue box, i.e., scaling of the axes, specification of the colour
and form of the surface, assign a title to the surface, etc.

- You can also*:*

  * *increase* or *decrease* the *size of the surface* by enclosing the *surface*
    with a *selection frame* (*selection rectangle*) while keeping the left
    mouse button pressed and then increase or decrease its size in the
    usual WINDOWS manner.

  * *rotate the surface* while keeping the mouse button pressed.

  ◆

$N := 25 \qquad i := 0..N \qquad k := 0..N$

$$u_i := \frac{2 \cdot \pi \cdot i}{N} \qquad v_k := \frac{\pi \cdot k}{N}$$

$$X_{i,k} := \cos(u_i) \cdot \sin(v_k) \qquad Y_{i,k} := 2 \cdot \sin(u_i) \cdot \sin(v_k)$$

$$Z_{i,k} := 3 \cdot \cos(v_k)$$

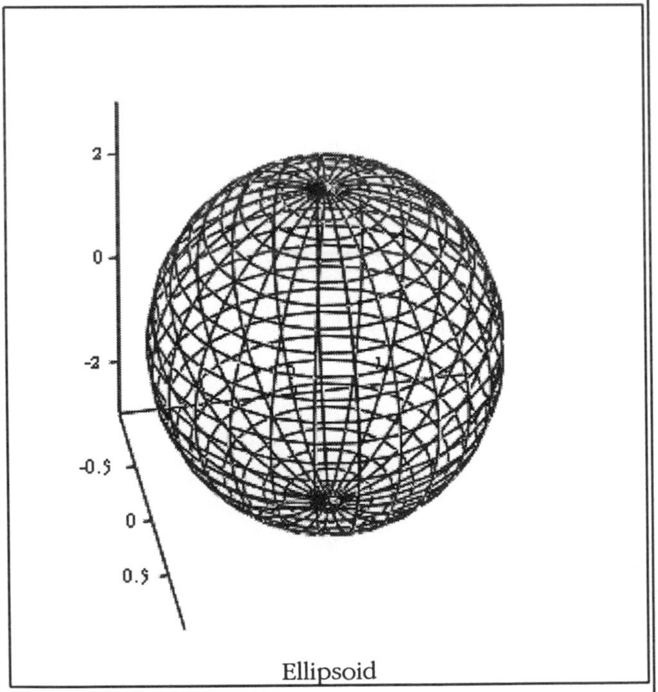

Ellipsoid

(X, Y, Z)

Figure 18.10. Ellipsoid from Example 18.5a

☞

Figures 18.10 and 18.11 can be used as *general templates* for the *graphical representation* of *surfaces* specified in *parametric form* or the form z=f(x,y).
I suggest that you store both figures as MATHCAD worksheets. These then can be loaded to create any graphic. You merely need to change the appropriate values.

◆

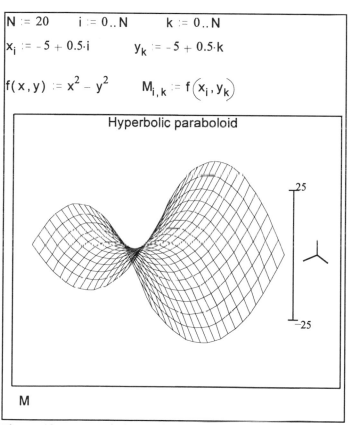

$$N := 20 \qquad i := 0..N \qquad k := 0..N$$

$$x_i := -5 + 0.5 \cdot i \qquad y_k := -5 + 0.5 \cdot k$$

$$f(x,y) := x^2 - y^2 \qquad M_{i,k} := f\left(x_i, y_k\right)$$

Hyperbolic paraboloid

M

Figure 18.11. Hyperbolic paraboloid from Example 18.5b

$N := 20 \qquad i := 0..N \qquad k := 0..N$

$x_i := -2 + 0.2 \cdot i \qquad y_k := -2 + 0.2 \cdot k$

$f(x,y) := \sin(x \cdot y) \qquad M_{i,k} := f(x_i, y_k)$

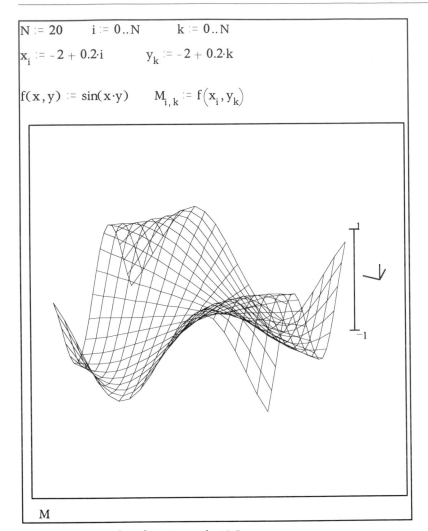

M

Figure 18.12. Surface from Example 18.5c

$$f(x, y) := \begin{vmatrix} x^2 + y^2 & \text{if} & x^2 + y^2 \leq 1 \\ 1 & \text{if} & 1 < x^2 + y^2 \leq 4 \\ \sqrt{x^2 + y^2} - 1 & \text{otherwise} \end{vmatrix}$$

$$N := 20 \qquad i := 0 .. N \qquad k := 0 .. N$$

$$x_i := -5 + 0.5 \cdot i \qquad y_k := -5 + 0.5 \cdot k$$

$$M_{i, k} := f(x_i, y_k)$$

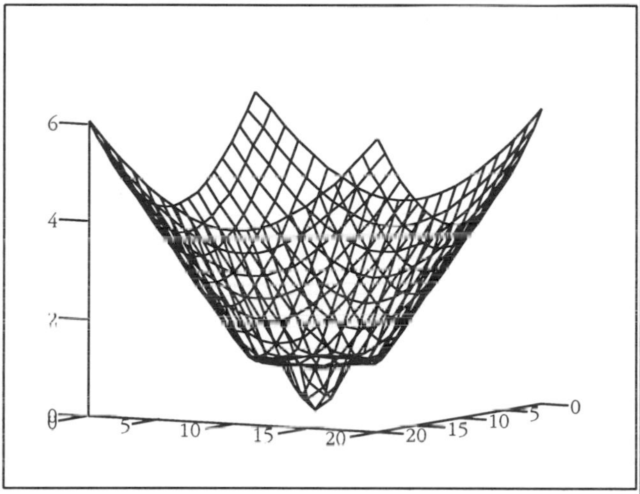

Figure 18.13. Surface from Example 18.5d

MATHCAD does not permit the *graphical representation* of *surfaces* specified in *implicit form*. You must revert to parametric representation in this case (see Example 18.5a).

♦

MATHCAD provides the *graphical representation* of *contour lines* (*level lines*) as an additional aid for the representation of surfaces; such displays are called *contour representations*. The following procedure is required:

• *First* calculate a *matrix* **M** analogue to the above discussed graphical representation of surfaces.

• *After* calculating the *matrix* **M** associated with the surface, perform one of the following actions

  * invoke the *menu sequence*

    Insert ⇒ Graph ⇒ Contour Plot

* click the

button in *operator palette no. 3*

to create a *graphic window*, in which you enter the designation of the matrix **M** in its lower placeholder (see Figure 18.14).

* *Finally*, if you are in *automatic mode, click* with the *mouse* outside the graphic window or press the ⏎-key, otherwise press the F9-function key, to get MATHCAD to draw the required *contour representation*.

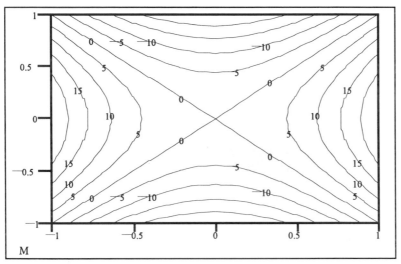

Figure 18.14. Contour representation of the hyperbolic paraboloid from Example 18.5b

You can also use the *3D Plot Wizard* to get MATHCAD to guide you through the capabilities of three-dimensional graphics; to use the wizard, activate the

Insert ⇒ Graph ⇒ 3D Plot Wizard

menu sequence.

◆

Starting with MATHCAD Version 8, it is also possible to draw several surfaces in one coordinate system. Figure 18.15 demonstrates this using the example of a *penetration* between a *sphere with radius 2:*

$x(u,v) = 2 \cdot \cos u \cdot \sin v$, $y(u,v) = 2 \cdot \sin u \cdot \sin v$, $z(u,v) = 2 \cdot \cos v$

and a *cylinder over the unit circle:*

$X(u,t) = \cos u$, $Y(u,t) = \sin u$, $Z(u,t) = t$

♦

$N := 40 \qquad i := 0 .. N \qquad k := 0 .. N$

$u_i := 2 \cdot \pi \cdot \dfrac{i}{N} \qquad v_k := 2 \cdot \pi \cdot \dfrac{k}{N} \qquad t_k := -4 + 8 \cdot \dfrac{k}{N}$

$x_{i,k} := 2 \cdot \cos(u_i) \cdot \sin(v_k) \qquad X_{i,k} := \cos(u_i)$

$y_{i,k} := 2 \cdot \sin(u_i) \cdot \sin(v_k) \qquad Y_{i,k} := \sin(u_i)$

$z_{i,k} := 2 \cdot \cos(v_k) \qquad Z_{i,k} := t_k$

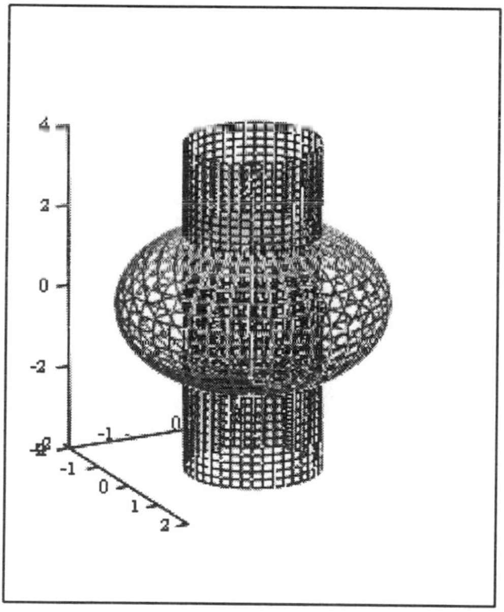

$(x, y, z) , (X, Y, Z)$

Figure 18.15. Surface penetration between sphere and cylinder

18.3 Scatter Plots

You frequently meet the situation in *practical problems* that a given *function* (functional relationship)

* *of one variable*

    y = f(x)

* *of two variables*

    z = f(x,y)

is not analytically given, but available only in the *form* of *n points (measured points)*, i.e., for *functions*

* of *one variable*

    by *n number-pairs* $(x_1, y_1), (x_2, y_2), ..., (x_n, y_n)$

* of *two variables*

    by *n number-triples* $(x_1, y_1, z_1), (x_2, y_2, z_2), ..., (x_n, y_n, z_n)$

Mathematics provides a number of *possibilities* to *approximate* these *points* using *functions* (e.g., *polynomials*), such as by *interpolation* or the *method of least squares*. We discuss these in Sections 17.2.4 and 27.2.

♦

MATHCAD also can also *graphically display* given *points*.

* The *graphical representation* of *n points (number-pairs)* in plane Cartesian coordinate systems is performed with the *following steps:*

    * *First* assign the *x-values* and the *y-values* to a *vector* **x** and *vector* **y** respectively; each of these vectors has n components.

    * *Then* use the

        button from *operator palette no. 1* to enter the *index domain* for the components of the two vectors in the *form*

        i := 1 .. n

        i.e., define the index i for the vector components as *range variable* (see Section 8.3).

    * *Then* use one of the following *two methods*

        – activate the *menu sequence*

            Insert ⇒ Graph ⇒ X-Y Plot

        – click the

graphic button in *operator palette no. 3*

to create a *graphic window* (*graphic frame*) for Cartesian coordinates in the worksheet at the cursor position; Figure 18.1 shows its layout.

* *Then* use in this *graphic window* the

button from *operator palette no. 1* to enter in the middle *placeholder* of the x-axis the *components* of the *vector* **x** in the form

x_i

and in the middle *placeholder* of the y-axis the *components* of the *vector* **y** in the form

y_i

when an index domain has been defined for i.
If you have not defined any domain for the index i (range variable), enter only the designation **x** or **y** for the vectors (see Example 18.6).

* *Finally, click outside* the *graphic window* or press the $\boxed{\leftarrow}$-key, if you are in automatic mode, otherwise press the $\boxed{F9}$-function key to obtain the required *scatter plot*.

• The *graphical representation* of *n number-triples* in spatial Cartesian coordinate systems is performed with the *following steps*:

* *First* assign the *x-values*, the *y-values* and the *z-values* to a *vector* **x**, *vector* **y**, and *vector* **z** respectively; each of these vectors has n components.

* *Then* use one of the following *two methods*

  – activate the *menu sequence*

    Insert ⇒ Graph ⇒ 3D Scatter Plot

  – click the

graphic button in *operator palette no. 3*

to create *graphic window* (*graphic frame*) in the worksheet at the cursor position; Figure 18.8 shows its layout.

* *Then* enter the designations of the three vectors (separated by commas) in the lower *placeholder* of this *graphic window* in the form

  x , y , z   (MATHCAD Version 7)

  (x , y , z)   (MATHCAD Version 8)

* *Finally, click outside* the *graphic window* or press the ⏎-key, if you
  are in automatic mode, otherwise press the F9-function key to obtain
  the required *scatter plot*.

As with all graphical representations in MATHCAD, you can also select various display modes for scatter plot.
For example, after a double mouse-click on the scatter plot graph you can
draw points

* *connected* by *straight lines*

* *isolated* in various forms

using the displayed dialogue box.

◆

☞

This procedure can also be used if you wish to display together two given
vectors with the same number n of components in plane Cartesian coordinates. This does not mean any more than the graphical representation of n
points whose coordinates are given by the vectors (see Example 18.6b).
You can also represent one vector graphically, i.e., display its components
as a function of the index (see Example 18.6b).

◆

Example 18.6:

a) The following five *measurement points* are given

   (1,2) , (2,4) , (3,3) , (4,6) , (5,5)

   Enter (or read) the x- and y-components for these points as vectors **x**
   and **y**

   $$x := \begin{bmatrix} 1 \\ 2 \\ 3 \\ 4 \\ 5 \end{bmatrix} \qquad y := \begin{bmatrix} 2 \\ 4 \\ 3 \\ 6 \\ 5 \end{bmatrix}$$

* We now display the points:

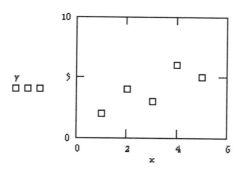

The same can be achieved using the *component notation* of vectors:

i := 1 .. 5

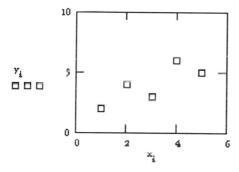

*   We now join the points with straight lines:

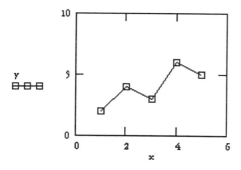

b) Display the components of the following vector as a function of the index i:

$$x := \begin{bmatrix} 1 \\ 3 \\ 2 \\ 4 \\ 5 \end{bmatrix}$$

$$i := 1 .. 5$$

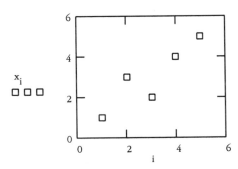

c) Display the following five *measurement points*

(1,2,8) , (2,4,5) , (3,3,6) , (4,6,4) , (5,5,9)

in a three-dimensional Cartesian coordinate system:

$$x := \begin{bmatrix} 1 \\ 2 \\ 3 \\ 4 \\ 5 \end{bmatrix} \qquad y := \begin{bmatrix} 2 \\ 4 \\ 3 \\ 6 \\ 5 \end{bmatrix} \qquad z := \begin{bmatrix} 8 \\ 5 \\ 6 \\ 4 \\ 9 \end{bmatrix}$$

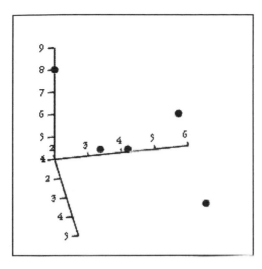

(x, y, z)

d) Display all the components in the given vector **x** against the components in the given vector **y** in the plane Cartesian coordinate system.

$$x := \begin{pmatrix} 3 \\ 6 \\ 8 \\ 2 \\ 5 \end{pmatrix} \qquad y := \begin{pmatrix} 9 \\ 7 \\ 5 \\ 3 \\ 1 \end{pmatrix}$$

This means that we display graphically the 5 points in the plane

(3 , 9) , (6 , 7) , (8 , 5) , (2 , 3) , (5 , 1)

We can use the same method as in Example a:

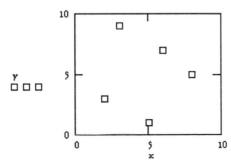

◆

18.4 Diagrams

MATHCAD can create so-called *3D bar charts*. The elements of a given *matrix* **M** are displayed as columns in these diagrams. *Proceed* as follows to create such charts:

- *First* create the required *matrix* **M**.
- *Then click the*

  graphic button for *3D bar diagrams* in *operator palette no. 3*

  to create a *graphic window* (*graphic frame*) for *bar charts* at the cursor position.
- *Then* enter the designation **M** for the *matrix* in the lower *placeholder* in this *graphic window.*
- *Finally, click outside* the *graphic window* or press the ⏎-key, if you are in automatic mode, otherwise press the F9-function key to obtain the required *bar chart.*

The *graphical representation* of a *function*

z=f(x,y)

of two variables also requires a matrix whose elements are formed from function values (see Section 18.2). Consequently, such functions can also be displayed as a bar chart; this is illustrated in Example 18.7b.

♦
Example 18.7:

a) Create a *matrix* **M** and display its elements as bars:

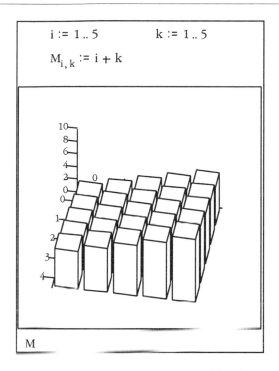

$i := 1 .. 5 \qquad k := 1 .. 5$

$M_{i,k} := i + k$

M

b) Display the *hyperbolic paraboloid* (*saddle surface*)

$z = x^2 - y^2$

from Example 18.5b over the square D

$-5 \leq x \leq 5 , \ -5 \leq y \leq 5$

as a *3D bar chart*:

$$f(x,y) := x^2 - y^2$$

$$N := 10 \qquad i := 1 .. N \qquad k := 1 .. N$$

$$x_i := -5 + i \qquad\qquad y_k := -5 + k$$

$$M_{i,k} := f\left(x_i, y_k\right)$$

M

♦

18.5 Animations

The *basic principle* for *animations* in computer algebra systems is realized by the successive display for a sequence of parameter values of functions that depend on this parameter; this produces animated graphics. This permits the user to see the effect of given parameters.

MATHCAD also supports, since Version 6, animated graphics:

- You proceed here analogue to the creation of the described graphics.

- The only change is the addition of a variable *parameter* with the designation **FRAME** in the function equation.

- *Then* activate the *menu sequence*

  View ⇒ Animate...

  When the

Animate

dialogue box appears, you can define here the *control range* for the
FRAME *parameter.*

- Finally, *enclose* the appropriate *graphic area* in the *worksheet* with a *selection rectangle* and click on

You can then initiate the *animation* by clicking on

in the **Playback** *window* that appears.

We will discuss the animation procedure by considering a simple example.

Example 18.8:

We use the *sine function* in the *interval* [2π,2π] with a *parameter* p, i.e.,

sin (x + p)

and display this as an animated graphic for the sequence of parameter values

p := 0 , 1 , 2 , ... , 9

Proceed as follows:

* *Define* the *function,* however using the *designation* **FRAME** for the *parameter* p. The display is then performed as described:

f(x) := sin(x + FRAME)

x := - 2 ·π , - 2 ·π + 0.001 .. 2 ·π

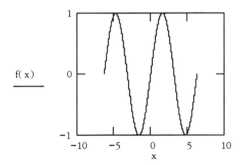

* *Then* use the *menu sequence*

View ⟹ Animate...

to open the

Animate

dialogue box in which we define the *control range* for the **FRAME** *parameter* from 0 to 9. We have also specified that 10 frames per second are to be displayed.

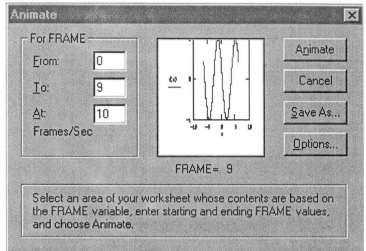

Enclose the previously drawn *graphic* with a *selection rectangle* and click on

* You can perform the *animation* by clicking on

in the
Playback

window that then appears.

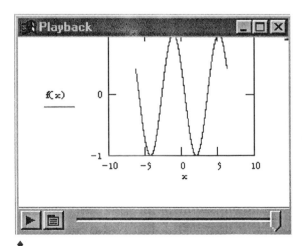

♦

18.6 Import and Export of Graphics

We discussed already in Chapter 9 the import and export of data and files (data exchange) as part of the *data management* that is principally *controlled* using the *menu sequence*

Insert ⇒ Component...

MATHCAD also provides similar capabilities for the *import* and *export* of *graphics* that can be *controlled* using the

- *menu sequence*

  Insert ⇒ Object...

- *functions* for *bitmap graphics* and other graphic formats

MATHCAD can use these facilities to also read, process and output pictures. Because this book is principally concerned with mathematical problems, we refer the reader to the manual or the integrated help for these tasks.

☞

If you wish to export the graphics created by MATHCAD (that we discussed in the previous chapters) to other WINDOWS applications, this can be done most simply using the usual WINDOWS method with copy and insert via the clipboard. This method was also used to take the graphics from MATHCAD into this book.

♦

19 Differentiation

As we already mentioned in the Introduction, a finite algorithm is available for the determination of the *derivatives* of differentiable functions that are formed from *elementary functions*. Consequently, MATHCAD can also exactly (symbolically) differentiate these functions without difficulty.

Thus, MATHCAD provides an effective tool for differentiation and so, as we will see during the course of this chapter, often avoids long-winded manual computations:

* We discuss in the following Section 19.1 the technique used to calculate derivatives.

* The subsequent Sections 19.2-19.5 handle important applications of the differential calculus that can also be performed using MATHCAD.

19.1 Calculation of Derivatives

We use MATHCAD to calculate

* *partial derivatives of arbitrary order*

$$f_{x_1} = \frac{\partial f}{\partial x_1} \ , \ f_{x_1 x_1} = \frac{\partial^2 f}{\partial x_1^{\,2}} \ , \ f_{x_1 x_2} = \frac{\partial^2 f}{\partial x_1 \partial x_2} \ , \ ...$$

for *functions*

$$z = f(x_1, x_2, ..., x_n)$$

of *n variables (n = 1, 2, ...)*

$$x_1, x_2, ..., x_n$$

and as *special cases*

* *derivatives of arbitrary order* (n = 1, 2, ...)

$$f'(x), f''(x), ..., f^{(n)}(x)$$

for *functions*

$$y = f(x)$$

of one variable

x

* *partial derivatives of arbitrary order*

$$f_x = \frac{\partial f}{\partial x} \ , \ f_{xx} = \frac{\partial^2 f}{\partial x^2} \ , \quad f_{xy} = \frac{\partial^2 f}{\partial x \partial y} \ , \ ...$$

for *functions*

$$z = f(x,y)$$

of *two variables*

x , y

MATHCAD provides one of the following *procedures* for the *exact (symbolic) differentiation* of a differentiable *function* of n variables

$$z = f(x_1, x_2, ..., x_n)$$

contained in the worksheet:

I   *Mark with editing lines* the *variable* in the function expression that is to be differentiated. Then activate the *menu sequence*

Symbolics ⇒ Variable ⇒ Differentiate

The *first derivative* of the function after the marked variable is returned. If you wish to calculate a higher derivative, you must repeat the described procedure.

II. The *first derivative* of a function for a variable is also returned when you click with the mouse on the button for the *differentiation operator*

from *operator palette no. 5*

and then fill both *placeholders* in the displayed *symbol*

$$\frac{d}{d \, \blacksquare} \blacksquare$$

as follows

$$\frac{d}{dx_i} f(x_1, x_2, ..., x_n)$$

when, in this case, you want to differentiate with respect to the variable x_i

If you require a higher derivative, nest the differentiation operator the appropriate number of times (see Example19.1d2).

III. The button for the *differentiation operator*

from *operator palette no. 5* can be used to calculate directly the *derivatives* of the *n-th order* (n = 1, 2, 3, ...) of a function when you fill the *placeholders* of the displayed *symbol*

$$\frac{d^{\blacksquare}}{d\,{\blacksquare}^{\blacksquare}}\,\blacksquare$$

as follows

$$\frac{d^n}{dx_i{}^n}f(x_1, x_2,..., x_n)$$

when, in this case, you want to differentiate n-times with respect to the variable

$$x_i$$

This operator must be nested for mixed derivatives of higher order (see Example 19.1j2).

For *methods II.* and *III.*, after *marking* the complete expression with *editing lines*, perform one of the following actions to initiate the *exact calculation* of *derivatives:*

*   activate the *menu sequence*

    Symbolics ⇒ Evaluate ⇒ Symbolically

*   activate the *menu sequence*

    Symbolics ⇒ Simplify

*   enter the *symbolic equal sign* → and then press the ⏎-key.

Methods I. and *II.* are recommended only for *derivatives* of the *first order*. *Method III.* is better for *higher derivatives*.
The differential operators from II. and III. must be nested appropriately to perform *mixed partial derivatives;* this is illustrated in Example 19.1j2.

♦

The differentiation operators from methods II. and III. should also be used for the *numerical calculation* of *derivatives* for given values; an assignment of the appropriate values must have been made to the variables beforehand. The subsequent input of the numerical equal sign = returns the result (see Example 19.1e).

However, MATHCAD has the *disadvantage* that only derivatives to the fifth order can be calculated numerically. Example 19.1f shows how higher derivatives can be calculated.

♦

Example 19.1:

a) Differentiate the following function using *Method I*.

$$\operatorname{asin}\left(\frac{x}{1+x}\right) \qquad \textit{by differentiation, yields}$$

$$\frac{1}{\sqrt{1-\dfrac{x^2}{(1+x)^2}}} \cdot \left[\frac{1}{(1+x)} - \frac{x}{(1+x)^2}\right]$$

This result can be simplified somewhat using the capabilities described in Section 13.2:

$$\frac{1}{\sqrt{1-\dfrac{x^2}{(1+x)^2}}} \left[\frac{1}{(1+x)} - \frac{x}{(1+x)^2}\right] \quad \textit{simplifies to} \quad \frac{1}{\left[\sqrt{\dfrac{(1+2x)}{(1+x)^2}} \cdot (1+x)^2\right]}$$

b) Differentiate the following function using

* *Method I.:*

$$x^{x^2} \quad \textit{by differentiation, yields} \quad x^{\left(x^2\right)} \cdot (2 \cdot x \cdot \ln(x) + x)$$

* *Method II.:*

The differentiation operator is used as follows and obviously also returns the same result:

$$\frac{d}{dx} x^{x^2} \rightarrow x^{x^2} \cdot (2 \cdot x \cdot \ln(x) + x)$$

c) If you have forgotten the *quotient rule* for the differentiation of the function

$$y(x) = \frac{f(x)}{g(x)}$$

MATHCAD returns it using *Method I.*:

$$\frac{f(x)}{g(x)} \quad \textit{by differentiation, yields} \quad \frac{\dfrac{d}{dx}f(x)}{g(x)} - \frac{f(x)}{g(x)^2} \cdot \frac{d}{dx}g(x)$$

and after *simplification*

simplifies to
$$\frac{\left[\left(\frac{d}{dx}f(x)\right)\cdot g(x) - f(x)\cdot\frac{d}{dx}g(x)\right]}{g(x)^2}$$

Method II. also returns the result:

$$\frac{d}{dx}\frac{f(x)}{g(x)} \rightarrow \frac{\frac{d}{dx}f(x)}{g(x)} - \frac{f(x)}{g(x)^2}\cdot\frac{d}{dx}g(x)$$

d) Calculate a fifth order derivative

d1) directly using *Method III.*:

$$\frac{d^5}{dx^5}\, x\cdot\ln(1+x^2) \rightarrow \frac{720}{\left(1+x^2\right)^3}\cdot x^2 - \frac{60}{\left(1+x^2\right)^2} - 1440\cdot\frac{x^4}{\left(1+x^2\right)^4} + 768\cdot\frac{x^6}{\left(1+x^2\right)^5}$$

d2) through five-level *nesting* of the *differentiation operator* from *Method II.*:

$$\frac{d\,d\,d\,d\,d}{dx\,dx\,dx\,dx\,dx}x\cdot\ln(1+x^2) \rightarrow \frac{720}{\left(1+x^2\right)^3}\cdot x^2 - \frac{60}{\left(1+x^2\right)^2} - 1440\cdot\frac{x^4}{\left(1+x^2\right)^4} + 768\cdot\frac{x^6}{\left(1+x^2\right)^5}$$

You can see that the nesting in Example d2 makes it more complex so that this differentiation operator can only be recommended for first derivatives.

e) *Calculate* the fifth derivative for the function from Example d *numerically* for the given value x=3 using the differentiation operator from *Method III.*:

$$x := 3$$

$$\frac{d^5}{dx^5}\, x\cdot\ln\left(1+x^2\right) = -0.185$$

f) Calculate a sixth derivative using the differentiation operator from *Method III.*:

*   *exactly*

$$\frac{d^6}{dx^6}\frac{1}{1-x} \rightarrow \frac{720}{\left(1-x\right)^7}$$

*   *numerically*

    for x:=2

Although the *result* –720 should appear, we receive the display:

$$x := 2$$

$$\frac{d^6}{dx^6}\ \frac{1}{1-x} =$$

Can only evaluate an nth order derivative when n=0, 1 .. 5.

because MATHCAD calculates derivatives numerically only to the fifth order.

You can solve this problem in two ways:

f1)*Nesting* the *differentiation operator:*

$$x := 2$$

$$\frac{d^3}{dx^3}\ \frac{d^3}{dx^3}\ \frac{1}{1-x} = -720$$

f2)*Assignment* of the *differentiation* using the *symbolic equal sign* → to a new function and the subsequent calculation of its function value for the required x-value:

$$g(x) := \frac{d^6}{dx^6}\ \frac{1}{1-x} \rightarrow \frac{720}{(1-x)^7}$$

$$g(2) = -720$$

g) *Defined functions,* such as

$$f(x) := \sin(x) + \ln(x) + x + 1$$

can be differentiated using the *symbolic equal sign* →, whereas the other methods can fail:

$$f(x) := \sin(x) + \ln(x) + x + 1$$

$$f(x) \quad by\ differentiations,\ yields \quad \frac{d}{dx}f(x)$$

$$\frac{d}{dx}f(x) \quad yields \quad \frac{d}{dx}f(x)$$

$$\frac{d}{dx}f(x) \quad simplifies\ to \quad \frac{(\cos(x)\cdot x + 1 + x)}{x}$$

$$\frac{d}{dx}f(x) \rightarrow \cos(x) + \frac{1}{x} + 1$$

h) Let us now consider the behaviour of MATHCAD when you attempt to differentiate a function that is not differentiable at all points of the domain of definition. We use the function

$$f(x) := |x|$$

whose derivative has the form

$$f'(x) = \begin{cases} 1 & \text{if } x > 0 \\ -1 & \text{if } x < 0 \end{cases}$$

and is not differentiable at the origin. MATHCAD returns the *incorrect result* for the *numerical differentiation* at the origin x=0

$$x := 0$$

$$\frac{d}{dx}f(x) = 0$$

MATHCAD returns the following result for the exact calculation:

$$\frac{d}{dx}f(x) \rightarrow \frac{|x|}{x}$$

that is correct except for the origin.

i) MATHCAD can also differentiate functions that are comprised of various expressions and defined with the **if**-statement/operator, as shown in the following example. The *function*

$$f(x) = \begin{cases} 1 & \text{if } x \leq 0 \\ x^2 + 1 & \text{if } x > 0 \end{cases}$$

with the first *derivative* $$f'(x) = \begin{cases} 0 & \text{if } x \leq 0 \\ 2 \cdot x & \text{if } x > 0 \end{cases}$$

is differentiable for all x.

You can define this function in MATHCAD using one of the following two methods (see Sections 10.3 and 17.2.3):

1. Using the **if**-statement

$$f(x) := \mathbf{if}\,(\,x \leq 0\,,\,1\,,\,x^2 + 1\,)$$

2. Using the **if**-operator from operator palette no. 6

$$f(x) := \begin{vmatrix} 1 & \text{if} & x \le 0 \\ x^2 + 1 & \text{if} & x > 0 \end{vmatrix}$$

The use of the *symbolic equal sign* → in MATHCAD *returns* the *derivative* f'(x) of the function f(x) if it was defined using method 1:

$$g(x) := \frac{d}{dx}f(x) \rightarrow \text{if}\,(x \le 0, 0, 2 \cdot x)$$

The *graphical representation* of f(x) and f'(x) displays the following figure

$$f(x) := \text{if}(x \le 0, 1, x^2 + 1) \qquad x := -2, -1.999 \ldots 2$$

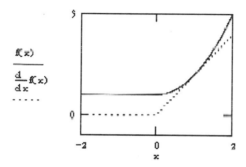

If the function has been defined using Method 2, MATHCAD does not return any result for the exact calculation of the derivative:

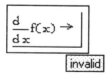

j) The *calculation* of *partial derivatives* is performed in a similar manner as for the differentiation of functions of one variable, as shown in the following examples. The only difference being that the differentiation operators from Method II. or III. must be nested for the mixed derivatives:

j1) There is no difference between the exact calculation of partial derivatives with respect to one variable for functions of several variables compared with functions of one variable, as shown in the following examples:

$$\frac{d^8}{dx^8}\ e^{x\cdot y} \rightarrow y^8\cdot\exp(x\cdot y) \qquad\qquad \frac{d^8}{dy^8}\ e^{x\cdot y} \rightarrow x^8\cdot\exp(x\cdot y)$$

j2) The exact calculation of a *mixed partial derivative* of higher order is possible by nesting the differentiation operator:

$$\frac{d^2}{dx^2}\frac{d^3}{dy^3}\ e^{x\cdot y} \rightarrow 6\cdot x\cdot\exp(x\cdot y) + 6\cdot x^2\cdot y\cdot\exp(x\cdot y) + x^3\cdot y^2\cdot\exp(x\cdot y)$$

j3) Calculate numerically the derivative from Example j2 at x=2 and y=1:

$$x := 2 \qquad\qquad y := 1$$

$$\frac{d^2}{dx^2}\frac{d^3}{dy^3}\ e^{x\cdot y} = 325.118$$

j4) If the numerical calculations meet partial derivatives above order 5, the same result as in Example f occurs, as shown below:

The calculation using

$$x := 2 \qquad y := 1$$

$$\boxed{\frac{d^7}{dy^7}\ e^{x\cdot y}} =$$

Can only evaluate an nth order derivative when n=0, 1 .. 5.

does not function, whereas the following trick is successful:

$$x := 2 \qquad y := 1$$

$$\frac{d^3}{dy^3}\frac{d^4}{dy^4}\ e^{x\cdot y} = 945.794$$

or

$$g(x,y) := \frac{d^7}{dy^7}\ e^{x\cdot y} \rightarrow x^7\cdot\exp(x\cdot y)$$

$$g(2,1) = 945.799$$

◆

☞

If you ignore the disadvantages of numerical calculations, MATHCAD works *effectively* for the *differentiation*, as shown in the given examples. MATH-CAD frees you from the intensive work often required for the differentiation of complicated functions and returns the result in seconds. However, because MATHCAD cannot always recognize whether a function is differentiable, you must check this before starting the differentiation (see Example 19.1h).

◆

19.2 Taylor Expansion

According to *Taylor's theorem*, a *function*

$f(x)$

that is (n+1)-times differentiable in the interval $(1 \rightarrow 0)$

$(x_0 - r, x_0 + r)$

has the representation (*Taylor expansion*)

$$f(x) = \sum_{k=0}^{n} \frac{f^{(k)}(x_0)}{k!}(x - x_0)^k + R_n(x)$$

for the *expansion point*

x_0

where the *remainder*

$R_n(x)$

has the *form* (Lagrange form of the remainder):

$$R_n(x) = \frac{f^{(n+1)}(x_0 + \vartheta \cdot (x - x_0))}{(n+1)!} \cdot (x - x_0)^{n+1} \qquad (0<\vartheta<1)$$

The *polynomial* of degree n

$$\sum_{k=0}^{n} \frac{f^{(k)}(x_0)}{k!}(x - x_0)^k$$

in the *Taylor expansion* is called the *Taylor polynomial* of degree n for $f(x)$ at the *expansion point*

x_0

If the following applies for all $x \in (x_0 - r, x_0 + r)$ for the *remainder*

$$\lim_{n \to \infty} R_n(x) = 0$$

the *function* f(x) can be expanded into the power series (*Taylor series*)

$$f(x) = \sum_{k=0}^{\infty} \frac{f^{(k)}(x_0)}{k!} \cdot (x - x_0)^k$$

with the *convergence interval* $|x - x_0| < r$.

It is usually difficult to prove that f(x) can be expanded into a Taylor series. As known from the theory, the existence of the derivative of arbitrary order for f(x) does not suffice here. However, for practical applications, the Taylor polynomial of degree n (for n=1,2,...) suffices in most cases to approximate a *function* f(x) in the neighbourhood of the expansion point with a polynomial of degree n.

♦

Because forming the *Taylor expansion* manually is tedious, MATHCAD uses one of the following procedures to calculate a *Taylor polynomial of degree n*:

I. Use of the **Symbolics** *menu*

- *First enter* the *function* f(x) to be expanded in the worksheet.
- *Then mark* with editing lines a *variable* x in the function expression.

  Finally, you can specify the required *degree n* of the *Taylor polynomial* (default value 6) in the

  Order of Approximation

  field of the *dialogue box* displayed by activating the menu sequence

  Symbolics ⇒ Variable ⇒ Expand to Series...

  If you now enter a positive integer n, the *(n–1)-th Taylor polynomial* is calculated at the *expansion point*

  $$x_0 = 0$$

II. Use of the **series** *keyword*

- *First* click with the mouse on the

  button from *operator palette no. 8* to insert the

  ■ series, ■ , ■ →

  symbol for the **series** *keyword* with three free placeholders in the worksheet at the cursor position.

- *Then* enter in the
  * left placeholder the function to be expanded
  * first right placeholder the coordinates for the expansion point separated by commas using the equal operator from operator palette no. 2
  * second right placeholder the required degree n of the Taylor polynomial (in the form n+1).
- *Then mark* the complete *expression* with *editing lines* and enter the symbolic equal sign →.
- *Finally,* press the ⏎-key to return the required Taylor expansion (see Examples 19.2 d2 and e).

Because MATHCAD can calculate the *Taylor expansion* with *Method I.* only at the *expansion point*

$$x_0 = 0$$

you must perform the *transformation*

$$x - u + x_0$$

beforehand to expand a function

$$f(x)$$

at an arbitrary expansion point

$$x_0$$

i.e., the function

$$F(u) = f(u + x_0)$$

is to be expanded at the point

u=0

with regard to the variable u (see Example 19.2d).

♦

MATHCAD can use *Method II.*, i.e., with the
series
keyword, to calculate the *Taylor polynomial* at some arbitrary expansion point. This can also be used to expand functions having several variables. Examples 19.2d and e show the procedure involved.

♦

MATHCAD returns the *Laurent expansion* if the function being expanded has a singularity at the expansion point (see Example 19.2f).

♦

Example 19.2:

a) Calculate the *Taylor polynomials* of degree 5 and 10 for the following function at the expansion point

$$x_0 = 0$$

i.e., we must enter for n the values 6 and 11 in the dialogue box:

$$\frac{1}{1 - x} \qquad \textit{converts to the series}$$

$$1 + x + x^2 + x^3 + x^4 + x^5$$

$$1 + x + x^2 + x^3 + x^4 + x^5 + x^6 + x^7 + x^8 + x^9 + x^{10}$$

It is known from the theory that the remainder converges to zero for $|x| < 1$ (for n tends to ∞), so the produced *geometric series* has the given function as sum.

b) For n=10, you obtain the following *Taylor expansion* for the *function* $\ln(1+x)$ at the expansion point

$$x_0 = 0$$

This expansion returns the known power series expansion for $|x| < 1$ (for n tends to ∞):

$$\ln(1 + x)$$

converts to the series

$$x - \frac{1}{2} \cdot x^2 + \frac{1}{3} \cdot x^3 - \frac{1}{4} \cdot x^4 + \frac{1}{5} \cdot x^5 - \frac{1}{6} \cdot x^6 + \frac{1}{7} \cdot x^7 - \frac{1}{8} \cdot x^8 + \frac{1}{9} \cdot x^9$$

c) Determine the *Taylor polynomial* for n=8 and n=12 for the following function and then display graphically the function and its Taylor polynomials. This graphic illustrates well the known behaviour that the Taylor polynomials provide a good approximation to the function only in the neighbourhood of the expansion point:

$$\frac{1}{1 + x^4} \qquad \textit{converts to the series} \qquad 1 - x^4$$

$$\frac{1}{1 + x^4} \qquad \textit{converts to the series} \qquad 1 - x^4 + x^8$$

$$x := -3, -2.99 .. 3$$

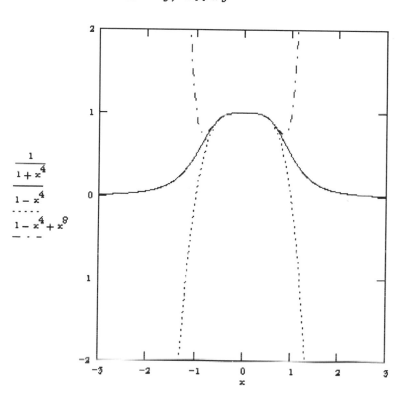

$\dfrac{1}{1+x^4}$

$1 - x^4$
....

$1 - x^4 + x^8$
— · —

d) Calculate the *Taylor expansion* for n=6 for the *function*

ln x

at the expansion point

$x_0 = 2$

   d1) Because MATHCAD can expand functions only at the point 0, we
   perform the *transformation*

   $x = u + 2$

   and expand the resulting function

   $F(u) = \ln(u + 2)$

   at the expansion point

   $u_0 = 0$:

   $\ln(u + 2)$ *converts to the series*

   $$\ln(2) + \frac{1}{2} \cdot u - \frac{1}{8} \cdot u^2 + \frac{1}{24} \cdot u^3 - \frac{1}{64} \cdot u^4 + \frac{1}{160} \cdot u^5$$

The **substitute** *keyword* from *operator palette no. 8* returns with the transformation

u = x – 2

the *required Taylor expansion* for the function ln x at the expansion point 2:

$$\ln(2) + \frac{1}{2} \cdot u - \frac{1}{8} \cdot u^2 + \frac{1}{24} \cdot u^3 - \frac{1}{64} \cdot u^4 + \frac{1}{160} \cdot u^5 \text{ substitute}, u = x - 2 \rightarrow$$

$$\ln(2) + \frac{1}{2} \cdot x - 1 - \frac{1}{8} \cdot (x-2)^2 + \frac{1}{24} \cdot (x-2)^3 - \frac{1}{64} \cdot (x-2)^4 + \frac{1}{160} \cdot (x-2)^5$$

d2) If you use the **series** *keyword* from *operator palette no. 8*, you obtain the expansion as follows:

$$\ln(x) \text{ series}, x = 2, 6 \rightarrow$$

$$\ln(2) + \frac{1}{2} \cdot x - 1 - \frac{1}{8} \cdot (x-2)^2 + \frac{1}{24} \cdot (x-2)^3 - \frac{1}{64} \cdot (x-2)^4 + \frac{1}{160} \cdot (x-2)^5$$

e) Expand the *function*

f(x,y) = sin x · sin y

of the two variables x and y at the expansion point

(π,π)

into a *Taylor polynomial* of degree 4 using the **series** *keyword* from *operator palette no. 8*:

$$\sin(x) \cdot \sin(y) \text{ series}, x = \pi, y = \pi, 5 \rightarrow$$

$$(x - \pi) \cdot (y - \pi) - \frac{1}{6} \cdot (x - \pi) \cdot (y - \pi)^3 - \frac{1}{6} \cdot (x - \pi)^3 \cdot (y - \pi)$$

f) If the function being expanded has a singularity at the expansion point, MATHCAD returns the *Laurent expansion*; the following example illustrates this:

$$\frac{1}{\sin(x)} \quad converts \ to \ the \ series \quad \frac{1}{x} + \frac{1}{6} \cdot x + \frac{7}{360} \cdot x^3 + \frac{31}{15120} \cdot x^5$$

The **series** *keyword* returns the same *result*:

$$\frac{1}{\sin(x)} \text{ series}, x = 0, 7 \rightarrow \frac{1}{x} + \frac{1}{6} \cdot x + \frac{7}{360} \cdot x^3 + \frac{31}{15120} \cdot x^5$$

19.3   Calculus of Errors

Because *measurements* are often used to obtain the quantities required for practical tests, the problem of *measurement errors* arises. The differential calculus can be used to estimate the effects of these measurement errors when the measured quantities represent the independent variables of a functional relationship, i.e., the measured quantities x_i represent the independent variables of a given function

$$z = f(x_1, x_2, ..., x_n)$$

The user needs to know how the *measurement errors* in x_i *affect* the value z calculated using the functional relationship (e.g., law of physics) f. We will consider this problem using a simple example.

Example 19.3:

As is well known, the *volume* V of a box with breadth b, height h and length l is calculated from

$$V = V(b, h, l) = b \cdot h \cdot l$$

i.e., the volume V is a function of the three variables b, h and l. If you wish to determine the volume of an existing box by measuring the length, breadth and height, these quantities are subject to *measurement errors* and, consequently, the calculation using the given formula provides an incorrect value for V. However, because bounds can be specified for the measurement errors, we are also interested in obtaining an *error bound* for the calculated volume V.

◆

This example shows the *problem* to be *solved*:

How does an *error* Δx_i in the quantities x_i, i.e., the $\tilde{x}_i$ obtained as approximation for the incorrect measurement for x_i,

$$\tilde{x}_i = x_i + \Delta x_i \qquad (\textit{vectorial}: \tilde{\mathbf{x}} = \mathbf{x} + \Delta \mathbf{x})$$

affect the resulting *error* Δz for the function

$$z = f(x_1, x_2, ..., x_n)$$

i.e., what is the accuracy of the obtained *approximate value*

$$\tilde{z} = z + \Delta z = f(x_1 + \Delta x_1, x_2 + \Delta x_2, ..., x_n + \Delta x_n) = f(\tilde{x}_1, \tilde{x}_2, ..., \tilde{x}_n)$$

Because you do not normally know the exact value for the *measurement errors* Δx_i, but rather the bounds for the *absolute errors*

$$|\Delta x_i| \leq \delta_i$$

it is only possible to give a bound for the absolute error for the obtained *approximate value* $\tilde{z} = z + \Delta z$:

$$|\Delta z| \leq \delta$$

You can use the Taylor expansion of first degree and ignore the remainder to calculate such bounds

$$\Delta z = f(x_1 + \Delta x_1, x_2 + \Delta x_2, ..., x_n + \Delta x_n) - f(x_1, x_2, ..., x_n)$$

$$= f(\tilde{\mathbf{x}}) - f(\mathbf{x})$$

and obtain the *approximate formula*

$$|\Delta z| \approx \left| \sum_{i=1}^{n} \frac{\partial f}{\partial x_i}(\tilde{\mathbf{x}}) \cdot \Delta x_i \right| \leq \sum_{i=1}^{n} \left| \frac{\partial f}{\partial x_i}(\tilde{\mathbf{x}}) \right| \cdot \delta_i \approx \delta$$

that can be used to calculate an *approximation* for the bound δ of the *absolute error* for z.

The equation can also be used to calculate a bound for the *relative error*

$$\frac{|\Delta z|}{\tilde{z}}$$

Examples for the *calculation* of the *absolute error* for *functions* of *two variables* (Example 19.4a) and *three variables* (Example 19.4b) follow. Because we defined the *error* as the

abs_error

function, you can use these two examples as a general model; you just need to set the argument values appropriately for the function call and re-define the functional relationship.

Example 19.4:

a) Use *Ohm's law* to calculate an *upper bound* for the *absolute error* for the determination of the *electrical resistance*, where (incorrect) measurements were used to determine the *voltage* U and the *current* I. We ignore the units and recommend the following *procedure:*

* We *first* define the *function* for which the error is to be calculated. This is Ohm's law in our example:

$$R(I, U) := \frac{U}{I}$$

* We *then* define a *bound* for the *absolute error* for a general function f of two variables as *function subroutine* **abs_error**:

$$\text{abs\_error}(f, x_1, x_2, \delta_1, \delta_2) := \left| \frac{d}{dx_1} f(x_1, x_2) \right| \cdot \delta_1 + \left| \frac{d}{dx_2} f(x_1, x_2) \right| \cdot \delta_2$$

* *Finally,* we calculate a *bound* for the R(I,U) function for the given *number values*

  I=20 , U=100 , $\delta_1 = \delta I = 0.02$, $\delta_2 = \delta U = 0.01$:

  abs_error(R , 20 , 100 , 0.02 , 0.01) = $5.5 \cdot 10^{-3}$ ∎

b) Calculate an *upper bound* for the *absolute error* of the *volume*

  $V = b \cdot h \cdot l$

  of a *box* when the *upper bound*

  δ=0.001m

  of the *measurement errors* for

  breadth b=10m, height h=5m and length l=15m
  is known; meters (m) are used here as unit. We adopt the same procedure as in Example a:

  $V(b, h, l) := b \cdot h \cdot l$

  $abs\_error(f, x_1, x_2, x_3, \delta_1, \delta_2, \delta_3) :=$

  $$\left| \frac{d}{dx_1} f(x_1, x_2, x_3) \right| \cdot \delta_1 + \left| \frac{d}{dx_2} f(x_1, x_2, x_3) \right| \cdot \delta_2 + \left| \frac{d}{dx_3} f(x_1, x_2, x_3) \right| \cdot \delta_3$$

  abs_error(V, 10 ·m, 5 ·m, 15 ·m, 0.001 ·m, 0.001 ·m, 0.001 ·m) = $0.275\ m^3$ ∎

In both Examples, it is better to use *literal indexes* for the *indexed variables.*
♦

19.4   Calculation of Limits

Calculate the *limits* of a *function* f(x) or an *expression* A(n) at x=a or n=a,
i.e.,

$$\lim_{x \to a} f(x) \qquad \text{or} \qquad \lim_{n \to a} A(n)$$

Indeterminate expressions of the form

$$\frac{0}{0} \ , \ \frac{\infty}{\infty} \ , \ 0 \cdot \infty \ , \ \infty - \infty \ , \ 0^0 \ , \ \infty^0 \ , \ 1^\infty \ , \ ...$$

can occur during the *calculation*. Subject to certain assumptions, *l'Hospital's rule* can be used in these situations. However, this rule does not always return a result. Consequently, MATHCAD cannot be expected to be always successful in calculating limits.

You use MATHCAD as follows to perform the *exact calculation* of a *limit*:

- *First* click with the mouse on the button for the *limit operator*

from *operator palette no. 5* to create the following *limit symbol*:

$$\lim_{\blacksquare \to \blacksquare} \blacksquare$$

at the cursor position in the worksheet in which f(x) or A(n) are entered in the *placeholders* following *lim* and in the *placeholders* below *lim* x and a or n and a, i.e.,

$$\lim_{x \to a} f(x) \quad \text{or} \quad \lim_{n \to a} A(n)$$

- *Finally, mark* the complete *expression* with *editing lines* and perform one of the following actions

  * Activate the *menu sequence*

    Symbolics ⇒ Evaluate ⇒ Symbolically

  * Activate the *menu sequence*

    Symbolics ⇒ Simplify

  * Enter the *symbolic equal sign* → and then press the ⏎-key.

  to initiate the *exact calculation* of a *limit*.

MATHCAD also supports the *calculation* of *one-sided* (i.e. *left-hand* or *right-hand*) *limits* through the use of the two buttons for the *limit operators*

    or

from *operator palette no. 5*

♦

MATHCAD does not support the *calculation* of *limits* using the numerical equal sign =.

♦

The

button from *operator palette no. 5* can be used in MATHCAD to enter ∞ (*infinity*) instead of a in the appropriate placeholder, which then permits *calculations of limits* for

$$x \to \infty \qquad \text{or} \qquad n \to \infty$$

◆

☞

If the *limit calculation fails* or you wish to check the returned *result*, it is better to get MATHCAD to draw f(x) or A(n).

◆

To demonstrate the effectiveness of MATHCAD, we calculate some limits in the following examples.

Example 19.5:

a) MATHCAD calculates the *left-hand* and *right-hand limit* for the following function:

$$\lim_{x \to 0^-} \frac{2}{1 + e^{\frac{-1}{x}}} \to 0 \qquad\qquad \lim_{x \to 0^+} \frac{2}{1 + e^{\frac{-1}{x}}} \to 2$$

You can see here that the two values are different, i.e., the *limit does not exist*, as MATHCAD correctly recognizes:

$$\lim_{x \to 0} \frac{2}{1 + e^{\frac{-1}{x}}} \to \text{undefined}$$

b) MATHCAD can calculate without difficulty the following limits that yield indeterminate expressions

b1)
$$\lim_{x \to 0} x^{\sin(x)} \to 1$$

b2)
$$\lim_{x \to \frac{\pi}{2}} \tan(x)^{\cos(x)} \to 1$$

b3)
$$\lim_{x \to \infty} \left(\frac{x + 1}{x - 1}\right)^{x + 3} \to \exp(2)$$

b4)

$$\lim_{x \to 1} \frac{x \cdot \ln(x) - x + 1}{x \cdot \ln(x) - \ln(x)} \to \frac{1}{2}$$

c) MATHCAD calculates the limit

$$\lim_{x \to \infty} \frac{3 \cdot x + \cos(x)}{x} \to 3$$

although it cannot be obtained using l'Hospital's rule, but only by transforming the function in

$$3 + \frac{\cos(x)}{x}$$

with the estimation for

$$\left| \frac{\cos(x)}{x} \right| \le \frac{1}{|x|}$$

d) MATHCAD calculates the following two limits, although these cannot be calculated using l'Hospital's rule, as you can readily verify.

$$\lim_{x \to \infty} \frac{5^x}{4^x} \to \infty \qquad\qquad \lim_{x \to \infty} \frac{4^x}{5^x} \to 0$$

e) MATHCAD even calculates limits for the principal value for *arctan* when the argument contains some arbitrary constant k.

$$\lim_{x \to \infty} \operatorname{atan}(k \cdot x) \to \frac{1}{2} \cdot \operatorname{csgn}(k) \cdot \pi$$

$$\lim_{x \to 0} \operatorname{atan}\left(\frac{k}{x}\right) \to \text{undefined}$$

$$\lim_{x \to 0^+} \operatorname{atan}\left(\frac{k}{x}\right) \to \frac{1}{2} \cdot \operatorname{csgn}(k) \cdot \pi$$

$$\lim_{x \to 0^-} \operatorname{atan}\left(\frac{k}{x}\right) \to \frac{-1}{2} \cdot \operatorname{csgn}(k) \cdot \pi$$

where *csgn* represents the *Signum function.*

f) The symbolic equal sign is recommended for the *calculation* of *limits* for *defined functions*:

Use the *symbolic equal sign* → to *calculate* the *limit*:

$$\lim_{x \to 0} f(x) \to \frac{3}{4}$$

for the *defined function*

$$f(x) := \frac{2 \cdot x + \sin(x)}{x + 3 \cdot \ln(x + 1)}$$

In contrast, the *menu sequences*

Symbolics ⇒ Evaluate ⇒ Symbolically

or

Symbolics ⇒ Simplify

do *not* produce any *result*:

$$\lim_{x \to 0} f(x) \qquad yields \qquad f(0)$$

or

$$\lim_{x \to 0} f(x) \qquad simplifies\ to \qquad f(0)$$

♦

These examples show that MATHCAD works very effectively for the calculation of limits.

♦

19.5 Discussion of Curves

Discussions of *Curves* (in the plane) are used to determine the *properties* of a given *function* f(x) and the form of its *curve*. These include

*   determination of the domain of definition and the range
*   investigation of symmetry properties
*   determination of discontinuities (poles, jumps, ...) and continuity intervals
*   investigation of differentiability

* determination of the intersection points with the x-axis (zeros) and the y-axis
* determination of the extreme values (maximum and minimum)
* determination of the points of inflection
* determination of the monotony and convexity intervals
* investigation of the behaviour at infinity (determination of the asymptotes)
* calculation of suitable function values.

☞

The previously discussed graphical properties of MATHCAD (see Section 18.1) can be used to *draw* the *curve* for arbitrary functions f(x), which then can be used to supply information on most of the specified properties for f(x). The graph can also be used to supply *starting values* for *numerical methods* that may need to be used, e.g., to determine the zeros, maximum, minimum and points of inflection should the exact (symbolic) calculation fail.

♦

However, you should not just limit yourself to a *discussion of curves* for the graphical display supplied by MATHCAD, but, because a graphic returned by MATHCAD may be erroneous, also investigate the listed *analytical properties* using the *functions/commands/menus* supplied by MATHCAD for the

* *solution of equations* (to determine zeros, extreme values, points of inflection)
* *differentiation* (to determine monotony and convexity intervals, to produce equations for the determination of extreme values and points of inflection).

Let us now use three examples to show the capabilities for performing *curve discussions* using MATHCAD.

Example 19.6:

We first draw the curves for the following functions and then investigate the properties analytically:

a) Investigate a *fractional rational function* of the form

$$f(x) = \frac{x+1}{x^3 + 6 \cdot x^2 + 11 \cdot x + 6}$$

for which MATHCAD uses the calculation of function values (with the step size 0.3) in the interval [−4, 4] to draw the following *function curve* in which straight lines join the function values (see Section 18.1):

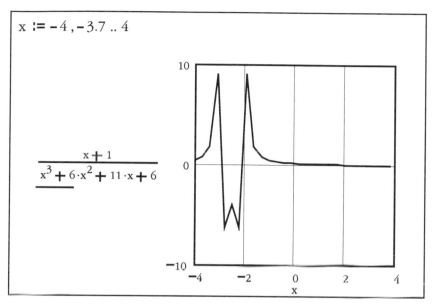

Because the poles are not apparent, this graph is inexact or incorrect. This effect is mainly caused by using a step size that is too large. Consequently, we redraw the function in the interval [−4, 4] using the smaller step size 0.1 and obtain the following function curve:

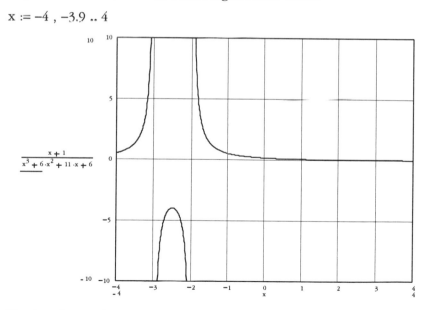

The last figure shows the function curve in a much better form, as confirmed by the following investigations, i.e., the displayed form of the curve depends mainly on the selection of the step size for the function value calculation.

We now use *analytical investigations* to test whether the drawn curve actually represents the required function curve:

* We first *factorize* (see Sections 13.6 and 16.2) to determine the *zeros* for the *denominator* of the function:

$$x^3 + 6 \cdot x^2 + 11 \cdot x + 6 \quad by\ factoring,\ yields \quad (x+3) \cdot (x+2) \cdot (x+1)$$

This shows that the given function has *discontinuities* at

$x = -1, -2$ and -3

We can use the result of the factorization to write the function in the *following form*:

$$f(x) = \frac{x+1}{(x+3)(x+2)(x+1)}$$

This shows that there is a *removable discontinuity* at $x = -1$, even though this cannot be recognized from the graph drawn by MATH-CAD. If you remove this discontinuity by dividing both the numerator and denominator of the function by x+1, the new function has the following simplified form:

$$f(x) = \frac{1}{(x+3)(x+2)}$$

It is obvious from this function expression that *poles* occur at -3 and -2 and no *zeros* exist. The equation also shows that the function is positive in the intervals $(-\infty, -3)$ and $(-2, +\infty)$ and negative in the interval $(-3, -2)$ and tends to zero asymptotically (for x tending to $\pm\infty$).

* To determine the *extreme values* (maximum and minimum), we must calculate the first derivative, set this to zero (*necessary optimization condition* - see Section 25.1) and obtain an exact solution for the resulting equation using the *menu sequence* (see Section 16.3):

Symbolics ⇒ Variable ⇒ Solve

$$\frac{d}{dx} \frac{1}{(x+3) \cdot (x+2)} \equiv 0 \quad has\ solution(s) \quad \frac{-5}{2}$$

We use the second derivative (*sufficient optimization condition* - see Section 25.1) to *check* the obtained *solution*:

$$x := \frac{-5}{2} \qquad \frac{d^2}{dx^2} \frac{1}{(x+3) \cdot (x+2)} = -32$$

This shows that a *maximum* with the function value f(−5/2)= −4 occurs at the point

x=−5/2

These tests confirm the correctness of the displayed graph for the considered function.

b) Consider a *curve* whose equation has the following *implicit form*:

$$y^2(1 + x) = x^2(1 - x)$$

Because MATHCAD does not permit the graphical display of implicitly given curves, you can obtain a solution in this case by drawing in the same coordinate system the curves for the two functions

$$y = \frac{x\sqrt{1-x}}{\sqrt{1+x}} \quad \text{und} \quad y = -\frac{x\sqrt{1-x}}{\sqrt{1+x}}$$

that result by solving the curve equation for y:

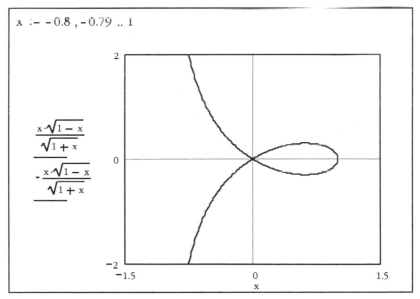

Because the graph and the equations show that the two branches of the *curve* are *symmetric*, you only need to analyse one of them:

* You can recognize the *domain of definition*
  −1< x ≤ 1
  and the *zeros*
  x=0 and x=1.

* The *extreme values* can be determined as in Example a:

$$\frac{d}{dx} \frac{x\cdot\sqrt{1-x}}{\sqrt{1+x}} \equiv 0 \qquad \textit{has solution(s)} \qquad \begin{bmatrix} \dfrac{-1}{2} + \dfrac{1}{2}\cdot\sqrt{5} \\[2ex] \dfrac{-1}{2} - \dfrac{1}{2}\cdot\sqrt{5} \end{bmatrix}$$

Thus a *maximum* for the upper branch is at the point

$$x = \frac{-1+\sqrt{5}}{2}$$

You can *check* this using the second derivative:

$$x := \frac{-1}{2} + \frac{1}{2}\cdot\sqrt{5} \qquad \frac{d^2}{dx^2} \frac{x\cdot\sqrt{1-x}}{\sqrt{1+x}} = -1.758$$

Because of the symmetry, this is a minimum for the lower branch.

Thus we have validated the principal properties of the curve and can confirm the curve displayed by MATHCAD.

c) Consider the *integral rational function (polynomial function)*

$$y = 0.025\,x^5 + 0.05\,x^4 - 0.6\,x^3 - 0.55\,x^2 + 2.575\,x - 1.5$$

Integral rational functions provide the least difficulties for a discussion of curves. They are continuous for all x-values. Consequently, MATHCAD also does not have any difficulties with the *graphical representation* of the function curve:

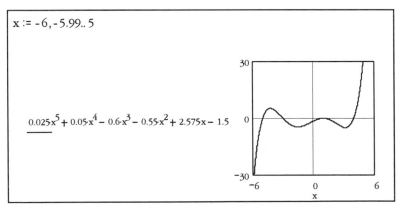

$$x := -6, -5.99 .. 5$$

$$0.025x^5 + 0.05\cdot x^4 - 0.6\cdot x^3 - 0.55x^2 + 2.575x - 1.5$$

Because no solution formula exists for polynomials above the fourth degree, the only *problem* occurs with the exact (symbolic*) determination* of the *zeros* and *extreme values*.

MATHCAD returns the following results for the given function:

* *Zeros:*

$$0.025 \cdot x^5 + 0.05 \cdot x^4 - 0.6 \cdot x^3 - 0.55 \cdot x^2 + 2.575 \cdot x - 1.5$$

by factoring, yields $\dfrac{1}{40} \cdot (x + 5) \cdot (x + 3) \cdot (x - 4) \cdot (x - 1)^2$

i.e., the function has the zeros −5, −3, 1, 4.

* *Extreme values:*

result as zeros for the first derivative of the function:

$$\frac{d}{dx}(0.025 \cdot x^5 + 0.05 \cdot x^4 - 0.6 \cdot x^3 - 0.55 \cdot x^2 + 2.575 \cdot x - 1.5) \equiv 0$$

has solution(s)
$$\begin{bmatrix} 1. \\ 3.1706844917212148503 + 2. \cdot 10^{-20} \cdot i \\ -4.2374443319589008082 - 1. \cdot 10^{-20} \cdot i \\ -1.5332401597623140424 \quad 1.\ 10^{-20} \cdot i \end{bmatrix}$$

The following figure shows that four real zeros exist:

$x := -5, -4.99 .. 5$

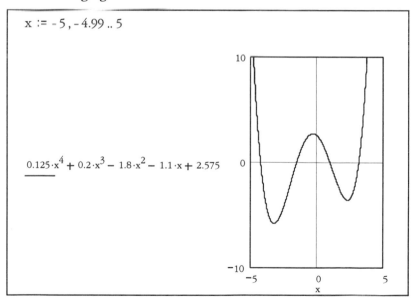

$0.125 \cdot x^4 + 0.2 \cdot x^3 - 1.8 \cdot x^2 - 1.1 \cdot x + 2.575 \qquad 0$

In addition to x = 1, you obtain satisfactory values for the remaining zeros if you ignore those imaginary parts calculated by MATHCAD that have a very small absolute value.

This provides the following approximations for the remaining extreme values

−4.24, −1.53, 3.17

* *Points of inflection*:

result from the zeros of the second derivative of the function

$$\frac{d^2}{dx^2}\ 0.025 \cdot x^5 + 0.05 \cdot x^4 - 0.6 \cdot x^3 - 0.55 \cdot x^2 + 2.575 \cdot x - 1.5 = 0$$

has solution(s)

$$\begin{pmatrix} 2.3170984433653077928 \\ -3.2224590675394149678 \\ -.294639375825892825 \end{pmatrix}$$

i.e., we obtain the following approximations for the points of inflection

−3.22, −0.29, 2.32.

♦

20 Integration

As we already mentioned in the Introduction, finite algorithms for the exact calculation exist only for some types of integrals. Consequently, we cannot expect MATHCAD to be able to calculate every integral. However, because MATHCAD can perform the often extensive calculations in a matter of seconds, as shown in the following section, it is very useful for those integrals that can be calculated.

If MATHCAD *cannot* exactly *calculate* an *integral*, it indicates this in one of the following ways:

* A message is displayed:
  No closed form found for integral

* The integral to be calculated is returned unchanged as the result (see Examples 20.11 and m.

* The calculation docs not complete. Press the (Esc)-key to terminate.

  ♦

Note that MATHCAD does not permit the use of *indexed variables* with an *array index* for integration.

♦

20.1 Indefinite Integrals

The determination of a function

F(x)

whose derivative

$F'(x)$

is the same as a given function

f(x)

produces the *integral calculus.*
A function F(x) calculated in this manner is called a *primitive* of f(x).

As *known* from the *integral calculus:*

*   All *primitives* F(x) that exist for a given function f(x) differ only by a constant.

*   The complete set of primitive functions for a function f(x) is called the *indefinite integral* and written in the form

$$\int f(x)\,dx$$

    where f(x) and x are called the *integrand* and *integration variable* respectively.
    ◆

The *integral calculus* must *answer* two main *questions*:

I.   Does every function f(x) have a *primitive* F(x)?

II. Which method can be used to determine a *primitive* F(x) for a given function f(x)?

The following answers can be provided:

*   The *first question*

    has a positive answer for many functions, because every *continuous function* f(x) has over a finite interval [a,b] a *primitive* F(x). However, this is only an *existence theorem* that occurs frequently in mathematics.
    Such theorems do not provide a *solution algorithm* to explicitly construct a primitive for a given continuous function f(x) consisting, for example, of known elementary functions

    x^n, e^x, $\ln x$, $\sin x,\dots$

    Although you know that a primitive F(x) exists, you do not know whether and how F(x) can be formed from elementary functions.

*   The *second question*

    has only a positive answer for specific classes of functions, because no generally applicable finite solution algorithm exists to determine a primitive.

Although no *finite algorithm* exists to *determine* a *primitive* F(x) for an *arbitrary continuous function* f(x), there are known methods (*integration methods*) that can be used to construct a *primitive* F(x) for special *functions* f(x). MATHCAD uses the best known of these

*   *partial integration*

*   *partial fraction decomposition* (for fractional rational functions)

*   *substitution*

If such methods are successful, MATHCAD is usually successful for the exact calculation of integrals and avoids extensive computational work. ◆

MATHCAD provides two possibilities for the *exact calculation* of *indefinite integrals*

$$\int f(x)\,dx$$

I. Enter the function f(x) to be integrated into the worksheet, mark a variable x with editing lines and then activate the *menu sequence*

   Symbolics ⇒ Variable ⇒ Integrate

II. Click the *integral button* for the *indefinite integration*

   in the *operator palette no. 5* to display the *integral symbol*

$$\int \blacksquare\,d\blacksquare$$

   at the cursor position in the worksheet. Then enter the *integrand* f(x) and the *integration variable* x in the two *placeholders*, i.e.,

$$\int f(x)\,dx$$

   and *mark* the complete *expression* with editing lines.
   Perform one of the following actions

   *   *Activate* the *menu sequence*

      Symbolics ⇒ Evaluate ⇒ Symbolically

   *   *Activate* the *menu sequence*

      Symbolics ⇒ Simplify

   *   Enter the *symbolic equal sign* → and then press the ⏎-key

   to initiate the exact calculation of the *indefinite integral*.
   Because it is easier to use the *symbolic equal sign*, this is the better method.

☞

In some situations you can avoid MATHCAD failing with the exact calculation of integrals if you *simplify* the *integrand* f(x) before starting the integration. We consider two of these cases:

*   You can use the procedure described in Section 13.3 to decompose *fractional rational functions* into partial fractions.

*   You can perform common *substitutions* beforehand (see Example 20.1l).
    ♦

Example 20.1:

The problems a, b, c and d can be solved using *partial integration* and do not present any difficulties for MATHCAD:

a)

$$\int x^3 \cdot e^{2 \cdot x} dx \rightarrow \frac{1}{2} \cdot x^3 \cdot \exp(2 \cdot x) - \frac{3}{4} \cdot x^2 \cdot \exp(2 \cdot x) + \frac{3}{4} \cdot x \cdot \exp(2 \cdot x) - \frac{3}{8} \cdot \exp(2 \cdot x)$$

b)

$$\int \operatorname{asin}(x) \, dx \rightarrow x \cdot \operatorname{asin}(x) + \sqrt{1 - x^2}$$

c)

$$\int x^3 \cdot \cos(x) \, dx \rightarrow x^3 \cdot \sin(x) + 3 \cdot x^2 \cdot \cos(x) - 6 \cdot \cos(x) - 6 \cdot x \cdot \sin(x)$$

d)

$$\int \sin(\ln(x)) \, dx \rightarrow \frac{1}{2} \cdot x \cdot (\sin(\ln(x)) - \cos(\ln(x)))$$

The problems e, f, g and h can be solved using *partial fraction decomposition*. Because of the difficulties caused by the partial fraction decomposition, MATHCAD can solve these problems only when the denominator polynomial is sufficiently simple (see Section 13.3):

e) MATHCAD can calculate the integral

$$\int \frac{2 \cdot x^2 + 2 \cdot x + 13}{x^5 - 2 \cdot x^4 + 2 \cdot x^3 - 4 \cdot x^2 + x - 2} \, dx \rightarrow$$

$$\ln(x - 2) - \frac{1}{2} \cdot \ln(x^2 + 1) - 4 \cdot \operatorname{atan}(x) - \frac{1}{4} \cdot \frac{(8 \cdot x - 6)}{(x^2 + 1)}$$

f) MATHCAD can calculate the integral

$$\int \frac{x^6 + 7 \cdot x^5 + 15 \cdot x^4 + 32 \cdot x^3 + 23 \cdot x^2 + 25 \cdot x - 3}{x^8 + 2 \cdot x^7 + 7 \cdot x^6 + 8 \cdot x^5 + 15 \cdot x^4 + 10 \cdot x^3 + 13 \cdot x^2 + 4 \cdot x + 4} \, dx \rightarrow$$

$$\ln(x^2 + 1) - \frac{3}{(x^2 + 1)} - \ln(x^2 + x + 2) + \frac{1}{(x^2 + x + 2)}$$

g) MATHCAD can calculate the following integral although it cannot decompose the integrand into partial fractions as we saw in Example
13.11a

$$\int \frac{1}{x^4 + 1}\,dx \rightarrow$$

$$\frac{1}{8}\cdot\sqrt{2}\cdot\ln\left[\frac{\left(x^2 + x\cdot\sqrt{2} + 1\right)}{\left(x^2 - x\cdot\sqrt{2} + 1\right)}\right] + \frac{1}{4}\cdot\sqrt{2}\cdot\mathrm{atan}\left(x\cdot\sqrt{2} + 1\right) + \frac{1}{4}\cdot\sqrt{2}\cdot\mathrm{atan}\left(x\cdot\sqrt{2} - 1\right)$$

h) MATHCAD cannot calculate the integral

$$\int \frac{1}{x^3 + 2\cdot x^2 + x + 1}\,dx$$

even though it has only a denominator polynomial of third degree. This
is caused by it having only one real non-integer zero. As already discussed in Section 13.3 as part of the partial fraction decomposition,
MATHCAD has difficulties with non-integer and complex zeros for denominator polynomials.

Although *substitutions* can be used to solve the problems i, j, k and l,
MATHCAD cannot always recognize this (see Example l). However, this is
not surprising, because there is no algorithm available to find a suitable
substitution:

i)

$$\int \frac{1}{x\cdot\sqrt{1 - x}}\,dx \rightarrow -2\cdot\mathrm{artanh}\left(\sqrt{1 - x}\right)$$

j) MATHCAD can calculate the integral

$$\int \frac{\sqrt[3]{1 + \sqrt[4]{x}}}{\sqrt{x}}\,dx \rightarrow \frac{12}{7}\cdot\sqrt[3]{\left(1 + \sqrt[4]{x}\right)^7} - 3\cdot\sqrt[3]{\left(1 + \sqrt[4]{x}\right)^4}$$

because the *substitution*

$$t = \sqrt[3]{1 + \sqrt[4]{x}} \qquad x = (t^3 - 1)^4$$

is successful.

k)

$$\int \frac{1}{2 + \cos(x)}\, dx \rightarrow \frac{2}{3}\sqrt{3} \cdot atan\left(\frac{1}{3} \cdot tan\left(\frac{1}{2} \cdot x\right) \cdot \sqrt{3}\right)$$

l) MATHCAD cannot calculate the following integral in the given form:

$$\int asin(x)^2 dx \rightarrow \int asin(x)^2 dx$$

However, if you use the manual *substitution* x = sin t to convert it to the form

$$\int t^2 \cos t\, dt$$

MATHCAD then can calculate it:

$$\int t^2 \cdot \cos(t)\, dt \rightarrow t^2 \cdot \sin(t) - 2 \cdot \sin(t) + 2 \cdot t \cdot \cos(t)$$

The *inverse substitution* t = arc sin x can also be performed with MATH-CAD using the **substitute** *keyword* (see Section 13.8):

$$t^2 \cdot \sin(t) - 2 \cdot \sin(t) + 2 \cdot t \cdot \cos(t) \text{ substitute}, t \blacksquare asin(x) \rightarrow$$

$$asin(x)^2 \cdot x - 2 \cdot x + 2 \cdot asin(x) \cdot \sqrt{1 - x^2}$$

m) MATHCAD cannot calculate the integral

$$\int x^x dx \rightarrow \int x^x dx$$

No integration method can be recognized to permit its calculation.

n) We can assign the result of an indefinite integration to a new function g(x), using the symbolic equal sign $\rightarrow$:

$$g(x) := \int x \cdot e^x dx \rightarrow x \cdot exp(x) - exp(x)$$

o) We consider the integration of defined functions:
    o1)The *symbolic equal sign* is successful for the integration of defined functions, such as

$$f(x) := \sin(x) + \ln(x) + x + 1$$

$$\int f(x)\,dx \rightarrow -\cos(x) + x \cdot \ln(x) + \frac{1}{2} \cdot x^2$$

o2)*Functions* that consist of *several expressions*, such as the continuous function

$$f(x) = \begin{cases} x & \text{if } x \le 0 \\ x^2 & \text{otherwise} \end{cases}$$

can be defined in MATHCAD in the following manner:

*   with the **if** statement

    $f(x) := \textbf{if} (x \le 0 , x , x\char`\^2)$

*   or with the **if** operator from *operator palette no. 6*

    $$f(x) := \begin{cases} x & \text{if } x \le 0 \\ x^2 & \text{otherwise} \end{cases}$$

Functions defined in this way cannot be integrated in MATHCAD with the given methods for the exact integration, although a *primitive* F(x) itself can be calculated without difficulty:

$$F(x) = \begin{cases} \frac{1}{2} \cdot x^2 & \text{if } x \le 0 \\ \frac{1}{3} \cdot x^3 & \text{otherwise} \end{cases}$$

The following figure shows the graphical display of the two functions f(x) and F(x):

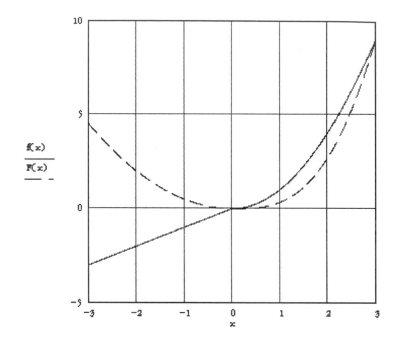

MATHCAD permits only the numerical calculation of integrals for such functions (see Example 20.5d).

♦

We have often used functions in the previous examples that can be easily integrated using the given methods. However, you should not conclude that MATHCAD can integrate all functions. Nevertheless, you can see that MATHCAD can calculate easily and correctly integrals of relatively complicated functions, provided this is possible using one of the given integration methods.

♦

20.2 Definite Integrals

In addition to *indefinite integrals*, integral calculus also investigates *definite integrals*

$$\int_a^b f(x)\,dx$$

with the *integrand* f(x), the *integration variable* x and the *integration limits* a and b, that are required for many application problems.

The *fundamental theorem* of the *differential and integral calculus*

$$\int_a^b f(x)\,dx = F(b) - F(a)$$

links indefinite and *definite integrals*, where F(x) represents a *primitive* for the integrand f(x).

♦

MATHCAD performs the *exact calculation* of *definite integrals*

$$\int_a^b f(x)\,dx$$

as follows:

- *First click the integral button* for the *definite integration*

  in the *operator palette no. 5* to create the *integral symbol* with four placeholders

  $$\int_■^■ ■\,d■$$

  in the worksheet at the cursor location.

- *Then* enter the *integration limits* a and b, the *integrand* f(x) and the *integration variable* x in the appropriate placeholders to produce

  $$\int_a^b f(x)\ dx$$

- *Then mark* the complete *expression* with *editing lines*.
- Perform one of the following *actions* to initiate the *exact calculation* of the *definite integral*
  - * *Activate* the *menu sequence*

    Symbolics ⇒ Evaluate ⇒ Symbolically
  - * *Activate* the *menu sequence*

    Symbolics ⇒ Simplify
  - * Enter the *symbolic equal sign* → and then press the ⏎-key.

Because it is easiest to use the *symbolic equal sign*, this is recommended.

☞

You can use *definite integrals* to *calculate* a *primitive* F(x) for a given function f(x) when you use the *formula*

$$F(x) \ = \ \int_a^x f(t)\,dt$$

that results directly from the fundamental theorem of the differential and integral calculus. The *primitive* calculated in this manner has the property F(a) = 0 (see Example 20.2c).

◆

Because MATHCAD has the same problems with the calculation of definite integrals as for indefinite integrals, we restrict ourselves in the following example to a few problems.

Example 20.2:

a) The result of the exact calculation of the following definite integral

$$\int_1^3 e^x \cdot \sin(x)\,dx \rightarrow$$

$$\frac{-1}{2}\cdot\exp(3)\cdot\cos(3) + \frac{1}{2}\cdot\exp(3)\cdot\sin(3) + \frac{1}{2}\cdot\exp(1)\cdot\cos(1) - \frac{1}{2}\cdot\exp(1)\cdot\sin(1)$$

is not particularly illustrative, because the contained real numbers could not be expressed in any other way. If you require a decimal approximation as the result, the contained solution must be calculated numerically (see Section 6.2). We perform this here by entering the numerical equal sign:

$$\frac{-1}{2}\cdot\exp(3)\cdot\cos(3) + \frac{1}{2}\cdot\exp(3)\cdot\sin(3) + \frac{1}{2}\cdot\exp(1)\cdot\cos(1) - \frac{1}{2}\cdot\exp(1)\cdot\sin(1)$$

$$= 10.95017031468552$$

b) Both the following definite integral and the associated indefinite integral are not calculated exactly

$$\int_2^3 x^x\,dx \rightarrow \int_2^3 x^x\,dx$$

This leaves only the numerical calculation (see Example 20.5a)

c) If you use the definite integral

$$F(x) := \int_1^x \sin(\ln(t)) \, dt \rightarrow \frac{-1}{2} \cdot x \cdot (-\sin(\ln(x)) + \cos(\ln(x))) + \frac{1}{2}$$

you also obtain a *primitive* for the *function*

f(x)=sin(ln(x))

that differs by the constant 1/2 from that calculated with the indefinite integral (see Example 20.1d):

$$F(x) := \int \sin(\ln(x)) \, dx \rightarrow \frac{1}{2} \cdot x \cdot (\sin(\ln(x)) - \cos(\ln(x)))$$

d) As with indefinite integrals, only the *symbolic equal sign* $\rightarrow$ can be used to calculate definite integrals for *defined functions*:

$$f(x) := \sin(x) + \ln(x) + x + 1$$

$$\int_1^2 f(x) \, dx \rightarrow -\cos(2) + \frac{3}{2} + 2 \cdot \ln(2) + \cos(1)$$

e) MATHCAD cannot calculate exactly either the definite integral nor the indefinite integral (see Example 20.1n2) for a *defined function* that consists of *several expressions*, such as

$$f(x) = \begin{cases} x & \text{if } x \leq 0 \\ x^2 & \text{otherwise} \end{cases}$$

♦

20.3 Improper Integrals

Consider the calculation of *improper integrals* that adopt a role in a number of practical applications. The following *forms* exist:

I. The integrand f(x) is bounded and the *integration interval* is *unbounded*, e.g.,

$$\int_a^\infty f(x) \, dx$$

II. The *integrand* f(x) is *unbounded* in the integration interval [a,b], e.g.,

$$\int_{-1}^{1} \frac{1}{x^2} \, dx$$

III. Both the *integration interval* and the *integrand* are *unbounded*, e.g.,

$$\int_{-1}^{\infty} \frac{1}{x} \, dx$$

Because the integration limit ∞ is permitted, MATHCAD can easily handle Case I of the unbounded integration interval.

☞

If MATHCAD does not produce a satisfactory result for the case of unbounded integration intervals, you can use the following trick:
instead of the *improper integral*

$$\int_{a}^{\infty} f(x) \, dx$$

calculate the *definite integral*

$$\int_{a}^{s} f(x) \, dx$$

with a fixed upper limit s and then determine the *limit* (see Section 19.4)

$$\lim_{s \to \infty} \int_{a}^{s} f(x) \, dx$$

as shown in Example 20.3e.

♦

Example 20.3:

a) MATHCAD calculates without difficulty the following convergent integral

$$\int_{1}^{\infty} \frac{1}{x^3} \, dx \to \frac{1}{2}$$

b) MATHCAD does not provide any result for the divergent integral

$$\int_{-\infty}^{\infty} \frac{1+x}{1+x^2} \, dx$$

that has the *principal value* π

$$\int_{-\infty}^{\infty} \frac{1+x}{1+x^2}\,dx \rightarrow \int_{-\infty}^{\infty} \frac{(x+1)}{(x^2+1)}\,dx$$

c) MATHCAD does not provide any result for the divergent integral

$$\int_{-\infty}^{\infty} x^3\,dx$$

that has the *principal value* 0

$$\int_{-\infty}^{\infty} x^3\,dx \rightarrow \int_{-\infty}^{\infty} x^3\,dx$$

d) MATHCAD calculates without difficulty the following convergent integral

$$\int_{1}^{\infty} e^{-x}\,dx \rightarrow \exp(-1)$$

e) Use the *limit calculation*

$$\lim_{s \to \infty} \int_{0}^{s} x^3 \cdot e^{-x}\,dx \rightarrow 6$$

to check the following result

$$\int_{0}^{\infty} x^3 \cdot e^{-x}\,dx \rightarrow 6$$

◆

It is more difficult to *calculate improper integrals* with *unbounded integrands* f(x). Because MATHCAD does not always recognize this case, incorrect results can occur.

Example 20.4:

a) If you symbolically integrate the *improper integral*

$$\int_{-1}^{1} \frac{1}{x^2}\,dx$$

without recognizing that the integrand is unbounded at x=0, you obtain −2 as incorrect result. The integral is actually divergent, which MATHCAD correctly recognizes:

$$\int_{-1}^{1} \frac{1}{x^2}\,dx \rightarrow \infty$$

b) MATHCAD does not provide any result for the *divergent integral*

$$\int_{-1}^{1} \frac{1}{x}\,dx$$

whose *Cauchy principal value* is 0:

$$\int_{-1}^{1} \frac{1}{x}\,dx \rightarrow \int_{-1}^{1} \frac{1}{x}\,dx$$

c) Both the integration interval and the integrand are unbounded for the following *improper integral* calculated by MATHCAD:

$$\int_{0}^{\infty} \frac{\ln(x)}{1+x^2}\,dx \rightarrow 0$$

d) MATHCAD calculates the following convergent integral

$$\int_{0}^{1} \frac{1}{\sqrt{x}}\,dx \rightarrow 2$$

◆

To summarize, it is desirable to check the *improper integrals calculated* by MATHCAD even when results are returned. It is also desirable to perform an additional calculation as definite integral with subsequent limit calculation (as in Example 20.3e).

◆

20.4 Numerical Methods

If the exact calculation of the *definite integral*

$$\int_a^b f(x)\,dx$$

fails, MATHCAD can *calculate* it *using approximations*, for which several numerical methods are provided.

MATHCAD can also *calculate primitives* F(x) (and thus indefinite integrals) as *approximations* for individual x-values if you use the *formula* shown in Section 20.2

$$F(x) = \int_a^x f(t)\,dt$$

and then numerically determine the contained definite integral for the required x values. Thus, MATHCAD provides a list of function values for a primitive F(x) with F(a)=0 that you then can display graphically to show the general function form (see Example 20.c).

♦

MATHCAD performs the *numerical calculation* of *definite integrals* as follows:

* *First* click the *integral button* for the *definite integration*

  in *operator palette no. 5* to create the following *integral symbol* with four placeholders

  $$\int_\blacksquare^\blacksquare \blacksquare \; d\blacksquare$$

  at the cursor location in the worksheet.

* Enter the *integration limits* a and b, the *integrand* f(x) and the *integration variable* x in the appropriate *placeholders* to obtain

  $$\int_a^b f(x)\,dx$$

  Place the mouse pointer on the integral sign and then press the right mouse button to display the following *dialogue box*

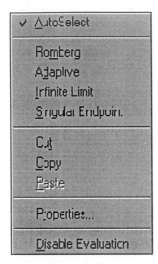

in which you can specify the numerical method to be used. MATHCAD
uses *AutoSelect* as standard, i.e., MATHCAD itself chooses the method to
be used. However, the user can click with the mouse to choose between, for example, a Romberg method (*Romberg*) or an adaptive quad-
rature method (*Adaptive*).

*   *Then mark* the complete *expression* with *editing lines*.
*   *Finally,* perform one of the following *activities*
    *   *Activate* the *menu sequence*
        Symbolics ⇒ Evaluate ⇒ Floating Point...
    *   Enter the *numerical equal sign* =
    to initiate the *numerical calculation* of the *definite integral*.
    The *built-in variable* **TOL** (see Sections 6.2 and 8.1) can be used to set
    the *precision* (*tolerance*).

☞

Because MATHCAD only differentiates between *exact* and *numerical calculation* for definite integrals when it has finished, I suggest you follow the
following procedure to calculate a given integral:

*   *First* enter the *symbolic equal sign* → and then press the ⏎-key to at-
    tempt the *exact calculation*.
*   If the exact calculation fails, replace the symbolic equal sign with the
    numerical =, which then initiates the *numerical calculation*.

◆

Example 20.5:

a) The following definite integral cannot be calculated exactly (see Exam-
    ple 20.2b) but only numerically:

$$\int_2^3 x^x \, dx = 11.675 \quad\blacksquare$$

b) The following definite integral can be calculated exactly. We compare
   the results of both calculation types:

   *   *exact calculation:*

$$\int_1^3 e^x \cdot \sin(x) \, dx \rightarrow$$

$$\frac{-1}{2} \cdot \exp(3) \cdot \cos(3) + \frac{1}{2} \cdot \exp(3) \cdot \sin(3) + \frac{1}{2} \cdot \exp(1) \cdot \cos(1) - \frac{1}{2} \cdot \exp(1) \cdot \sin(1) = 10.95 \quad\blacksquare$$

   *   *numerical calculation:*

$$\int_1^3 e^x \cdot \sin(x) \, dx = 10.95 \quad\blacksquare$$

   You can see that the numerical calculation here works correctly and
   supplies the same result as the exact calculation.

c) MATHCAD *does not calculate* the following *indefinite integral exactly*

$$\int \frac{e^x}{\sin(x) + 2} \, dx \rightarrow \int \frac{\exp(x)}{(\sin(x) + 2)} \, dx$$

   The *formula*

$$y = F(x) = \int_a^x f(t) \, dt \quad (x \geq a)$$

   can be used to *approximate* a *primitive* F(x) with F(a)=0 using the numerical calculation of a definite integral in a given number of x-values.
   The calculated points can be displayed graphically using the method described in Section 18.3.

   We *approximate* the *primitive* in the *interval* [0,2] in the x-values

   0, 0.1, 0.2, ... , 2

   The calculation can be performed easily in MATHCAD if you define x as
   range variable.

x := 0, 0.1 .. 2

$$F(x) := \int_0^x \frac{e^t}{\sin(t) + 2} \, dt$$

This causes the *primitive* F(x) (with F(0)=0) in the following x-values to be approximated as:

x	F(x)
0	0
0.1	0.051
0.2	0.105
0.3	0.163
0.4	0.223
0.5	0.288
0.6	0.356
0.7	0.43
0.8	0.509
0.9	0.594
1	0.686
1.1	0.785
1.2	0.894
1.3	1.012
1.4	1.142
1.5	1.285
1.6	1.442
1.7	1.616
1.8	1.809
1.9	2.023
2	2.264

The graphical display of the calculated approximation values for the primitive function F with F(0)=0 in the x-values 0, 0.1, 0.2, ...2:

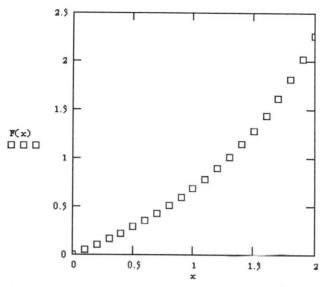

Connect the calculated approximation values with straight lines:

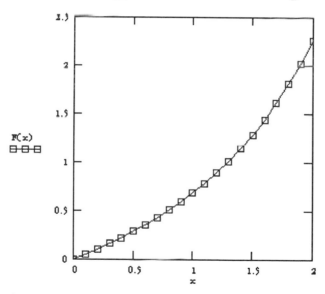

d) As with indefinite integrals, MATHCAD cannot calculate exactly the definite integral for *functions* that consist of *several expressions,* such as,

$$f(x) := \begin{cases} x & \text{if } x \le 0 \\ x^2 & \text{otherwise} \end{cases}$$

(see Examples 20.1n2 and 20.2e).
It is only possible to *calculate* the *definite integral numerically.*

$$\int_{-1}^{2} f(x)\,dx = 2.167 \quad \blacksquare$$

e) Calculate the definite integral

$$\int_{0}^{1} f(x)\,dx$$

for a function f(x) given in the interval [0,1] by the following points (number pairs)

$$vx := \begin{bmatrix} 0 \\ 0.1 \\ 0.2 \\ 0.3 \\ 0.4 \\ 0.5 \\ 0.6 \\ 0.7 \\ 0.8 \\ 0.9 \\ 1 \end{bmatrix} \qquad vy := \begin{bmatrix} 3 \\ 5 \\ 7 \\ 5 \\ 9 \\ 8 \\ 7 \\ 1 \\ 4 \\ 6 \\ 8 \end{bmatrix}$$

where the **vx** and **vy** *vectors* contain the x and y-coordinates respectively.

We approximate the given points using a polygonal arc (linear interpolation) or cubic splines that we discussed in Section 17.2.4, and

*   display the result graphically:

vs := cspline(vx , vy)   x := 0 , 0.01 .. 1

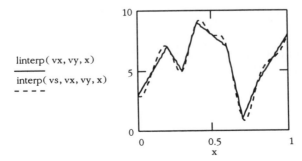

linterp(vx, vy, x)

interp(vs, vx, vy, x)

- - - -

* calculate an approximation for the definite integral using the interpolation functions **linterp** and **interp** of MATHCAD:

$$\int_0^1 \text{linterp}(vx, vy, x)\, dx = 5.74997540742708 \quad \blacksquare$$

$$\int_0^1 \text{interp}(vs, vx, vy, x)\, dx = 5.70618513630292 \quad \blacksquare$$

♦

If the numerical integration method used by MATHCAD does not provide any satisfactory results for the integral to be calculated, you can write your own programs using the tools described in Chapter 10.

♦

20.5 Multiple Integrals

Now that we have discussed in detail the integrals for functions of one variable (known as *simple integrals*), we use some examples in this section to discuss the *calculation* of *multiple integrals* in the plane and in the space, i.e., *double integrals* and *triple integrals* of the form

$$\iint_D f(x, y)\, dx\, dy \quad \text{and} \quad \iiint_G f(x, y, z)\, dx\, dy\, dz$$

(D and G are bounded domains in the plane or in the space).

The calculation of multiple integrals can be transformed into the calculation of repeated (two or three) simple integrals if the D and G domains represent so-called normal domains.

MATHCAD can perform *exact* and *numerical calculations* of *multiple integrals* by *nesting* the *integral operators* for definite integrations from *operator palette no. 5* in the similar manner as in Sections 20.2 and 20.4. A manually performed coordinate transformation can often improve the effectiveness of the calculation (see Examples 20.6b and c).

♦

We now use some examples to consider the problems involved with the calculation of multiple integrals.

Example 20.6:

a) The calculation of the *volume* of a three-dimensional domain bounded by the planes

$$z = x + y, \ z = 6 \ , \ x = 0 \ , \ y = 0 \ , \ z = 0$$

produces the following *triple integral* that MATHCAD can calculate without difficulty using the *triple nesting* of the *integral operator* for definite Integrals:

* *exact calculation*

$$\int_0^6 \int_0^{6-x} \int_{x+y}^6 1 \, dz \, dy \, dx \rightarrow 36$$

* *numerical calculation*

$$\int_0^6 \int_0^{6-x} \int_{x+y}^6 1 \, dz \, dy \, dx = 36 \ \blacksquare$$

b) The *volume calculation* for the domain in the first octant bounded by the surfaces

$$z = 0 \ (xy \ plane) \ , \ z = x^2 + y^2 \ (parabola) \ , \ x^2 + y^2 = 1 \ (cylinder)$$

produces the following exactly calculated integral by MATHCAD:

$$\int_0^1 \int_0^{\sqrt{1-x^2}} \int_0^{x^2+y^2} 1 \, dz \, dy \, dx \rightarrow \frac{1}{8} \cdot \pi$$

when you use *cylinder coordinates* to perform a manual *coordinate transformation* to produce the *simpler integral:*

$$\int_0^{\frac{\pi}{2}} \int_0^1 \int_0^{r^2} r \, dz \, dr \, d\phi \rightarrow \frac{1}{8} \cdot \pi$$

c) MATHCAD *cannot provide* an *exact solution* for the following double integral in this form

$$\int_0^3 \int_0^{\sqrt{9-x^2}} \sqrt{x^2+y^2} \, dy \, dx$$

The *numerical calculation* using MATHCAD yields

$$\int_0^{^{\bullet}3} \int_0^{^{\bullet}\sqrt{9 - x^2}} \sqrt{x^2 + y^2}\, dy\, dx = 14.135 \quad \blacksquare$$

You must *manually* perform a *coordinate transformation* (polar coordinates) to provide an integral that MATHCAD can *calculate exactly*.

$$\int_0^{^{\bullet}\frac{\pi}{2}} \int_0^{^{\bullet}3} r^2\, dr\, d\phi \rightarrow \frac{9}{2}\cdot\pi = 14.137 \quad \blacksquare$$

$\blacklozenge$

These examples show that the *calculation* of *multiple integrals* in MATH-CAD does not cause any additional problems provided that the contained simple integrals can be calculated.

If a *coordinate transformation* exists that simplifies the integral, you should perform this manually before starting the calculation with MATHCAD.

$\blacklozenge$

21 Infinite Series and Products

Whereas we discussed in Chapter 14 the case of *finite sums/series* and *products* that do not provide MATHCAD with any difficulties, this chapter handles the infinite case. We consider MATHCAD's facilities for

- the *calculation* of *infinite number series* and *products*,
- the *expansion* of *functions* in *function series*, where we discuss the *special cases* that are important for applications
  * *power series* (Section 21.2)
  * *Fourier series* (Section 21.3)

21.1 Number Series and Products

Chapter 14 discussed the use of MATHCAD to calculate *finite sums* (*series*) and *products* of the form

$$\sum_{k=m}^{n} a_k = a_m + a_{m+1} + \ldots + a_n \qquad \text{bzw.} \qquad \prod_{k=m}^{n} a_k = a_m \cdot a_{m+1} \cdot \ldots \cdot a_n$$

where the *terms*

$$a_k \qquad (k = m, \ldots, n)$$

are *real numbers:*

- The associated *sum operator* or *product operator* is selected with a mouse click from *operator palette no. 5*
- In the *placeholders* of the displayed *symbols*

$$\sum_{\blacksquare = \blacksquare}^{\blacksquare} \blacksquare \qquad \text{or} \qquad \prod_{\blacksquare = \blacksquare}^{\blacksquare} \blacksquare$$

enter

  * the *general term*
    $$a_k$$

after the sum sign or product sign
* k and m below the sum sign or product sign
* n above the sum sign or product sign

(see Example 21.1).

MATHCAD can calculate *infinite sums* of *real numbers*, which are called *infinite series* (*number series*) and *infinite products* of *real numbers* (*number products*) of the *form*

$$\sum_{k=m}^{\infty} a_k = a_m + a_{m+1} + \dots \qquad \text{bzw.} \qquad \prod_{k=m}^{\infty} a_k = a_m \cdot a_{m+1} \cdot \dots$$

in a similar manner as for finite sums and products. Rather than n, enter *infinity* in the upper placeholder of the sum operator or product operator by clicking with the mouse on the *infinity* button

from *operator palette no. 5.*

♦

The *two major difficulties* that occur during the calculation of infinite *number series* or *number products* come from the theory and so also apply to MATHCAD:

* No universally applicable *convergence criteria* exist to determine the convergence/divergence for an arbitrary series/product.

* However, even when convergence can be proven, a finite calculation algorithm *does not* normally *exist*, so that MATHCAD soon reaches its limits. You can sometimes achieve an approximation by specifying a suitably large value for n instead of ∞:

    * Because of the Leibniz criteria, this is always successful for *alternating series*. A lower boundary for n that can be calculated here provides an approximation with given accuracy (see Example 21.1e).

    * As known from the theory, because the obtained values can be completely incorrect, care should be exercised for products and non-alternating series.

♦

Note that MATHCAD can *only* perform the *exact calculation* of infinite number series and products performing one of the following actions

* *Activate* the *menu sequence*

    Symbolics ⇒ Evaluate ⇒ Symbolically

* *Activate* the *menu sequence*

Symbolics ⇒ Simplify

* Enter the *symbolic equal sign* → and then press the ⏎-key

MATHCAD does not provide any functions for numerical calculations, which, for example, would be useful for alternating series.

♦

Let us now use a number of examples.

Example 21.1:

a) MATHCAD recognizes the *divergence* of the following *series*:

$$\sum_{k=1}^{\infty} \frac{1}{\sqrt{k}} \to \infty$$

b) MATHCAD does not make any decision on the convergence or divergence for the *series*

$$\sum_{k=2}^{\infty} \frac{1}{k \cdot \ln(k)}$$

whose divergence can be tested using the integral test. The series is returned unchanged:

$$\sum_{k=2}^{\infty} \frac{1}{k \cdot \ln(k)} \to \sum_{k=2}^{\infty} \frac{1}{(k \cdot \ln(k))}$$

c) MATHCAD assigns the value 1/2 in accordance with the Abelian summation method to the *series*

$$\sum_{k=0}^{\infty} (-1)^k \to \frac{1}{2}$$

that apparently diverges, because the necessary convergence test is not satisfied.

d) MATHCAD calculates the following *convergent series*:

d1)

$$\sum_{k=1}^{\infty} \frac{1}{k^2} \to \frac{1}{6} \cdot \pi^2$$

d2)

$$\sum_{k\,=\,1}^{\infty} \frac{1}{(2 \cdot k - 1) \cdot (2 \cdot k + 1)} \rightarrow \frac{1}{2}$$

e) Although the *alternating series*

$$\sum_{k=1}^{\infty} (-1)^{k+1} \cdot \frac{k}{k^2 + 1}$$

satisfies the Leibniz criterion and is thus convergent, MATHCAD does not find any solution. MATHCAD reacts as for Example b.
You can obtain a sequence of approximate values for the series through the numerical calculation of finite sums for increasing n:

$$\sum_{k\,=\,1}^{100} (-1)^{(k+1)} \cdot \frac{k}{k^2 + 1} = 0.264635993910762 \quad \blacksquare$$

$$\sum_{k\,=\,1}^{1000} (-1)^{(k+1)} \cdot \frac{k}{k^2 + 1} = 0.269110753207134 \quad \blacksquare$$

$$\sum_{k\,=\,1}^{10000} (-1)^{(k+1)} \cdot \frac{k}{k^2 + 1} = 0.269560505208509 \quad \blacksquare$$

In accordance with the *Leibniz theorem*, the inequality

$$\left| a_{n+1} \right| = \frac{n+1}{(n+1)^2 + 1} < \varepsilon$$

can be used to determine the number n in order to use the *finite sum*

$$\sum_{k=1}^{n} (-1)^{k+1} \cdot \frac{k}{k^2 + 1}$$

to *approximate* the given alternating series with the given accuracy ε. Because MATHCAD does not solve this inequality for n, you can obtain a solution in the following manner:
Define the function

$$f(n) := \frac{n+1}{(n+1)^2 + 1}$$

and calculate for n = 100 , 1000 , 10000 , ... the function values

$$f(100) = 9.9 \cdot 10^{-3}$$

$$f(1000) = 9.99 \cdot 10^{-4}$$

$$f(10000) = 9.999 \cdot 10^{-5}$$

that provide an error bound when the appropriate number of terms of the series are accumulated.

f) MATHCAD calculates the following *alternating series*:

$$\sum_{k=1}^{\infty} (-1)^{(k-1)} \cdot \frac{1}{k} \rightarrow \ln(2)$$

g) MATHCAD can *solve* the *following problems* for *infinite products*:

g1) The *product*

$$\prod_{k=1}^{\infty} \left(1 + \frac{1}{k}\right) \rightarrow \infty$$

diverges to ∞.

g2) The *product*

$$\prod_{k=2}^{\infty} \left(1 - \frac{1}{k}\right) \rightarrow 0$$

diverges to 0.

g3) The *product*

$$\prod_{k=2}^{\infty} \left(1 - \frac{1}{k^2}\right) \rightarrow \frac{1}{2}$$

converges to 1/2.

♦

Example 21.1e shows how you can calculate *alternating series* when MATHCAD does not return any solution.

♦

21.2 Power Series

In Section 19.2 we used MATHCAD to calculate the *Taylor expansion*

$$f(x) = \sum_{k=0}^{n} \frac{f^{(k)}(x_0)}{k!} \cdot (x - x_0)^k + R_n(x)$$

for functions of one real variable x.

If for the *remainder*

$$R_n(x) = \frac{f^{(n+1)}(x_0 + \vartheta \cdot (x - x_0))}{(n+1)!} \cdot (x - x_0)^{n+1}$$

the following apply

$$\lim_{n \to \infty} R_n(x) = 0 \text{ for all } x \in (x_0 - r, x_0 + r)$$

then the *power series*

$$f(x) = \sum_{k=0}^{\infty} \frac{f^{(k)}(x_0)}{k!} \cdot (x - x_0)^k$$

can be used to represent the function f(x) with the *convergence interval*

$$|x - x_0| < r \qquad (r - convergence\ radius)$$

that is known as the *Taylor series*. This is also known as a *Taylor series expansion* or *power series expansion* for the given *function*
f(x)
at the *expansion point*
x_0

It is normally difficult to prove that a *function* f(x) can be *expanded* into a *Taylor series*. As known from the theory, the existence of the derivatives of arbitrary order of f(x) does not suffice here.

As we have already seen from Section 19.2, MATHCAD cannot calculate the general term of the Taylor series and thus the remainder, and so cannot provide any help here.

However, the *Taylor polynomial* of *degree n* (for n=1,2,...) suffices for many practical applications to provide a *polynomial of degree n* as approximation for a *given function* f(x). We have already shown this in Section 19.2.

♦

21.3 Fourier Series

Although the *Fourier series expansion* plays a large role for many practical problems (in particular in electronics, acoustics and optics), MATHCAD, in contrast to other computer algebra systems, does not provide any integrated functions for this expansion.

For a *periodic function* f(x) with the *period* 2p or for a given function f(x) over the interval [–p, p], the expansion into a *Fourier series* has the form

$$f(x) = \frac{a_0}{2} + \sum_{k=1}^{\infty} (a_k \cdot \cos \frac{k \cdot \pi \cdot x}{p} + b_k \cdot \sin \frac{k \cdot \pi \cdot x}{p})$$

with the *Fourier coefficients*

$$a_k = \frac{1}{p} \int_{-p}^{p} f(x) \cdot \cos \frac{k \cdot \pi \cdot x}{p} dx \quad \text{and} \quad b_k = \frac{1}{p} \int_{-p}^{p} f(x) \cdot \sin \frac{k \cdot \pi \cdot x}{p} dx$$

The pointwise convergence of the Fourier series formed in this manner is guaranteed for most functions f(x) that occur in practice (Dirichlet theorem).

☞

If you wish to expand a *function* f(x) in the interval

$[-\pi , \pi]$

into a *Fourier series*, we have a special case of our general expansion. The series in this case is

$$f(x) = \frac{a_0}{2} + \sum_{k=1}^{\infty} (a_k \cdot \cos k \cdot x + b_k \cdot \sin k \cdot x)$$

and the *Fourier coefficients* now have the form:

$$a_k = \frac{1}{\pi} \int_{-\pi}^{\pi} f(x) \cdot \cos k \cdot x \; dx \qquad b_k = \frac{1}{\pi} \int_{-\pi}^{\pi} f(x) \cdot \sin k \cdot x \; dx$$

♦

However, MATHCAD can return the *Fourier series* for a given *function* f(x) when you use the *integration methods* from Section 20.2 to calculate the *Fourier coefficients*.

We explain this method in the following Example 21.2 that can be used to calculate the *Fourier series* (to the N-th term)

$$F_N(x) = \frac{a_0}{2} + \sum_{k=1}^{N} (a_k \cdot \cos \frac{k \cdot \pi \cdot x}{p} + b_k \cdot \sin \frac{k \cdot \pi \cdot x}{p})$$

for *any functions* f(x). You only need to make the appropriate changes to

*   the function f(x)

* the value p for the interval [-p, p]
* the number N for the terms to be calculated

Example 21.2:

a) *Expand* the *function*

$$f(x) = x^2$$

in the *interval* [−1,1] into a *Fourier series* with 5 terms:
We must calculate the two *Fourier coefficients*

$$a_k = \int_{-1}^{1} x^2 \cdot \cos k \cdot \pi \cdot x \; dx \quad \text{and} \quad b_k = \int_{-1}^{1} x^2 \cdot \sin k \cdot \pi \cdot x \; dx$$

for k = 0, 1, ... , 10.
MATHCAD can solve the problem using the following method. We proceed in such a manner that we can expand any functions. You only need to make the appropriate changes to the function f(x), p and N:

$$f(x) := x^2 \quad p := 1$$

$$N := 5 \quad k := 0 .. N$$

$$a_k := \frac{1}{p} \cdot \int_{-p}^{p} f(x) \cdot \cos\left(k \cdot \pi \cdot \frac{x}{p}\right) dx$$

$$b_k := \frac{1}{p} \cdot \int_{-p}^{p} f(x) \cdot \sin\left(k \cdot \pi \cdot \frac{x}{p}\right) dx$$

$$F_N(x) := \frac{a_0}{2} + \left[\sum_{k=1}^{N} \left(a_k \cdot \cos\left(k \cdot \pi \cdot \frac{x}{p}\right) + b_k \cdot \sin\left(k \cdot \pi \cdot \frac{x}{p}\right)\right) \right]$$

Graph of the function f(x) and its Fourier series (for N=5):

$$x := -p, -p + 0.001 .. p$$

The graphic shows that the Fourier series provides a good approximation to the given function.

b) We use the formulae from Example a to calculate the *Fourier series* with
10 terms for the *function*

$$f(x) = x^3$$

in the *interval* [−2,2] :

$$f(x) := x^3 \qquad p := 2$$

$$N := 10 \quad k := 0..N$$

$$a_k := \frac{1}{p} \cdot \int_{-p}^{p} f(x) \cdot \cos\left(k \cdot \pi \cdot \frac{x}{p}\right) dx$$

$$b_k := \frac{1}{p} \cdot \int_{-p}^{p} f(x) \cdot \sin\left(k \cdot \pi \cdot \frac{x}{p}\right) dx$$

$$F_N(x) := \frac{a_0}{2} + \left[\sum_{k=1}^{N} \left(a_k \cdot \cos\left(k \cdot \pi \cdot \frac{x}{p}\right) + b_k \cdot \sin\left(k \cdot \pi \cdot \frac{x}{p}\right) \right) \right]$$

Graph of the function f(x) and its Fourier series (for N=10):

$$x := -p, -p + 0.001 .. p$$

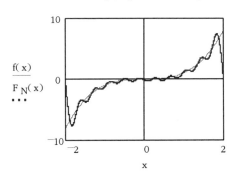

The graphic shows that the Fourier series provides a good approximation to the given function within the considered interval but with the known deviation at the endpoints of the interval. This is caused by the continuation of this function, which is discontinuous at the endpoints of the interval (in contrast to the function from Example a) and the associated Fourier series converges here to the mean value 0.

♦

22 Vector Analysis

The *vector analysis* considers *vectors* in the plane R^2 and in the space R^3 that *depend* on *variables* (i.e., *vector functions*) using methods from differential and integral calculus.

The discussions in this chapter show that MATHCAD can still be *improved* with regard to vector analysis:

* No functions exist to calculate the most important operators gradient, rotation and divergence. Consequently, we discuss means of defining these.

* No capabilities are provided for the graphical display of three-dimensional vector arrays.

* Curvilinear and surface integrals cannot be calculated directly. Consequently, we discuss how MATHCAD can be used to calculate these.

Other computer algebra systems provide better facilities here.

♦

22.1 Fields and their Graphical Display

Scalar and *vector fields* adopt a major role in *vector analysis*, where
* a *scalar quantity* (numerical value) u is assigned to each point P of the plane/space for a *scalar field*,
* a *vector* **v** is assigned to each point P of the plane/space for a *vector field*.

This permits the mathematical *description* in a Cartesian coordinate system in the *plane* R^2 or in the *space* R^3 of
* *scalar fields* using a (scalar) *function* of the form

  $u = u(x, y) = u(\mathbf{r})$ in the *plane* R^2

  $u = u(x, y, z) = u(\mathbf{r})$ in the *space* R^3

* *vector fields* using a *vector function* of the form

* $\mathbf{v} = \mathbf{v}\,(x,\,y) = \mathbf{v}\,(\,\mathbf{r}\,) = v_1(x,y)\cdot\mathbf{i} + v_2(x,y)\cdot\mathbf{j}$

   in the *plane* R^2 (*two-dimensional vector fields*)

* $\mathbf{v} = \mathbf{v}\,(x,\,y,\,z) = \mathbf{v}\,(\,\mathbf{r}\,) = v_1(x,y,z)\cdot\mathbf{i} + v_2(x,y,z)\cdot\mathbf{j} + v_3(x,y,z)\cdot\mathbf{k}$

   in the *space* R^3 (*three-dimensional vector fields*)

where

* $\mathbf{r} = x\cdot\mathbf{i} + y\cdot\mathbf{j}$ in the *plane* R^2

   $\mathbf{r} = x\cdot\mathbf{i} + y\cdot\mathbf{j} + z\cdot\mathbf{k}$ in the *space* R^3

   designate the *radius vector* with the *length* (*magnitude*)

* $r = |\mathbf{r}| = \sqrt{x^2 + y^2}$ in the *plane* R^2

* $r = |\mathbf{r}| = \sqrt{x^2 + y^2 + z^2}$ in the *space* R^3

* **i j k**

   designate the *basis vectors* of the coordinate system.

As we have seen, you obtain the special case of a *two-dimensional vector field* when you set

$v_3(x,y,z) = 0$

for *three-dimensional fields* and omit the third coordinate z.
MATHCAD can *graphically display two-dimensional fields*; Figures 22.1 and 22.2 show examples.

♦

Example 22.1:

a) Figure 22.1 contains the graph of the two-dimensional *vector field*

$$\mathbf{v} = \mathbf{v}(x,y) = \frac{x - y}{\sqrt{x^2 + y^2}}\cdot\mathbf{i} + \frac{x + y}{\sqrt{x^2 + y^2}}\cdot\mathbf{j}$$

b) The three-dimensional *vector field*

   $\mathbf{v} = \mathbf{v}\,(\,\mathbf{r}\,) = x\cdot\mathbf{i} + y\cdot\mathbf{j} + z\cdot\mathbf{k}$

   is a *potential field*. Figure 22.2 shows the *graphical display* of this field in the *plane*, i.e., for z=0.

   ♦

MATHCAD requires the following procedure for the *graphical display* of *two-dimensional vector fields*

$\mathbf{v} = \mathbf{v}\,(x,\,y) = v_1(x,y)\;\mathbf{i} + v_2(x,y)\;\mathbf{j}$

* *First calculate the* **V1** *and* **V2** *matrices* for the two functions

$v_1(x, y)$, $v_2(x, y)$

of the *vector field* in the same manner as for surfaces (see Figures 22.1 and 22.2).

* *Then* activate the *menu sequence*

Insert $\Rightarrow$ Graph $\Rightarrow$ Vector Field Plot

or click with the mouse the

button in *operator palette no. 3* to create a *graphic window*.

* Then enter the *designation* of the *calculated matrices* **V1** and **V2** in the form

V1 , V2   (Version 7)

(V1 , V2)   (Version 8)

in the placeholders of the *graphic window* (see Figure 22.1).

* *Then* click with the mouse outside the graphic window or press the ⎘- key to obtain the *graphical display* of the vector field if you are in automatic mode.

Figures 22.1 and 22.2 show the *graphical displays* of the *two-dimensional vector fields* from Example 22.1.

You can use the displayed figures as *general templates* for the *graphical display* of two-dimensional vector fields.

♦

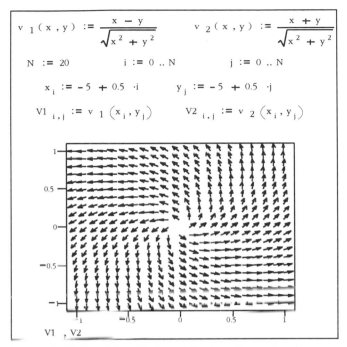

Figure 22.1. Graphical display of the two-dimensional
vector field from Example 22.1a

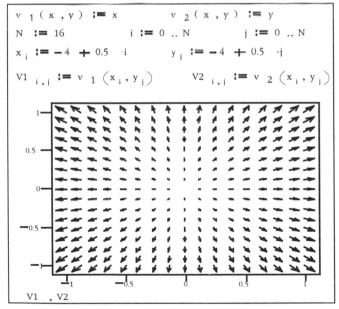

Figure 22.2. Graphical display of the two-dimensional
vector field from Example 22.1b

22.2 Gradient, Rotation and Divergence

The following differential operators play a fundamental role in the vector analysis for the characterization of fields:

- The *gradient*

  grad

  assigns to each *scalar field*

  u(**r**)

  a *vector field* of the form

  $$\mathbf{grad}\ u(\mathbf{r}) = u_x(\mathbf{r}) \cdot \mathbf{i} + u_y(\mathbf{r}) \cdot \mathbf{j} + u_z(\mathbf{r}) \cdot \mathbf{k}$$

  that is designated as *gradient field* provided that the u(**r**) function has partial derivatives

  $$u_x = \frac{\partial u}{\partial x}\ ,\quad u_y = \frac{\partial u}{\partial y}\ ,\quad u_z = \frac{\partial u}{\partial z}$$

- The *rotation* of a *vector field*

  $$\mathbf{v} = \mathbf{v}\,(x, y, z) = \mathbf{v}\,(\mathbf{r}) = v_1(x,y,z) \cdot \mathbf{i} + v_2(x,y,z) \cdot \mathbf{j} + v_3(x,y,z) \cdot \mathbf{k}$$

  is calculated from

  $$\mathbf{rot}\ \mathbf{v}(\mathbf{r}) = \begin{vmatrix} \mathbf{i} & \mathbf{j} & \mathbf{k} \\ \dfrac{\partial}{\partial x} & \dfrac{\partial}{\partial y} & \dfrac{\partial}{\partial z} \\ v_1 & v_2 & v_3 \end{vmatrix} =$$

  $$\left(\frac{\partial}{\partial y} v_3 - \frac{\partial}{\partial z} v_2\right) \cdot \mathbf{i} + \left(\frac{\partial}{\partial z} v_1 - \frac{\partial}{\partial x} v_3\right) \cdot \mathbf{j} + \left(\frac{\partial}{\partial x} v_2 - \frac{\partial}{\partial y} v_1\right) \cdot \mathbf{k}$$

- The *divergence* of the *vector field*

  $$\mathbf{v} = \mathbf{v}\,(x, y, z) = \mathbf{v}\,(\mathbf{r}) = v_1(x,y,z) \cdot \mathbf{i} + v_2(x,y,z) \cdot \mathbf{j} + v_3(x,y,z) \cdot \mathbf{k}$$

  is calculated from

  $$\mathrm{div}\ \mathbf{v}(\mathbf{r}) = \frac{\partial v_1}{\partial x} + \frac{\partial v_2}{\partial y} + \frac{\partial v_3}{\partial z}$$

Vector fields

v(r)

that can be represented as *gradient field* of a *scalar field* (designated as *potential*) u(**r**), i.e.,

$\mathbf{v}(\mathbf{r}) = \mathbf{grad}\ u(\mathbf{r})$

adopt an important role for practical applications.

This property can be tested using the *rotation* with the *condition*

$\mathbf{rot}\ \mathbf{v}(\mathbf{r}) = 0$

which under certain assumptions is necessary and sufficient for the *existence* of a *potential*. Such vector fields $\mathbf{v}(\mathbf{r})$ are called *potential fields*.

◆

In contrast with other computer algebra systems, MATHCAD does not provide *any functions* for the *calculation* of *gradient, rotation* and *divergence*. You can determine these in MATHCAD only by calculating the given formulae. We demonstrate this procedure in Example 22.2, where the *symbolic equal sign* → must be used for the *exact calculation*.

◆

Example 22.2:

a) MATHCAD can calculate the *gradients* of a given *function*

$u(x,y,z) = u(\mathbf{r})$

using the *matrix operator* and the *differentiation operator* from *operator palette no. 4 or 5*

$$\mathbf{grad}\,(u,x,y,z) := \begin{pmatrix} \dfrac{d}{dx}u(x,y,z) \\[2ex] \dfrac{d}{dy}u(x,y,z) \\[2ex] \dfrac{d}{dz}u(x,y,z) \end{pmatrix}$$

This *vector function*

$\mathbf{grad}\,(u,\,x,\,y,\,z)$

is generally applicable; it only requires that you have defined the function $u(x,y,z)$ appropriately, such as

$u(x,y,z) := x\cdot y\cdot z$

$$\mathbf{grad}\,(u,x,y,z) \rightarrow \begin{pmatrix} y\cdot z \\ x\cdot z \\ x\cdot y \end{pmatrix}$$

b) MATHCAD can calculate the *rotation* of a *vector field*

$\mathbf{v} = \mathbf{v}\,(x,\,y,\,z) = \mathbf{v}\,(\,\mathbf{r}\,) = v_1(x,y,z)\cdot\mathbf{i} + v_2(x,y,z)\cdot\mathbf{j} + v_3(x,y,z)\cdot\mathbf{k}$

using the *matrix operator* and the *differentiation operator* from *operator palette no. 4* or *5*, e.g.

$$
\mathbf{rot\_v}(x,y,z) := \begin{bmatrix} \dfrac{d}{dy}v_3(x,y,z) - \dfrac{d}{dz}v_2(x,y,z) \\[3mm] \dfrac{d}{dz}v_1(x,y,z) - \dfrac{d}{dx}v_3(x,y,z) \\[3mm] \dfrac{d}{dx}v_2(x,y,z) - \dfrac{d}{dy}v_1(x,y,z) \end{bmatrix}
$$

This *defined vector function*

rot_v(x, y, z)

is generally applicable if you have previously entered the components of the given vector field. This applies for both two and three-dimensional fields.

In the case of *two-dimensional fields*, just set

$v_3(x,y,z)=0$

Calculate the rotation for the following two and three-dimensional fields

$v_1(x,y,z) := x$ $v_1(x,y,z) := x$

$v_2(x,y,z) := y$ $v_2(x,y,z) := y$

$v_3(x,y,z) := 0$ $v_3(x,y,z) := z$

$$
\mathbf{rot\_v}(x,y,z) \rightarrow \begin{pmatrix} 0 \\ 0 \\ 0 \end{pmatrix}
$$

i.e., both fields are *potential fields*.

c) MATHCAD can calculate the *divergence* of a *vector field*

$\mathbf{v} = \mathbf{v}(x, y, z) = \mathbf{v}(\mathbf{r}) = v_1(x,y,z)\cdot\mathbf{i} + v_2(x,y,z)\cdot\mathbf{j} + v_3(x,y,z)\cdot\mathbf{k}$

using the *differentiation operator* from *operator palette no. 5*, e.g.

$\mathbf{div\_v}(x,y,z) := \dfrac{d}{dx}v_1(x,y,z) + \dfrac{d}{dy}v_2(x,y,z) + \dfrac{d}{dz}v_3(x,y,z)$

This *function*

div_v(x, y, z)

is generally applicable. You must only have entered the appropriate *components* of the given *vector field*, such as:

$$v_1(x,y,z) := x \cdot y \qquad v_2(x,y,z) := x \cdot z \qquad v_3(x,y,z) := y \cdot e^z$$

You then obtain the *result*

$$\text{div\_v}(x,y,z) \rightarrow y + y \cdot \exp(z)$$

◆

Because the preferred array index did not function in the MATHCAD version available to the author, the literal index was used in Examples 22.2b and c. It was also not possible to define rotation, divergence analogously to the gradient as a function that contains the name of the vector field as a parameter.

◆

If a *potential*

$$u(\mathbf{r})$$

is available, i.e.,

rot v (r)–0

applies for a *vector field*

$$\mathbf{v}(\mathbf{r})$$

it is generally difficult to calculate this potential using the integration of the relationships

$$\frac{\partial u}{\partial x} = v_1(\mathbf{r})$$

$$\frac{\partial u}{\partial y} = v_2(\mathbf{r})$$

$$\frac{\partial u}{\partial z} = v_3(\mathbf{r})$$

In contrast to other computer algebra systems, MATHCAD does not provide any functions to calculate the *potential*. You can only attempt here to integrate the three equations using the integration methods specified in Section 20.1.

Example 22.3:

MATHCAD calculates with the **rot_v** function, defined in Example 22.2b

$$\mathbf{rot\_v}(x,y,z) \rightarrow \begin{pmatrix} 0 \\ 0 \\ 0 \end{pmatrix}$$

for the *Coulomb field*

$$\mathbf{v}(x,y,z) = \mathbf{v}(\mathbf{r}) = C \cdot \frac{\mathbf{r}}{r^3} \quad (C \text{ constant})$$

i.e., it is a *potential field*. You obtain the associated *potential* u(**r**) by integrating

$$u(\mathbf{r}) = -\frac{C}{r}$$

We leave this as an exercise for the reader to calculate using MATHCAD.
♦

22.3 Curvilinear and Surface Integrals

Although the *calculation* of *curvilinear* and *surface integrals* is another important part of vector analysis, MATHCAD does not provide *any functions*. However, you can use MATHCAD to calculate such integrals if you first manually transform them to single or double integrals and then use MATHCAD to solve these using the appropriate integration methods.
We illustrate the procedure in the following example that calculates both a curvilinear and a surface integral.
♦

Example 22.4:

a) Calculate the *curvilinear integral*:

$$\int_C 2xy \, dx + (x - y) \, dy$$

along the parabola

$$y = x^2$$

between the points

(0,0) and (2,4)

Using the *computational formula* for *curvilinear integrals*, you obtain the *definite integral*

$$\int_0^2 (2 \cdot x \cdot x^2 + (x - x^2) \cdot 2 \cdot x) \, dx$$

that MATHCAD can calculate without difficulty:

$$\int_0^2 2 \cdot x \cdot x^2 + (x - x^2) \cdot 2 \cdot x \, dx \rightarrow \frac{16}{3}$$

b) The surface area of the *cone* K

$$z = \sqrt{x^2 + y^2}$$

lying between the planes z = 0 and z = 1 must be calculated.

The *computational formula* transforms the surface integral of the first kind to be calculated into a double integral

$$\iint\limits_{K} dS = \int\limits_{-1}^{1} \int\limits_{-\sqrt{1-x^2}}^{\sqrt{1-x^2}} \sqrt{1 + z_x^2 + z_y^2}\, dy\, dx = \int\limits_{-1}^{1} \int\limits_{-\sqrt{1-x^2}}^{\sqrt{1-x^2}} \sqrt{2}\, dy\, dx$$

that MATHCAD can calculate

$$\int_{-1}^{1} \int_{-\sqrt{1-x^2}}^{\sqrt{1-x^2}} \sqrt{2}\, dy\, dx \rightarrow \pi \cdot \sqrt{2}$$

◆

23 Differential Equations

Differential equations are equations that contain *functions* and their *derivatives*. These unknown functions must be determined so that a given differential equation is satisfied identically.

Ordinary differential equations differ from *partial differential equations* in the following manner. The *solution functions* for

* *ordinary differential equations*
  depend only on one (independent) variable

* *partial differential equations*
  depend on several (independent) variables.

♦

As for algebraic equations (see Chapter 16), *methods* are provided for the *exact solution*, in particular for *linear differential equations*, whereas only special cases of nonlinear differential equations can be calculated exactly.

♦

In contrast to other computer algebra systems, MATHCAD does *not provide any functions* for the *exact solution* of *differential equations.*
MATHCAD has not included the corresponding functions from the MAPLE *computer algebra system.* The only way of performing the exact solution in MATHCAD lies in the use of the Laplace and Fourier transformations that we discuss in Chapter 24.

♦

23.1 Ordinary Differential Equations

Solution algorithms to determine *general solutions* exist for *linear differential equations* of the *n-th order* of the form

$$a_n(x) \cdot y^{(n)} + a_{n-1}(x) \cdot y^{(n-1)} + \dots + a_1(x) \cdot y' + a_0(x) \cdot y = f(x)$$

when the *coefficients*

$$a_k(x)$$

satisfy certain *conditions*, for example

* $a_k(x)$ constant,

    i.e., an *equation* with *constant coefficients* is available.

* $a_k(x) = b_k \cdot x^k$ (b_k – constant),

    i.e., an *Euler equation* is available.

For *non-homogeneous differential equations* (i.e., $f(x) \neq 0$), the function $f(x)$ on the right-hand side must not be too complicated in order to be able to exactly determine the general solution.

Theorems to determine an exact solution also exist for special nonlinear ordinary differential equations, such as Bernoulli, Riccati's, Lagrange, Clairaut's, Bessel's, Legendre's equations.

♦

MATHCAD cannot determine general solutions for differential equations. It provides only various *functions for the numerical solution*, where *initial* and *boundary value problems* for *differential equation systems* of the *first order* can be *solved*. These functions are discussed in the next two sections.

23.1.1 Initial Value Problems

MATHCAD can *numerically solve systems* of *n differential equations* of the *first order* of the form

y ′ (x) = **f** (x , **y**(x))

with the *initial conditions* (for x = a)

y (a) = **y**^{a}

where the *solution vector* **y** (x) and **f** (x, y) represent the n-dimensional vectors

$$y(x) = \begin{pmatrix} y_1(x) \\ y_2(x) \\ \vdots \\ y_n(x) \end{pmatrix} \qquad \text{resp.} \qquad f(x,y) = \begin{pmatrix} f_1(x,y) \\ f_2(x,y) \\ \vdots \\ f_n(x,y) \end{pmatrix}$$

Because you can reduce every *n-th order differential equation* of the form

$$y^{(n)} = f(x, y, y', ..., y^{(n-1)})$$

to a *system* of *n differential equations* of the *first order* of the form

$$y_1' = y_2$$
$$y_2' = y_3$$
$$\vdots$$
$$y_{n-1}' = y_n$$
$$y_n' = f(x, y_1, y_2, \ldots, y_n)$$

by setting

$$y = y_1$$

MATHCAD can also be used to numerically solve *differential equations* of the *n-th order*.

♦

The *numerical function*

rkfixed (y, *a, b, points,* **D)**

most often applied in MATHCAD for the solution uses a *Runge-Kutta method* of the fourth order; the following *arguments* must be entered:

* **y** designates the *vector* of the *initial values* $\mathbf{y^a}$ at the point x = *a*, to which a previous assignment of the form

  $$\mathbf{y} := \mathbf{y^a}$$

  has been made.

* *a* and *b* are the *end points* of the *solution interval* [*a,b*] on the x-axis, where *a* is the *initial value* for x, for which the function value

  $$\mathbf{y}(a) = \mathbf{y^a}$$

  of the *solution vector* $\mathbf{y}(x)$ is given.

* *points* designates the *number* of *equidistant x-values* between *a* and *b*, in which approximate values for the solution vector are to be determined.

* **D** designates the *vector* of the *right-hand side* of the *system* of *differential equations*, to which these were previously assigned in the form

  $$\mathbf{D}(x, y) := \mathbf{f}(x, y)$$

The **rkfixed** *numerical function* can be most easily used for one *differential equation* of the *first order* (i.e., n=1)

$$y'(x) = f(x, y(x))$$

with the *initial condition*

$$y(a) = y^a$$

The function in this case returns a *result matrix* with two columns. This contains

* in the first column, the x-values
* in the second column, the approximate values of the *solution function* y(x) calculated for the x-values of the first column.

The *points* argument determines the *number* of *rows* for this *result matrix* (i.e., the x-values).

If you have *only one differential equation* and so only one *initial condition,* a component must be used in MATHCAD to assign this initial condition to a vector **y**. The next Example 23.1 shows the exact method involved.

◆

Example 23.1:

This example shows the *result matrix* in *matrix form.* We have set this as described in Chapter 15.

The *indexing* of the associated vectors and matrices starts with 0, i.e., **ORIGIN** := 0.

a) The solution of the *first order linear differential equation*

$$y' = - 2\,x{\cdot}y + 4{\cdot}x$$

with the *initial condition*

$$y(0) = 3$$

can be calculated exactly and is

$$y = 2 + e^{-x^2}$$

MATHCAD can provide the *numerical solution* in the interval [0,2] as follows:

$$y_0 := 3 \quad D(x,y) := -2{\cdot}x{\cdot}y + 4{\cdot}x$$

$$u := \text{rkfixed}(y,0,2,10,D)$$

The *result matrix* **u** calculated by MATHCAD contains in the first column the *x-values* for which the associated *function values* y(x) of the *solution* of the differential equation in the second column have been *calculated* as *approximations:*

$$
u = \begin{pmatrix}
0 & 3 \\
0.2 & 2.96 \\
0.4 & 2.85 \\
0.6 & 2.7 \\
0.8 & 2.53 \\
1 & 2.37 \\
1.2 & 2.24 \\
1.4 & 2.14 \\
1.6 & 2.08 \\
1.8 & 2.04 \\
2 & 2.02
\end{pmatrix}
$$

The graphical display of the exact and numerical solution shows how well the two match. Because the indexing starts with 0, the *result matrix* **u** contains the x-values in column 0 and the associated approximate values of the solution function y(x) in column 1:

$$x := 0, 0.001 .. 2$$

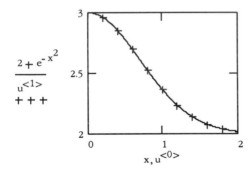

b) The first order nonlinear differential equation

$$y' = y^2 \cdot (\cos x - \sin x) - y$$

with the *initial condition*

$$y(0) = 1/2$$

has the *exact solution*

$$y = \frac{1}{2 \cdot e^x - \sin x}$$

MATHCAD returns in the interval [0,4] the following *numerical solution*, which we compare with the exact solution:

$$y_0 := \frac{1}{2} \qquad D(x,y) := y^2 \cdot (\cos(x) - \sin(x)) - y$$

$$i := 0 .. 10 \qquad x_i := 0.4 \cdot i$$

$$u := rkfixed(y, 0, 4, 10, D) \qquad y_i := \frac{1}{\left(2 \cdot e^{x_i} - \sin(x_i)\right)}$$

The *result matrix* **u** returned by the **rkfixed** *numerical function* contains the x-values in the first column and the calculated approximate values for the solution y(x) in the second column.

The left-hand output table shows the exact values of the solution y(x) in the appropriate x-values. The comparison with the result matrix **u** shows that the approximate values calculated by MATHCAD provide a good match with the exact solution values:

x_i	y_i
0	0.5
0.4	0.385
0.8	0.268
1.2	0.175
1.6	0.112
2	0.072
2.4	0.047
2.8	0.031
3.2	0.02
3.6	0.014
4	0.009

$$u = \begin{pmatrix} 0 & 0.5 \\ 0.4 & 0.385 \\ 0.8 & 0.268 \\ 1.2 & 0.175 \\ 1.6 & 0.112 \\ 2 & 0.072 \\ 2.4 & 0.047 \\ 2.8 & 0.031 \\ 3.2 & 0.02 \\ 3.6 & 0.014 \\ 4 & 0.009 \end{pmatrix}$$

c) The *non-homogeneous linear differential equation* of the first order

$$y' = \frac{1+y}{1+x}$$

with the *initial condition*

$$y(0) = 1$$

has the straight line

$$y = 2 \cdot x + 1$$

as the *exact solution.*

MATHCAD returns the *numerical solution* in the interval [0,2] using

$$y_0 := 1 \quad D(x,y) := \frac{1 + y}{1 + x} \quad u := rkfixed(y,0,2,10,D)$$

The *graphical display* shows the good match of the approximation solution calculated by MATHCAD with the exact solution:

$$x := 0, 0.001 .. 2$$

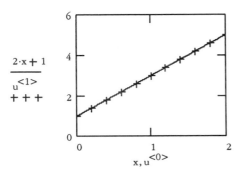

$2 \cdot x + 1$

$\overline{u^{<1>}}$

$+ + +$

$$x, u^{<0>}$$

d) An exact solution is not known for the nonlinear differential equation of the first order

$$y' = y^2 + x^2 \quad , \quad y(0) = -1$$

MATHCAD returns in the interval [0,1] the following *numerical solution:*

$$y_0 := -1 \quad D(x,y) := y^2 + x^2$$

$$u := rkfixed(y,0,1,10,D)$$

$$u = \begin{pmatrix} 0 & -1 \\ 0.1 & -0.91 \\ 0.2 & -0.83 \\ 0.3 & -0.76 \\ 0.4 & -0.7 \\ 0.5 & -0.63 \\ 0.6 & -0.57 \\ 0.7 & -0.49 \\ 0.8 & -0.42 \\ 0.9 & -0.33 \\ 1 & -0.23 \end{pmatrix}$$

The *graphical display* of the calculated approximate values for the solution function (connected by straight lines) has the following form:

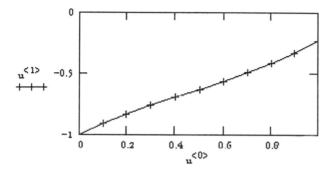

e) The nonlinear differential equation of the first order

$$y' = e^y \quad , \quad y(0) = -1$$

has the following *exact solution*:

$$y(x) = -\ln(e - x)$$

that can be easily obtained using the method of separation of variables. MATHCAD calculates the following *numerical solution*:

$$y_0 := -1 \qquad D(x, y) := e^y$$

$$u := \text{rkfixed}(y, 0, 1, 10, D)$$

$$u = \begin{bmatrix} 0 & -1 \\ 0.1 & -0.963 \\ 0.2 & -0.924 \\ 0.3 & -0.883 \\ 0.4 & -0.841 \\ 0.5 & -0.797 \\ 0.6 & -0.751 \\ 0.7 & -0.702 \\ 0.8 & -0.651 \\ 0.9 & -0.598 \\ 1 & -0.541 \end{bmatrix}$$

The graphical display shows that the approximate solution calculated by MATHCAD provides a good match to the exact solution:

$$x := 0, 0.001 .. 1$$

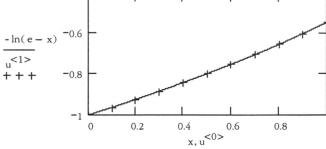

♦

If you want to solve *initial value problems* for *systems* of n *differential equations* of the first order in the interval [a, b]

$$y_1' = f_1(x, y_1, ..., y_n)$$
$$y_2' = f_2(x, y_1, ..., y_n)$$
$$\vdots$$
$$y_n' = f_n(x, y_1, ..., y_n)$$

having the *initial conditions*

$$y_1(a) = y_1^a \quad , \quad y_2(a) = y_2^a \quad , \quad ... \quad , \quad y_n(a) = y_n^a$$

proceed in the same manner as for the above Example 23.1 for one differential equation of first order and use the *numerical function*

rkfixed

in which you must now assign the vector of the initial conditions and the right-hand side of the system to the two arguments **y** and **D**, i.e.,

$$\mathbf{y} := \begin{pmatrix} y_1^a \\ y_2^a \\ \vdots \\ y_n^a \end{pmatrix} \qquad \mathbf{D(x, y)} := \begin{pmatrix} f_1(x, y_1, ..., y_n) \\ f_2(x, y_1, ..., y_n) \\ \vdots \\ f_n(x, y_1, ..., y_n) \end{pmatrix}$$

The **rkfixed** *function* returns a *result matrix* with n+1 columns, in which

* the *first column* contains the x-values
* the remaining n columns contain the calculated function values for the solution functions

$$y_1, ..., y_n$$

Let us now consider two examples for the numerical solution of differential equations of higher (n-th) order and systems of the first order.

Example 23.2:

a) The *system* of *differential equations*

$$y_1' = \frac{y_1}{2 \cdot y_1 + 3 \cdot y_2}$$

$$y_2' = \frac{y_2}{2 \cdot y_1 + 3 \cdot y_2}$$

with the *initial conditions*

$$y_1(0) = 1 \ , \ y_2(0) = 2$$

has the exact solution

$$y_1(x) = \frac{x}{8} + 1 \quad , \quad y_2(x) = \frac{x}{4} + 2$$

The *numerical solution* using MATHCAD with the **rkfixed** *numerical function* is performed as follows:

$$\mathbf{y} := \begin{pmatrix} 1 \\ 2 \end{pmatrix} \qquad \mathbf{D(x, y)} := \begin{pmatrix} \dfrac{y_1}{2 \cdot y_1 + 3 \cdot y_2} \\ \dfrac{y_2}{2 \cdot y_1 + 3 \cdot y_2} \end{pmatrix}$$

$$\mathbf{Y} := \mathbf{rkfixed}\,(\mathbf{y}\,,\,0\,,\,2\,,\,10\,,\,\mathbf{D})$$

$$\mathbf{Y} = \begin{pmatrix}
0 & 1 & 2 \\
0.2 & 1.025 & 2.05 \\
0.4 & 1.05 & 2.1 \\
0.6 & 1.075 & 2.15 \\
0.8 & 1.1 & 2.2 \\
1 & 1.125 & 2.25 \\
1.2 & 1.15 & 2.3 \\
1.4 & 1.175 & 2.35 \\
1.6 & 1.2 & 2.4 \\
1.8 & 1.225 & 2.45 \\
2 & 1.25 & 2.5
\end{pmatrix}$$

MATHCAD calculates in the *result matrix* **Y** in the columns 2 and 3 the function values for the solution functions

$$y_1(x) \quad \text{and} \quad y_2(x)$$

for the x-values 0 , 0.2 , 0.4 , ... , 2 of the first column.

b) The *initial value problem* for the second order differential equation

$$y'' - 2 \cdot y = 0 \quad , \quad y(0) = 2 \quad , \quad y'(0) = 0$$

that has the *exact solution*

$$y(x) = e^{\sqrt{2}x} + e^{-\sqrt{2}x}$$

can be *transformed* into the following *initial value problem* for *first order systems*:

$$y_1' = y_2 \quad , \quad y_1(0) = 2$$
$$y_2' = 2 \cdot y_1 \quad , \quad y_2(0) = 0$$

The function

$$y_1(x)$$

of this system returns the solution of the given second order differential equation.

We use the **rkfixed** *numerical function* in the *interval* [0,4] with the *step size* 0.4 in MATHCAD to perform the calculation and draw the exact solution and the found numerical solution in one coordinate system.

Because the indexing in this example started with 1 (i.e., **ORIGIN := 1**), the *result matrix* **Y** calculated by MATHCAD contains in the first column the x-values from the interval [0,4] and in the second and third columns the calculated *approximate values* for $y_1(x)$ and $y_2(x)$.

$$y := \begin{pmatrix} 2 \\ 0 \end{pmatrix} \qquad D(x,y) := \begin{pmatrix} y_2 \\ 2 \cdot y_1 \end{pmatrix}$$

$$Y := \textbf{rkfixed}\,(y\,,\,0\,,\,4\,,\,10\,,\,D)$$

$$Y = \begin{pmatrix} 0 & 2 & 0 \\ 0.4 & 2.329 & 1.685 \\ 0.8 & 3.421 & 3.924 \\ 1.2 & 5.637 & 7.452 \\ 1.6 & 9.702 & 13.426 \\ 2 & 16.953 & 23.807 \\ 2.4 & 29.768 & 42.003 \\ 2.8 & 52.355 & 73.987 \\ 3.2 & 92.128 & 130.258 \\ 3.6 & 162.144 & 229.289 \\ 4 & 285.386 & 403.587 \end{pmatrix}$$

The graphical display of the exact and numerical solution shows how well the two match:

x := 0 , 0.001 .. 4

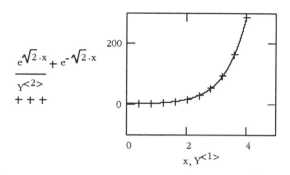

$\dfrac{e^{\sqrt{2}\cdot x} + e^{-\sqrt{2}\cdot x}}{Y^{<2>}}$

+ + +

♦

In addition to the *standard function* **rkfixed** that uses a Runge-Kutta method, MATHCAD has *further functions* for the *numerical solution* of first order differential equation systems that we will now discuss. These functions have the same arguments as for **rkfixed**:

- **Bulstoer**

  The function uses the Bulirsch-Stoer method.

- **Rkadap**

  In contrast to the Runge-Kutta method **rkfixed**, the solution is not calculated using equidistant x-values. The step size used in the function calculation is selected to be dependent on the function change. However, the result is output with equidistant x-values.

- **Stiffb** and **Stiffr**

  These functions use the Bulirsch-Stoer or Rosenbrock method to solve *stiff differential equations*. The matrix designation **J** appears in the last position as additional argument for these functions. A *matrix*

  J(x,y)

  of the type (n,n+1) that contains as

  *   *first column* $\dfrac{\partial \mathbf{D}}{\partial x}$

  *   *remaining n columns* $\dfrac{\partial \mathbf{D}}{\partial y_k}$ (k = 1, 2, ... , n)

  must have been previously assigned to this designation **J**.

If you only require the value of the *solution* at the *end point* b of the solution interval, you can use the **bulstoer, rkadapt, stiffb** and **stiffr** *functions* with the *arguments*

(**y** , a , b , acc , **D** , *pointsmax* , *min*)

The new arguments have the following significance:

*   *acc*

    controls the *accuracy* of the solution by changing the step size. The value 0.001 is set as default.

*   *pointsmax*

    specifies the maximum number of points in the interval [a,b] in which the approximate values are calculated.

*   *min*

    determines the minimum distance between the x-values in which the approximate values are calculated.

We investigate the use of all MATHCAD numerical functions in the following example.

Example 23.3:

We solve an example of a *stiff differential equation* that often appears in literature

$$y' = -10\ (y - \arctan x) + \frac{1}{1 + x^2}$$

with the *initial condition*

$$y(0) = 1$$

that has the *exact solution*

$$y\ =\ e^{-10x} + \arctan x$$

In the following example, we use all the discussed *functions* for the *numerical solution* and compare the results:

a) Use of the **rkfixed** *function*

$$y_0 := 1 \quad D(x,y) := -10 \cdot (y - atan(x)) + \frac{1}{1 + x^2}$$

$$u1 := rkfixed(y, 0, 5, 20, D)$$

$$x := 0, 0.01 .. 5$$

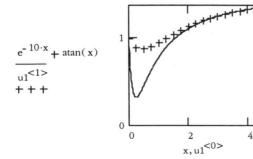

$$e^{-10 \cdot x} + atan(x)$$

$$\frac{}{u1^{<1>}}$$

$$+ + +$$

b) Use of the **Bulstoer** *function*

$$y_0 := 1 \qquad D(x,y) := -10 \cdot (y - \text{atan}(x)) + \frac{1}{1 + x^2}$$

u2 := Bulstoer(y, 0, 5, 20, D)

x := 0, 0.01 .. 5

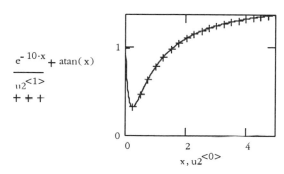

$$\frac{e^{-10 \cdot x} + \text{atan}(x)}{\text{u2}^{<1>}}$$
$$+ + +$$

c) Use of the **Rkadapt** *function*

$$y_0 := 1 \qquad D(x,y) := -10 \cdot (y - \text{atan}(x)) + \frac{1}{1 + x^2}$$

u3 := Rkadapt(y, 0, 5, 20, D)

x := 0, 0.01 .. 5

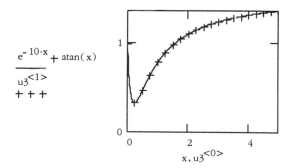

$$\frac{e^{-10 \cdot x} + \text{atan}(x)}{\text{u3}^{<1>}}$$
$$+ + +$$

d) Use of the **Stiffb** *function*

$$y_0 := 1 \qquad D(x,y) := -10 \cdot (y - atan(x)) + \frac{1}{1 + x^2}$$

$$J(x,y) := \left[\frac{10}{1 + x^2} - 2 \cdot \frac{x}{(1 + x^2)^2} \quad -10 \right]$$

$$u4 := Stiffb(y, 0, 5, 20, D, J) \qquad x := 0, 0.01 .. 5$$

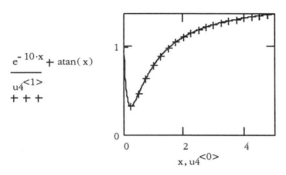

$$e^{-10 \cdot x} + atan(x)$$

$$\overline{u4^{<1>}}$$
$$+ + +$$

e) Use of the **Stiffr** *function*

$$y_0 := 1 \qquad D(x,y) := -10 \cdot (y - atan(x)) + \frac{1}{1 + x^2}$$

$$J(x,y) := \left[\frac{10}{1 + x^2} - 2 \cdot \frac{x}{(1 + x^2)^2} \quad -10 \right]$$

$$u5 := Stiffr(y, 0, 5, 20, D, J)$$

$$x := 0, 0.01 .. 5$$

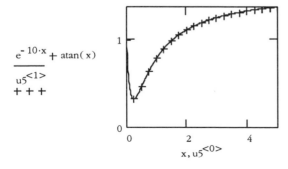

$$e^{-10 \cdot x} + atan(x)$$

$$\overline{u5^{<1>}}$$
$$+ + +$$

Note that J(x,y) is a row vector with two components in Examples d and e. The use of the five *numerical functions* shows that only the **rkfixed** standard algorithm, which caused no problems in other situations, does not produce a satisfactory result for *stiff differential equations.*

♦

Because you do not have much advance knowledge of the properties of the solution for a given differential equation, it is better to perform the numerical solution of the same problem using several of the given numerical functions.

♦

Because the use of the numerical functions is somewhat complicated, it is better to *write* and *save* your *own worksheets*. The previous examples can be used as templates. Because you only need to change the differential equation, the initial conditions and the solution interval, this saves you much work.

The

* **Differential Equations Function Pack**

* **Numerical Recipes**

* **Numerical Methods**

electronic books also contain worksheets for the solution of differential equations.

23.1.2 Boundary Value Problems

In addition to the previously considered *initial value problems*, for which the values for the solution functions and its derivatives are given for just one x-value, *boundary value problems* play an important role in practical applications for which the values for the solution functions and its derivatives are given for several x-values. *Boundary values* in the two *end points a* and *b* of the *solution interval* [a, b] are frequently given.

Whereas those initial value problems, subject to only weak assumptions, have a unique solution, such a solution is much more difficult to obtain for boundary value problems. As the following example demonstrates, even simple problems may have no solution.

Example 23.4:

Consider the simple second order linear differential equation:

$$y'' + y = 0$$

It has the *general solution*

$$y(x) = A \cdot \cos x + B \cdot \sin x \qquad \text{(A and B arbitrary constants)}$$

However, this general solution is of no interest for practical problems. We search for solutions here that satisfy certain conditions (initial values or boundary values).

We supply so-called *two-point boundary values* in the following example, i.e., values for the solution function for two different x-values. We change these values to show the three possibilities that can occur for the solution of *boundary value problems*. For the *boundary conditions*

a) $y(0) = y(\pi) = 0$

in addition to the *trivial solution*

$y(x) = 0$

there exist *additional solutions* of the form

$y(x) = B \cdot \sin x$ (B arbitrary constant)

b) $y(0) = 2$, $y(\pi/2) = 3$

there exists the unique *solution*

$y(x) = 2 \cdot \cos x + 3 \cdot \sin x$

c) $y(0) = 0$, $y(\pi) = -1$

there exist *no solutions.*

◆

MATHCAD provides two *numerical functions* **sbval** and **bvalfit** to calculate *boundary value problems,* however, these assume that a system of first or-der differential equations is present. We only use the *function*

sbval (v , a , b , D , load , score)

whose *arguments* have the following meaning:

*   **v**

vector (column vector) for the *estimations* of the *initial values* in point a that are not given

* a , b

end points of the *solution interval* $[a,b]$

*   **D (x , y)**

This vector has the same meaning as for initial value problems and contains the right-hand side of the *differential equations*

*   **load (a , v)**

This vector (column vector) initially contains the *given initial values* and then the estimate values from the vector **v** for the *missing initial values* in point a.

*   **score** (b , **y**)

This vector has the same number of components as the estimate vector **v** and contains the differences between those functions y_i for which the boundary values are given in point b and their *given* values in point b.

☞

Whereas the given numerical values must be entered in the **sbval** function for the *end points* a and b of the solution *interval* $[a,b]$, these are entered only symbolically for **load** and **score**, e.g., as a and b.

Consequently, it is better to assign the given numerical values to the end points a and b before starting the calculation and then use only the designations a and b in all functions (see Example 23.6b).

♦

The **sbval** function returns a *vector* as the *result* that contains the *missing initial values*. This permits us to treat the given problem as an initial value problem and then use the appropriate numerical functions.

To permit the use of the **sbval** function for a higher order equation, you must transform this equation to a first order system in the same manner as for initial value problems.

The *following example* shows the *use* of the **sbval** *function*.

Example 23.5:

The *boundary value problem*

$$y''' + y'' + y' + y = 0$$

$$y(0) = 1 , y(\pi) = -1 , y'(\pi) = -2$$

has the *unique solution*

$$y(x) = \cos x + 2 \sin x.$$

To use the **sbval** function here, we must transform this third-order differential equation to the *first order system*

$$y_1' = y_2$$
$$y_2' = y_3$$
$$y_3' = - y_3 - y_2 - y_1$$

with the *boundary values*

$$y_1(0) = 1 , \quad y_1(\pi) = -1 , \quad y_2(\pi) = -2$$

The **sbval** *function* can be used to determine the *missing initial values*

$$y_2(0) \quad \text{und} \quad y_3(0)$$

required to permit the solution as *initial values problem:*

To determine the missing two initial values, in each case we assign 1 as *estimate value* in the vector **v** and start the indexing at 1, i.e., **ORIGIN** := 1:

$$v := \begin{pmatrix} 1 \\ 1 \end{pmatrix} \qquad \text{load}(a,v) := \begin{pmatrix} 1 \\ v_1 \\ v_2 \end{pmatrix}$$

$$D(x,y) := \begin{pmatrix} y_2 \\ y_3 \\ -y_3 - y_2 - y_1 \end{pmatrix}$$

$$\text{score}(b,y) := \begin{pmatrix} y_1 + 1 \\ y_2 + 2 \end{pmatrix}$$

$$S := \text{sbval} (v, 0, \pi, D, \text{load}, \text{score})$$

$$S = \begin{pmatrix} 2 \\ -1 \end{pmatrix}$$

After using **sbval** function, the **S** vector contains the *approximate values* for the missing *initial values*

$$y_2(0) \quad \text{and} \quad y_3(0)$$

Because you now have the required initial values:

$$y_1(0) = 1, \ y_2(0) = 2, \ y_3(0) = -1$$

you can now use the numerical functions for initial value problems to solve the given problem

$$y := \begin{pmatrix} 1 \\ 2 \\ -1 \end{pmatrix} \qquad D(x,y) := \begin{pmatrix} y_2 \\ y_3 \\ -y_3 - y_2 - y_1 \end{pmatrix}$$

$$Y := \text{rkfixed}(y, 0, 1, 10, D)$$

Because the indexing of the *result matrix* **Y** starts with 1, column 1 and column 2 contain the x-values and the calculated approximate values for the *solution* $y_1(x)$ respectively. The following *graph* shows both the calculated approximate solution and the exact solution:

$x := 0, 0.001 .. 1$

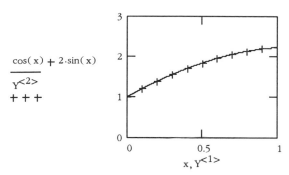

$$\frac{\cos(x) + 2 \cdot \sin(x)}{Y^{<2>}}$$
$+++$

x, Y$^{<1>}$

We can see from the graph that the approximate solution provides a good match to the exact solution.

♦

MATHCAD requires that the *estimations* for the *initial values* must be entered as vector **v**, even when only one value is present. We describe the procedure for this case in the following example.

♦

Example 23.6:

To solve the boundary value problems from Example 23.4b, we must first transform the *differential equation*

$y'' + y = 0$

with the *boundary value conditions*

$y(0) = 2 , y(\pi/2) = 3$

into the *first order system* of the form

$y_1' = y_2$
$y_2' = -y_1$

with the *boundary value conditions*

$y_1(0) = 2 , y_1(\pi / 2) = 3$

We can then use the **sbval** function. Because only one initial condition $y_2(0)$ is missing (i.e., y'(0)), the vector **v** used to *estimate* the *initial values* contains only one value.

sbval does not function, if, as in Example a, v is not used as the vector:

a)

$$v := 0 \qquad load(a, v) := \begin{pmatrix} 2 \\ v \end{pmatrix} \qquad D(x, y) := \begin{pmatrix} y_2 \\ -y_1 \end{pmatrix}$$

$$\text{score}(b,y) := y_1 - 3$$

$$S := \text{sbval}\left(\underset{\uparrow}{v}, 0, \frac{\pi}{2}, D, \text{load}, \text{score}\right)$$

| must be vector |

v must be defined as vector if you want to use the **sbval** function success-fully. Because MATHCAD does not permit the definition of vectors (matri-ces) with only one element using the method described in Section 15.1, you must use the following trick:

b) In the following example we use the recommended variant in which we assign the given values to the *end points a* and *b* at the start:

$$a := 0 \qquad b := \pi/2$$

$$v_1 := 0 \qquad \text{load}(a,v) := \begin{pmatrix} 2 \\ v_1 \end{pmatrix} \qquad D(x,y) := \begin{pmatrix} y_2 \\ -y_1 \end{pmatrix}$$

$$\text{score}(b,y) := y_1 - 3$$

$$S := \text{sbval}(v,a,b,D,\text{load},\text{score})$$

$$S = 3$$

The missing *initial value returned* by **sbval**

$$y_2(0) = y'(0) = 3$$

permits the solution of the problem as the initial value problem using the **rkfixed** *numerical function*:

$$y := \begin{pmatrix} 2 \\ 3 \end{pmatrix} \qquad D(x,y) := \begin{pmatrix} y_2 \\ -y_1 \end{pmatrix}$$

$$Y := \text{rkfixed}(y, a, b, 10, D)$$

The *graphical display* shows that the returned approximate solution provides a good match with the exact solution:

$$x := 0, 0.001 .. 2$$

$$\frac{2 \cdot \cos(x) + 3 \cdot \sin(x)}{Y^{<2>}}$$
$$+ + +$$

23.2 Partial Differential Equations

Because the methods for the exact solution are much more difficult for *partial differential equations,* it is not surprising that MATHCAD does not provide any functions. However, you can use the expansion into Fourier series and Fourier and Laplace transformations to provide exact solutions for special linear partial differential equations using MATHCAD.

MATHCAD provides only the following two *functions* for the *numerical solution*

relax and **multigrid**

for the *Poisson differential equation* for the plane

$$\frac{\partial^2 u(x,y)}{\partial x^2} + \frac{\partial^2 u(x,y)}{\partial y^2} = f(x,y)$$

over a *square domain.* In this case, you must use the difference method to transform the differential equation into a difference equation, which the two functions then can solve numerically. The MATHCAD help explains the procedure involved if you enter the two function names **relax** and **multigrid** as terms.

You can avoid this method if you use the MATHCAD *worksheets* provided for *partial differential equations,* such as those contained in Section 15.4 *"Relaxation Methods for Boundary Value Problems"* of the electronic book

Numerical Recipes Extension Pack

The following example shows two such worksheets used to solve the Poisson differential equation.

Example 23.7:

a) Section 15.4 of the electronic book **Numerical Recipes Extension Pack** contains the following worksheet for the numerical solution of the Poisson differential equation:

sor   **N R**   C: 869
              F: 860

solves a partial differential equation by performing successive overrelaxation to find a solution to the equation

$$a_{j,L} \cdot u_{j+1,L} + b_{j,L} \cdot u_{j-1,L} + c_{j,L} \cdot u_{j,L+1} + d_{j,L} \cdot u_{j,L-1} + e_{j,L} \cdot u_{j,L} = f_{j,L}$$

Its arguments are:

- square matrices a, b, c, d, e, f o f coefficients, all the same size
- an initial guess at the solution u, a square matrix of the same size
- rjac, the spectral radius of the Jacobi iteration matrix

The output is the solution matrix u.

$$R := 16 \qquad j := 0..R \qquad L := 0..R$$

When solving Poisson's equation with a single charge at the center of grid, the coefficients are

$$a_{j,L} := 1 \qquad b_{j,L} := 1 \qquad c_{j,L} := 1 \qquad d_{j,L} := 1$$

$$e_{j,L} := -4 \qquad f_{j,L} := 0 \qquad f_{\frac{R}{2},\frac{R}{2}} := 1$$

$$rjac := \cos\left(\frac{\pi}{R+1}\right)$$

$$u_{j,L} := 0$$

$$S := sor(a,b,c,d,e,f,u,rjac)$$

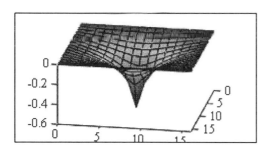

S

b) The problem solved in the following worksheet affects the stationary (time-independent) *temperature distribution* T(x,y) in a square plate that contains a heat source. A constant temperature of 0 degrees is maintained at the plate edge. The **multigrid** *numerical function can solve* and draw the solution function T(x,y) as solution for the partial differential equation (Laplace or Poisson equation with boundary conditions 0 at the plate edge) that describes this problem.

The Heat Equation

Problem

Find the temperature **T(x,y)** of a square plate with an internal heat source. The boundary of the source is pinned at zero degrees. We assume that the source does not change with time, and solve for the steady-state temperature distribution.

Algorithm

This application uses the partial differential equation solver   **multigrid** .

At any point free of the heat source, the heat equation reduces to Laplace's equation in two variables

$$T_{xx} + T_{yy} = 0$$

$R := 32$ Size of the grid is **R + 1**

$\rho_{R,R} := 0$ Sets the dimensions of the source ρ

$c := 8 \qquad d := 14$

$heat := 1545$ position and strength of source

$\rho_{c,d} := heat$

$G := multigrid\,(\rho, 2)$

We show the solution as a surface plot and a contour plot showing the lines of constant temperature.

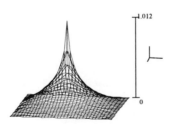

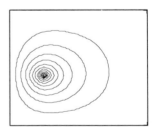

- G   G

◆

24 Transformations

We discuss the *integral transformations*

* *Laplace transformation*
* *Fourier transformation*

and other *transformations* such as

* *Z-transformation*
* *Wavelet transformation*

in this chapter that provide a wide application spectrum.
We show in the following Sections 24.1 24.4 how MATHCAD is used to perform these transformations.
In Section 24.5 we use the *Laplace* and *Z-transformation* to solve

* *differential equations*
* *difference equations.*

24.1 Laplace Transformation

The *Laplace transform*

L [f] = F(s)

of a *function* (*original function*)
f(t)
is determined under certain assumptions from the *integral transformation*

$$L [f] = F(s) = \int_{0}^{\infty} f(t) e^{-st} dt$$

The *inverse Laplace transformation*, provided certain assumptions are satisfied, is determined from

$$f(t) = \frac{1}{2\pi i} \int_{c-i\infty}^{c+i\infty} e^{st} F(s) ds$$

MATHCAD is used as *follows* to perform *Laplace transformations* and *inverse Laplace transformations:*

- *Calculation* of the *Laplace transformation:*
  - * *First* enter the *function* to be *transformed* (*original function*) f(t) in the worksheet.
  - * *Then mark* a *variable* t with *editing lines*.
  - * The *menu sequence*

    Symbolics ⇒ Transform ⇒ Laplace

    then returns the *Laplace transform* (*image function*) F(s).

- *Calculation* of the *inverse Laplace transformation:*
  - * *First mark* with *editing lines* a *variable* s in the *Laplace transform* (*image function*) F(s).
  - * The *menu sequence*

    Symbolics ⇒ Transform ⇒ Inverse Laplace

    then returns the *inverse transformation*.

The newer version of MATHCAD also provides *keywords*

                    and

from *operator palette no.8* to realize the *Laplace transformation* and its *inverse transformation:*

Enter the function to be transformed in the left-hand placeholder and the variable in the right-hand placeholder in the displayed symbols

 ▪ laplace, ▪ →   or   ▪ invlaplace, ▪ →

and then press the ⎵-key (see Examples 24.1 and 24.4).

♦

Note that MATHCAD for the *Laplace transformation* displays the *Laplace transform* (*image function*) F(s) as a function of s and the *original function* f(t) as a function of t.

♦

As we saw in Chapter 20, no finite algorithms exist for the calculation of the improper integrals for the Laplace transformation and its inverse transformation. Consequently, MATHCAD cannot always be expected to find a solution for Laplace transformations.

♦

Let us now consider some examples for Laplace transformations and their inverse transformations.

Example 24.1:

a) Calculate the *Laplace transforms* for several *elementary functions*

$\cos(t)$ *has Laplace transform* $\dfrac{s}{(s^2 + 1)}$

$\sin(t)$ *has Laplace transform* $\dfrac{1}{(s^2 + 1)}$

t^n *has Laplace transform* $\text{laplace}(t^n, t, s)$

t *has Laplace transform* $\dfrac{1}{s^2}$

$e^{-a \cdot t}$ *has Laplace transform* $\dfrac{1}{(s + a)}$

$t \cdot e^{-a \cdot t}$ *has Laplace transform* $\dfrac{1}{(s + a)^2}$

1 *has Laplace transform* $\dfrac{1}{s}$

You obtain the same transformations when you use the *keyword* **laplace**, e.g.,

$\cos(t) \ \text{laplace}, t \ \rightarrow \dfrac{s}{(s^2 + 1)}$

With the exception of the transformation of the t^n function, MATHCAD can calculate all these Laplace transforms.

b) We calculate the *inverse transformations* in the same order as for the Laplace transforms in a):

$\dfrac{s}{s^2 + 1}$ *has inverse laplace transform* $\cos(t)$

$\dfrac{1}{s^2 + 1}$ *has inverse laplace transform* $\sin(t)$

$\dfrac{1}{s^2}$ *has inverse laplace transform* t

$$\frac{1}{s+a} \qquad \textit{has inverse laplace transform} \qquad \exp(-a \cdot t)$$

$$\frac{1}{(s+a)^2} \qquad \textit{has inverse laplace transform} \qquad t \cdot \exp(-a \cdot t)$$

$$\frac{1}{s} \qquad \textit{has inverse laplace transform} \qquad 1$$

You obtain the same *inverse transformations* when you use the *keyword* **invlaplace**, e.g.,

$$\frac{s}{s^2+1} \quad \text{invlaplace}, s \;\; \rightarrow \; \cos(t)$$

c) The solution of differential equations raises the question how the derivatives of a function are to transform:
The *Laplace transforms* for the *derivatives*

y'(t) and y''(t)

of the *function*

y(t)

yield

$$s \cdot Y(s) - y(0)$$

or

$$s^2 \cdot Y(s) - s \cdot y(0) - y'(0)$$

where

Y(s)

designates the *Laplace transform* of y(t).

MATHCAD calculates these transforms in the following form:

c1) *Transformation* of the *first derivative:*

$$\frac{d}{dt} y(t) \quad \textit{has Laplace transform} \qquad \text{laplace}(y(t),t,s) \cdot s - y(0)$$

or using the **laplace** *keyword*

$$\frac{d}{dt} y(t) \; \text{laplace}, t \;\; \rightarrow \; s \cdot \text{laplace}(y(t),t,s) - y(0)$$

MATHCAD returns for the *Laplace transform* of the function y(t) the notation

laplace (y(t) , t , s)

which is impractical for applications, such as the solution of differential equations. So it is desirable to replace *laplace* (y(t) , t , s) with a new function, such as

Y(s)

(see Example 24.4).

c2) *Transformation* of *second derivatives:*

$$\frac{d^2}{dt^2} y(t) \quad \text{has Laplace transform}$$

$$s \cdot (s \cdot laplace(y(t), t, s) - y(0)) - \left(\left| \begin{array}{l} t \leftarrow 0 \\ \frac{d}{dt}y(t) \end{array} \right. \right)$$

or using the **laplace** *keyword*

$$\frac{d^2}{dt^2} y(t) \text{ laplace, } t \;\rightarrow\; s \cdot (s \cdot laplace(y(t), t, s) - y(0)) - \left(\left| \begin{array}{l} t \leftarrow 0 \\ \frac{d}{dt}y(t) \end{array} \right. \right)$$

The same comment as for c1 applies for the representation of the Laplace transforms.

♦

24.2 Fourier Transformation

The *Fourier transformation* also belongs to the *integral transformations* and is closely related to the Laplace transformation. It can also be used to solve differential equations.

MATHCAD performs *Fourier transformations* and *inverse Fourier transformations* in a similar manner as Laplace transformations:

* *Calculation* of the *Fourier transformation:*
  * *First* enter the *function* to be *transformed* (*original function*) in the worksheet.
  * *Then mark* a *variable* with *editing lines.*
  * *Finally,* the *menu sequence*
    Symbolics ⇒ Transform ⇒ Fourier

returns the *Fourier transform*.

- *Calculation* of the *inverse Fourier transformation:*
  * *First mark* a *variable* in the *Fourier transform* (*image function*) with *editing lines*.
  * *Finally*, the *menu sequence*

    Symbolics ⇒ Transform ⇒ Inverse Fourier

    returns the *inverse transformation*.

The *Fourier transformation* and its *inverse transformation* can also be realized in the newer versions of MATHCAD using the *keywords*

              or

from the *operator palette no.8,* when you enter the function to be transformed in the left-hand placeholder and the variable in the right-hand placeholder in the displayed symbols

▪ fourier, ▪ →   or   ▪ invfourier, ▪ →

and then press the (⏎)-key.
♦

Although MATHCAD also has *functions* for the *discrete Fourier transformation*, we do not discuss these in this book. The *Fourier* word in the integrated help provides detailed information.
♦

24.3 Z-Transformation

In many application areas the complete form of a function f(t), in which t normally represents the time, is either not known or is not of interest, rather just its values at certain points t_n (n = 0 , 1 , 2 , 3 , ...) are available.
Thus, you have only a *number sequence* for the function f(t)

$$\{f_n\} = \{f(t_n)\} \qquad n = 0 , 1 , 2 , 3 , ...$$

You obtain such a number sequence in practice, for example, by measuring at a series of times or through the discrete sampling of continuous signals.
You can write this number sequence as a function f of the index n if t_n takes only integer values (e.g., t_n=n), i.e.,

$$\{f_n\} = \{f(n)\}$$

The *Z-transformation* assigns the infinite series

$$Z[f_n] = F(z) = \sum_{n=0}^{\infty} f_n \cdot \left(\frac{1}{z}\right)^n$$

to each *number sequence* (*original sequence*)

$$\{f_n\}$$

which, if it converges, is called the *Z-transform* F(z) (*image function*).
We can see that the *Z-transformation* is the discrete analogue of the Laplace transformation.
The *Z-transformation* and the *inverse Z-transformation* are performed using MATHCAD in the same manner as for Laplace and Fourier transformations:

- *Calculation* of the *Z-transformation:*
    * *First* enter the *number sequence* f (n) to be *transformed* in the work sheet.
    ⁜ *Then mark* an *index n* with *editing lines*
    ⁜ *Finally*, the *menu sequence*

    Symbolics ⇒ Transform ⇒ Z

    returns the *Z-transform* (*image function*).

- *Calculation* of the *inverse Z-transformation:*
    * *First mark* a *variable* in the *Z-transform* (*image function*) F(z) with *editing lines*
    * *Finally,* the *menu sequence*

    Symbolics ⇒ Transform ⇒ Inverse Z

    returns the *inverse transformation.*

☞

The newer versions of MATHCAD can also use the

                    and

keywords from the *operator palette no. 8* to realize the *Z-transformation* and its *inverse Z-transformation:*
Enter the number sequence/function to be transformed in the left-hand placeholder and the index/variable in the right-hand placeholder of the displayed symbols

▪ ztrans, ▪ →   or   ▪ invztrans, ▪ →

and then press the ⏎-key to return the required transformation. ♦

☞

Note that MATHCAD for the *Z-transformation* represents the *Z-transform* (*image function*) F(z) as a function of z and the *original sequence* f(n) as a function of n.

♦

We now consider some examples for the Z-transformation and their inverse transformation.

Example 24.2:

a) Calculate the *Z-transformation* and its *inverse transformation* for some number sequences using the **ztrans** or **invztrans** keyword:

Z-transformation   *inverse Z-transformation*

$$1 \text{ ztrans, n} \rightarrow \frac{z}{(z-1)} \qquad\qquad \frac{z}{(z-1)} \text{ invztrans, z} \rightarrow 1$$

$$n \text{ ztrans, n} \rightarrow \frac{z}{(z-1)^2} \qquad\qquad \frac{z}{(z-1)^2} \text{ invztrans, z} \rightarrow n$$

$$n^2 \text{ ztrans, n} \rightarrow z \cdot \frac{(z+1)}{(z-1)^3} \qquad\qquad z \cdot \frac{(z+1)}{(z-1)^3} \text{ invztrans, z} \rightarrow n^2$$

$$a^n \text{ ztrans, n} \rightarrow \frac{-z}{(-z+a)} \qquad\qquad \frac{-z}{(-z+a)} \text{ invztrans, z} \rightarrow a^n$$

$$\frac{a^n}{n!} \text{ ztrans, n} \rightarrow \exp\left(\frac{1}{z} \cdot a\right) \qquad\qquad \exp\left(\frac{1}{z} \cdot a\right) \text{ invztrans, z} \rightarrow \frac{a^n}{n!}$$

b) You require the *Z-transformation* of

$$y(n+1), y(n+2), \dots$$

to solve *difference equations*. MATHCAD yields

$$y(n+1) \text{ ztrans, n} \rightarrow z \cdot \text{ztrans}(y(n), n, z) - y(0) \cdot z$$

$$y(n+2) \text{ ztrans, n} \rightarrow z^2 \cdot \text{ztrans}(y(n), n, z) - y(0) \cdot z^2 - y(1) \cdot z$$

Because the *notation ztrans* (y(n) , n , z) returned by MATHCAD for the *Z-transform* of y(n) is cumbersome for applications such as the solution of differences equations, it is better to substitute the new function Y(z) (see Example 24.5).

♦

24.4 Wavelet Transformation

The wavelet transformation is used extensively in linear signal analysis. Consequently, the newer versions of MATHCAD have included the **wave** and **iwave** functions for the discrete wavelet transformation and its inverse. The following *application example* that shows the use of these functions has been taken from the **QuickSheets** *electronic book* of the **Resource Center**.

Example 24.3:

This QuickSheet illustrates the application of wavelet transforms

Given a signal, for instance a single square wave:

$$N := 256 \qquad\qquad S_{N-1} := 0$$

$$n := \frac{3 \cdot N}{8}, \frac{3 \cdot N}{8} + 1 .. \frac{5 \cdot N}{8} \qquad S_n := 1 \qquad i := 0, 1 .. 255$$

Wavelet transform:

$$W := \text{wave} \ (S)$$

The number of levels contained in this transform is

$$\text{Nlevels} := \frac{\ln(N)}{\ln(2)} - 1 \qquad \text{Nlevels} = 7 \qquad k := 1, 2 .. \text{Nlevels}$$

To obtain a sense of the relative importance of each level, expand
as follows:

$$\text{coeffs (level) := submatrix } \left(W, 2^{\text{level}}, 2^{\text{level}+1} - 1, 0, 0 \right)$$

$$C_{i,k} := \text{coeffs}(k)_{\text{floor}\left[\frac{i}{\left(\frac{N}{2^k}\right)}\right]}$$

Plot several levels of
coefficients simultaneously
this way.

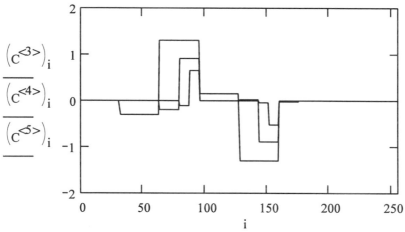

We wish to represent signal with less data; this is done by

- zeroing out the higher level coefficients, and

- computing the inverse wavelet transform of the new
  coefficient vectors.

First level at which coefficients are set to zero
($L \leq 7$ may be varied)

$L := 5$

$j := 2^L .. N - 1$ $W_j := 0$ $S' := \text{iwave}(W)$

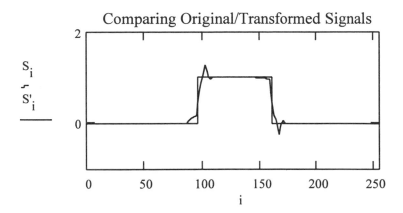

If acceptable, this offers a way of compressing the data needed to represent a signal.

♦

24.5 Solution of Difference and Differential Equations

The basic principle using transformations to solve equations consists of the following steps:

I.   *First* transform the equation to be solved (*original equation*) into an *image equation* that is normally simpler.

II. *Then solve the image equation* for the *image function*.

III. *Finally,* the *inverse transformation* of the *image function* returns the *solution* of the *original equation*.

The following section shows the use of the *Laplace transformation* and *Z-transformation* to solve *differential equations* and *difference equations*.

We start with the use of the *Laplace transformation* that can be successfully used for the exact solution of initial and boundary value problems for linear *differential equations* with constant coefficients. The *principle* follows:

*   *First* use *Laplace transformation* to convert the *differential equation* (*original equation*) for the function (*original function*) y(t) into an *algebraic equation* (*image equation*) for the *Laplace transform* (*image function*) Y(s).

*   *Then solve* the returned *image equation* for the *image function* Y(s).

*   *Finally,* use the *inverse Laplace transformation* on the *image function* Y(s) to obtain the solution y(t) of the differential equation.

☞

Examples 24.4a and b show that it is possible to successfully handle *initial value problems*, because you have here the function values for the solution function and their derivatives at the initial point t=0.

If the initial conditions are not given at the point t=0, you must use a transformation beforehand to bring the problem into this form. However, you normally have the initial conditions at point t=0 for *time-dependent problems* (e.g., in *electrical engineering*).

Example 24.4c demonstrates how a *Laplace transformation* can also be used to *solve* simple *boundary value problems*.

◆

☞

Because the results returned by MATHCAD for the Laplace transform for higher derivatives of a function y(t) are cumbersome when used to solve differential equations (see Example 24.1c), you can transform higher order differential equations into a system of first order differential equations. Example 24.4a1 illustrates this procedure.

◆

Example 24.4:

The differential equations in the following examples are entered without the equal sign, for example, the form

$$\frac{d}{dt} y_1(t) - y_2(t)$$

is used rather than

$$\frac{d}{dt} y_1(t) = y_2(t)$$

Whereas this method functions for all versions of MATHCAD, the use of the equal sign was not accepted in the Version 8 of MATHCAD available to the author.

a) Use the *Laplace transformation* to solve the homogeneous *second order differential equation* (*harmonic oscillator*)

y" + y = 0

with the *initial conditions*

y(0) = 2 , y'(0) = 3

a1) Transform the given second order differential equation into the first order *system* of the form

$$y_1' - y_2 = 0$$
$$y_2' + y_1 = 0$$

with the *initial conditions*

$$y_1(0) = 2 , y_2(0) = 3$$

The following *method* can be used to *solve* this *initial value problem* using the *Laplace transformation*:

* Use the **laplace** keyword to apply the *Laplace transformation* on the *differential equation system*.

* Then *solve* the resulting *system of linear algebraic equations (image equations)* for the *Laplace transforms (image functions)*.

* Finally, use the **invlaplace** *keyword* for the *inverse Laplace transformation* to calculate from the *image functions* the solution y(t) for the given differential equation.

MATHCAD displays this as the following worksheet (we have used the literal index for the indexing):

$$\frac{d}{dt}y_1(t) - y_2(t) \ \text{laplace}, t \ \rightarrow$$

$$s \cdot \text{laplace}(y1(t), t, s) - y1(0) - \text{laplace}(y2(t), t, s)$$

$$\frac{d}{dt}y_2(t) + y_1(t) \ \text{laplace}, t \ \rightarrow$$

$$s \cdot \text{laplace}(y2(t), t, s) - y2(0) + \text{laplace}(y1(t), t, s)$$

This produces a system of linear algebraic equations (image equations) for the unknown Y_1 *and* Y_2, *when you set*

$$Y_1 := \text{laplace}(y1(t), t, s) \qquad\qquad Y_2 := \text{laplace}(y2(t), t, s)$$

and use the given initial values for y1(0), y2(0).
MATHCAD can solve this system without difficulty:

given

$$Y_1 \cdot s - 2 - Y_2 \blacksquare 0$$

$$Y_2 \cdot s - 3 + Y_1 \blacksquare 0$$

$$\text{find}(Y_1, Y_2) \rightarrow \begin{pmatrix} \dfrac{(2 \cdot s + 3)}{(s^2 + 1)} \\ \dfrac{(3 \cdot s - 2)}{(s^2 + 1)} \end{pmatrix}$$

The use of the inverse Laplace transformation on Y_1 *returns the exact solution of the given differential equation:*

$$\frac{2 \cdot s + 3}{s^2 + 1} \text{ invlaplace, s } \rightarrow 2 \cdot \cos(t) + 3 \cdot \sin(t)$$

a2) The Laplace transformation is applied directly on the given second-order differential equation:

* The direct use of the **laplace** *keyword* on the given second-order differential equation (without equal sign) returns:

$$\frac{d^2}{d t^2} y(t) + y(t) \text{ laplace, t } \rightarrow$$

$$s \cdot (s \cdot laplace(y(t), t, s) - y(0)) - \left(\left| \begin{array}{c} t \leftarrow 0 \\ \\ \frac{d}{d t} y(t) \end{array} \right| \right) + laplace(y(t), t, s)$$

* Then replace *Laplace transform*

laplace(y(t),t,s)

by the function Y(s), replace the initial values y(0) and y'(0) with the actual values 2 and 3 and use the **solve** *keyword* (see Chapter 16) to solve the expression for Y(s):

$$s \cdot (s \cdot Y(s) - 2) - 3 + Y(s) \text{ solve, } Y(s) \rightarrow \frac{(2 \cdot s + 3)}{(s^2 + 1)}$$

* Finally, the *inverse Laplace transformation* with the **invlaplace** *keyword* returns the *solution* of the *differential equation*

$$\frac{(2 \cdot s + 3)}{(s^2 + 1)} \text{ invlaplace, s } \rightarrow 2 \cdot \cos(t) + 3 \cdot \sin(t)$$

b) Consider the differential equation with the initial conditions from Example a with the additional *nonhomogenity* cos t, i.e., the equation

y" + y = cos t

and use the method from a2. MATHCAD displays the following worksheet:

$$\frac{d^2}{d t^2} y(t) + y(t) - \cos(t) \text{ laplace, t } \rightarrow$$

$$s \cdot (s \cdot laplace(y(t), t, s) - y(0)) - \left(\left| \begin{array}{c} t \leftarrow 0 \\ \\ \frac{d}{d t} y(t) \end{array} \right| \right) + laplace(y(t), t, s) - \frac{s}{(s^2 + 1)}$$

The same method as used in Example a2 yields the following *image equation* for the *Laplace transform* Y(s), for which the **solve** *keyword* calculates the following solution:

$$s \cdot (s \cdot Y(s) - 2) - 3 + Y(s) - \frac{s}{s^2 + 1} \quad \text{solve, } Y(s) \quad \rightarrow \frac{(2 \cdot s^3 + 3 \cdot s + 3 \cdot s^2 + 3)}{(s^4 + 2 \cdot s^2 + 1)}$$

The *solution* of the *given differential equation* is obtained with the *inverse Laplace transformation* using the **invlaplace** *keyword*:

$$\frac{(2 \cdot s^3 + 3 \cdot s + 3 \cdot s^2 + 3)}{(s^4 + 2 \cdot s^2 + 1)} \quad \text{invlaplace, } s \quad \rightarrow \frac{1}{2} \cdot t \cdot \sin(t) + 2 \cdot \cos(t) + 3 \cdot \sin(t)$$

c) Solve the *boundary value problem* from Example 23.6.

$$y'' + y = 0 \ , \ \ y(0) = 2 \ , \ \ y(\pi/2) = 3$$

using *Laplace transformation*.
MATHCAD displays this as the following worksheet in which we have used the method from Example a2 and have written the missing initial condition y' (0) as parameter a·

$$s \cdot (s \cdot \text{laplace}(y(t), t, s) - y(0)) - \left| \left| \begin{matrix} t \leftarrow 0 \\ \\ \dfrac{d}{dt} y(t) \end{matrix} \right. \right| + \text{laplace}(y(t), t, s)$$

$$(s \cdot (s \cdot Y(s) - 2) - a) + Y(s) \quad \text{solve, } Y(s) \quad \rightarrow \frac{(2 \cdot s + a)}{(s^2 + 1)}$$

$$\frac{(2 \cdot s + a)}{(s^2 + 1)} \quad \text{invlaplace, } s \quad \rightarrow 2 \cdot \cos(t) + a \cdot \sin(t)$$

The replacement of the given boundary condition in the solution calculates the unknown parameter a

$$2 \cdot \cos\left(\frac{\pi}{2}\right) + a \cdot \sin\left(\frac{\pi}{2}\right) \equiv 3 \qquad \text{has solution(s)} \quad 3$$

Thus, the *solution*

$$2 \cdot \cos(t) + 3 \cdot \sin(t)$$

is calculated for the given *boundary value problem*.

♦

Example 24.4c shows the *method* used for the *solution* of *boundary value problems* using *Laplace transformation*:

The given boundary value problem is solved as an initial value problem with unknown initial conditions using Laplace transformation, where the parameters a , b , ... are substituted for the unknown initial conditions. Finally, the given boundary conditions are used to determine the still unknown parameters a , b , ...

♦

Finally, consider the use of the *Z-transformation* for the exact *solution* of linear *difference equations* with constant coefficients that occur in many practical problems both in engineering and science, as well as economics, such as the description of electrical networks.

Before we discuss the method used in MATHCAD required for the Z-transformation, we provide a short discussion of the *properties* of *linear difference equations*.

Linear difference equations of *order m*

- with *constant coefficients* have the *form*

$$y_n + a_1 \cdot y_{n-1} + a_2 \cdot y_{n-2} + ... + a_m \cdot y_{n-m} = b_n \qquad (n \geq m)$$

  or without using indices

$$y(n) + a_1 \cdot y(n-1) + a_2 \cdot y(n-2) + ... + a_m \cdot y(n-m) = b_n$$

  In this equation:

  * $a_1 , a_2 , ... , a_m$

    mean given constant real *coefficients.*

  * $\{b_n\}$ \hfill $(n = m , m+1 , ...)$

    mean the sequence for the given *right-hand side.*

    If all elements b_n of the sequence are zero, analogue to differential equations, these are called *homogeneous difference equations.*

  * $\{y_n\}$ \qquad or \qquad $\{y(n)\}$ \hfill $(n = 0 , 1 , 2 , ...)$

    mean the sequence of the *solutions.*

- have similar properties as *linear algebraic equations* and *differential equations:*

  * The *general solution* of a *linear difference equation* of the m-th order depends on m arbitrarily chosen real constants.

  * If you specify the initial values

    $$y_0 , y_1 , ... , y_{m-1}$$

    for a linear difference equation of the m-th order

    the *solution sequence*

    $$y_m , y_{m+1} , y_{m+2} , ...$$

is *determined uniquely*.

* The *general solution* of a *nonhomogeneous linear difference equation* results as the *sum* of the *general solution* of the *homogeneous* and a *special solution* of the *nonhomogeneous equation*.

   ◆

Instead of the *index* n, difference equations also use the *index* t to indicate that *time* is used.

◆

As with the Laplace transformation, the *principle* behind the use of the *Z-transformation* for the *solution* of *difference equations* consists of the following:

* *First* use *Z-transformation* to transform the *difference equation* (*original equation*) for the function (*original function*) y(n) into an *algebraic equation* (*image equation*) for the *Z- transform* (*image function*) Y(z).

* *Then solve* the *image equation* for Y(z).

* *Finally*, use the *inverse Z-transformation* on the *image function* Y(z) to obtain the solution y(n) of the given difference equation.

When the Z-transformation is used in MATHCAD to solve difference equations, note that indexes cannot be used as input in the worksheet, rather items must be written in the form y(n).

◆

Example 24.5:

A simple electrical network can, for example, be described with a *second-order difference equation* of the form

u(n+2) − 3·u(n+1) + u(n) = 0

for the *voltage* u. We use the **ztrans** *keyword* from *operator palette no.8* to solve this difference equation for the *initial conditions*

u(0) = 0 , u(1) = 1

* use the **ztrans** *keyword* directly on the given difference equation (without equal sign) to obtain:

   $u(n + 2) - 3 \cdot u(n + 1) + u(n)$ ztrans, n →

$z^2 \cdot \text{ztrans}(u(n), n, z) - u(0) \cdot z^2 - u(1) \cdot z - 3 \cdot z \cdot \text{ztrans}(u(n), n, z) + 3 \cdot u(0) \cdot z + \text{ztrans}(u(n), n, z)$

* Then, replace the *Z-transform* (*image function*)
   ztrans (u(n) , n , z)

with the function U(z), and the initial values u(0) and u(1) with the given values 0 and 1 and then use the **solve** *keyword* to solve the expression for U(z) (see Chapter 16):

$$z^2 \cdot U(z) - z - 3 \cdot z \cdot U(z) + U(z) \ \text{solve}, U(z) \ \rightarrow \ \frac{z}{(z^2 - 3 \cdot z + 1)}$$

* Finally, use the *inverse Z-transformation* with the **invztrans** *keyword* to return the *solution* u(n) for the *difference equation*

$$\frac{z}{(z^2 - 3 \cdot z + 1)} \ \text{invztrans} \ z \ \rightarrow \ \frac{1}{5} \cdot \frac{\left[-2^n \cdot \sqrt{5} + (-2)^n \cdot \left[\dfrac{1}{\left(-3 + \sqrt{5}\right)} \right]^n \cdot \left(3 + \sqrt{5}\right)^n \cdot \sqrt{5} \right]}{\left(3 + \sqrt{5}\right)^n}$$

◆

If we have a difference equation in the form (n = m , m+1 , ...)

$$y(n) + a_1 \cdot y(n-1) + a_2 \cdot y(n-2) + \dots + a_m \cdot y(n-m) \ = \ b_n$$

we must transform this equation into the form (n = 0 , 1 , ...)

$$y(n+m) + a_1 \cdot y(n+m-1) + a_2 \cdot y(n+m-2) + \dots + a_m \cdot y(n) = b_{n+m}$$

in order to apply MATHCAD.

◆

25 Optimization

A wide range of *mathematical optimization problems* exist. These include problems concerned with

* *linear programming*
* *nonlinear programming*
* *integer programming*
* *dynamic programming*
* *stochastic programming*
* *vectorial (multiobjective) programming*
* *variational calculus*
* *optimal control.*

Optimization problems are increasingly gaining in importance for practical problem situations in all scientific areas. This is caused by the wish to use an optimum strategy, for example, to achieve maximum profit for minimum expenditure.

In the mathematical *optimization theory* we have an *objective function* (*cost function*) that provides a *minimum* (smallest) or *maximum* (largest) *value* while taking account of certain *constraints* provided as equations or inequalities.

♦

The individual areas of the *mathematical optimization* (*mathematical programming*) differ through the form of the

* *objective function*
* *constraints.*

♦

Although, MATHCAD does not include any integrated functions to return the exact solution of optimization problems, it provides functions for an *approximate* (*numerical*) *solution:*

* The **minimize** and **maximize** *numerical functions*

   for the approximate solution of extreme value problems, and linear and nonlinear programming problems.

- *Functions* for *numerical methods*

  from the **Numerical Recipes** *electronic book.*

Thus, the availability of optimization functions and the use of existing functions/commands/menus enables MATHCAD to successfully solve a wide range of optimization problems for real functions. We demonstrate these in

- Section 25.1 for *extreme value problems.*

  These are understood to be problems in which

  * a *function* of *n variables* specifies the *objective function*
  * *equations* of *n variables* specify the *constraints*

  which, with simple structure, can be solved in MATHCAD exactly using the methods provided for differentiation and solution of equations or numerically using the **minimize** and **maximize** *functions.*

- Section 25.2 for *linear programming* problems

  that can be solved numerically using the

  * **minimize** or **maximize**
  * **simplx**

    function from the **Numerical Recipes** *electronic book.*

- Section 25.3 for *nonlinear programming* problems

  that can be solved numerically using the

  minimize or **maximize**

  function.

We also show in Section 25.4 how you can use the MATHCAD *programming facilities* to create your own simple *programs* in order to solve optimization problems using *numerical methods.*

25.1 Extreme Value Problems

An optimization problem for a *function* f with or one or more variables involves determining a *local minimum* or *maximum*, i.e.,

$$y = f(x) \underset{x}{\rightarrow} \text{Minimum} / \text{Maximum}$$

or

$$z = f(x_1, x_2, \ldots, x_n) \underset{x_1, x_2, \ldots, x_n}{\rightarrow} \text{Minimum} / \text{Maximum}$$

If you do not explicitly differentiate between minimum and maximum, this
is called an *extremum* or *optimum*. Problems of this type are known as *extreme value problems.*

The *necessary optimality conditions* (set the first order derivatives to zero)
required for the *exact solution* of *extreme value problems* can be achieved
for *functions*

* of *one variable* $y = f(x)$:

  $f'(x) = 0$

* of *two variables* $z = f(x,y)$:

  $f_x(x, y) = 0$
  $f_y(x, y) = 0$

* general for *n variables* $z = f(\mathbf{x}) = f(x_1, x_2, \ldots, x_n)$:

  $f_{x_1}(x_1, x_2, \ldots, x_n) = 0$
  $f_{x_2}(x_1, x_2, \ldots, x_n) = 0$
  $$\vdots$$
  $f_{x_n}(x_1, x_2, \ldots, x_n) = 0$

The *solutions* of these *equations* of the *optimality conditions* are called *stationary points*. The calculated stationary points must be checked for *optimi-
mality* using *sufficient conditions*

* $f''(x) \neq 0$ ($f''(x) > 0$ Minimum, $f''(x) < 0$ Maximum)

  for *functions of one variable* $y = f(x)$

* $f_{xx} \cdot f_{yy} - (f_{xy})^2 > 0$ ($f_{xx}(x,y) > 0$ Minimum, $f_{xx}(x,y) < 0$ Maximum)

  for *functions of two variables* $z = f(x,y)$

Because you must test the n-row Hessian matrix formed from the second-
order derivatives of the objective function f for positive definiteness, the use
of sufficient conditions becomes more difficult for $n \geq 3$ (i.e., starting at three
variables).

☞

If *extreme value problems* also make use of *constraints* in the form of *equations* (*equality constraints*), there are *two ways* of obtaining the *exact solution:*

* *solve* the *equations* for certain *variables* (if possible) and *substitute* in the
  objective function

* use of the *method of Lagrange multipliers.*

Both methods produce a problem without constraints, the exact solution for which we discussed at the start of this section (also refer to Example 25.1b).

◆

Thus, the following *method* is used in MATHCAD for the *exact solution* of *extreme value problems:* Problems of the form

$$f(x_1, x_2, ..., x_n) \rightarrow \underset{x_1, x_2, ..., x_n}{\text{Minimum / Maximum}}$$

where only *equality constraints*

$$g_j(x_1, x_2, ..., x_n) = 0, \quad j = 1, 2, ..., m$$

can occur, can be handled stepwise:

*   *First* calculate the *partial derivatives* of the *objective function* f or of the *Lagrange function*

$$L(\mathbf{x}; \lambda) = L(x_1, x_2, ..., x_n; \lambda_1, \lambda_2, ..., \lambda_m) =$$

$$f(x_1, x_2, ..., x_n) + \sum_{j=1}^{m} \lambda_j g_j(x_1, x_2, ..., x_n)$$

    using the method from Section 19.1, and then set this to zero (*necessary optimality condition*).

*   *Finally,* you can attempt to solve these *equations* from the *optimality conditions* using the method for the solution of equations discussed in Chapter 16. Because these equations are normally nonlinear, difficulties can occur here.

Because the solved optimality conditions are only necessary conditions, the *solutions* must still be tested to determine whether they are *optimum*.

◆

The following example, in which we *solve* simple practical problems *exactly* and use the **minimize** and **maximize** *numerical functions* to return *approximate solutions,* shows the procedure for *solving extreme value problems* using MATHCAD.

Example 25.1:

a) A *company's storage costs* for a *certain article are to be minimized.* The company requires 400 units of this article per week. The transport costs for the delivery of the required article are $100 per delivery, irrespective of the delivered quantity. The company's storage costs for this article are $2 per unit and week. The problem to minimize the storage costs for the article has the following mathematical model:

If x units are supplied per transport delivery, we require a total of L = 400/x deliveries per week and the transport costs are 100/x per unit. Assuming that the article stock level decreases linearly, we can assume that each unit has the same storage time. This is half the time between two deliveries, i.e., 1/2L weeks. Because it has $2 storage costs per week, L deliveries cause unit costs of $1/L. This produces the *cost function*

$$f(x) = \frac{100}{x} + \frac{x}{400}$$

that is to be *minimized* for x > 0. The minimum of f(x) is attained for x = 200. Thus, 200 units are to be supplied per transport delivery, i.e., we need two deliveries to minimize the storage costs.

We now use MATHCAD to *calculate* this *solution*.

a1)*Using* the *necessary optimality condition:*

To *determine* the *exact zeros* for the *first derivative* of the *objective function* f(x), we use the *differentiation operator* and the **solve** *keyword* to *solve* the *equation:*

$$\frac{d}{dx}\left(\frac{100}{x} + \frac{x}{400}\right) = 0 \text{ solve}, x \rightarrow \begin{bmatrix} 200 \\ -200 \end{bmatrix}$$

a2)Use of the **minimize** *numerical function* :

Enter the following *solution block:*

$$f(x) := \frac{100}{x} + \frac{x}{400} \qquad x := 20$$

$$\text{minimize}(f, x) = 200$$

i.e., the *function* to be minimized must be *defined* and a *starting value* (*initial estimate*) *assigned* to the *variable* x (taken as 20) before using **minimize**.

The *sufficient condition* H(x) = f "(x) > 0 for a *minimum* returns:

$$H(x) := \frac{d^2}{dx^2}\left(\frac{100}{x} + \frac{x}{400}\right)$$

$$H(200) = 2.5 \cdot 10^{-5}$$

Thus x = 200 is a *local minimum*, which the graph of the function f(x) also confirms:

x := 150 .. 250

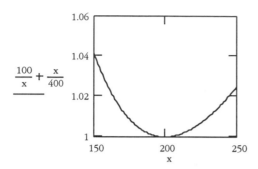

$$\frac{\frac{100}{x} + \frac{x}{400}}{}$$

b) The problem to select the rectangle with the largest *area* F from all *rectangles* with the same perimeter U reduces to the *extreme value problem*

$$F(x,y) = x \cdot y \rightarrow \underset{x,y}{\text{Maximum}}$$

with the *equality constraint*

$$2 \cdot (x + y) = U$$

where the variables x and y represent the length and height of the rectangle. The problem specification indicates that only positive values for x and y are sensible. We ignore these additional constraints

$$x \geq 0 \text{ and } y \geq 0$$

in the following.
MATHCAD provides the *following capabilities* to *solve* this problem:

b1)A *first solution possibility* is to solve the equality constraint for a variable, e.g., for y

$$y = \frac{U}{2} - x$$

and then substitute in the objective function F(x,y). This returns an extreme value problem without constraints for a function f(x) that only depends on the variable x:

$$f(x) = x \cdot (\frac{U}{2} - x) \rightarrow \underset{x}{\text{Maximum}}$$

We can now use MATHCAD to solve the *necessary optimality condition* for the function f(x), for which we use the *differentiation operator* and the **solve** *keyword*:

$$\frac{d}{dx}x \cdot \left(\frac{U}{2} - x\right) \equiv 0 \text{ solve}, x \rightarrow \frac{1}{4} \cdot U$$

We have obtained here a square with edge length x = y = U/4 as the solution.

b2)A *second solution possibility* is to use the *method of Lagrange multipliers:*

First form the associated *Lagrangian*

L (x, y, λ) = x· y + λ·(2· (x + y) − U)

for the problem and then maximize this for the variables x, y and λ. This is a problem without constraints that we can calculate by solving the necessary optimality conditions, for which we again use the *differentiation operator* and the **solve** *keyword*:

$$
\begin{bmatrix}
\dfrac{d}{dx}(x \cdot y + \lambda \cdot (2 \cdot (x + y) - U)) = 0 \\[2mm]
\dfrac{d}{dy}(x \cdot y + \lambda \cdot (2 \cdot (x + y) - U)) = 0 \\[2mm]
\dfrac{d}{d\lambda}(x \cdot y + \lambda \cdot (2 \cdot (x + y) - U)) = 0
\end{bmatrix}
\text{ solve, } \begin{bmatrix} x \\ y \\ \lambda \end{bmatrix} \rightarrow \begin{bmatrix} \dfrac{1}{4} \cdot U & \dfrac{1}{4} \cdot U & \dfrac{-1}{8} \cdot U \end{bmatrix}
$$

We obtain here the same result U/4 for x and y as for the first solution b1. The value obtained for the *Lagrange multiplier* λ is not of interest for the given problem.

b3)We can use the **maximize** *function* only when a value has been given for the perimeter U. We set U=50.

We use zero as the starting values for the variables x and y. Because we now have an equality constraint, we must write the *solve block* in the following form:

f(x, y) := x·y

x := 0   y := 0

given

2·(x + y) − 50 = 0

$$
\text{maximize}(f, x, y) = \begin{bmatrix} 12.5 \\ 12.5 \end{bmatrix}
$$

MATHCAD can calculate the solution x = y = U/4 = 12.5 here, even though nonfeasible starting values have been used.

c) Calculate solutions for the *extreme value problem*

$f(x,y) = \sin x + \sin y + \sin(x+y) \rightarrow \underset{x,y}{\text{Minimum}}$

for a function of two variables.

The *graphical display* shows that the given function over the x,y-plane has an infinite number of minimum and maximum values. Naturally MATHCAD cannot calculate all of these:

$N := 40 \qquad i := 0..N \qquad k := 0..N$

$x_i := -10 + 0.5 \cdot i \qquad y_k := -10 + 0.5 \cdot k$

$f(x,y) := \sin(x) + \sin(y) + \sin(x + y)$

$M_{i,k} := f(x_i, y_k)$

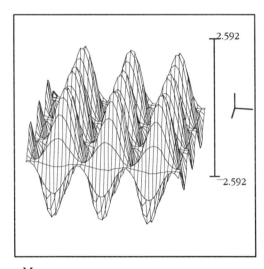

2.592

-2.592

M

To obtain a solution with MATHCAD, we use the *necessary optimality conditions* and the **minimize** or **maximize** *numerical functions*:

c1) *Solution* of the *necessary optimality conditions*:

    *   *exact solution*

$$\begin{bmatrix} \dfrac{d}{dx}(\sin(x) + \sin(y) + \sin(x + y)) = 0 \\[2mm] \dfrac{d}{dy}(\sin(x) + \sin(y) + \sin(x + y)) = 0 \end{bmatrix} \text{solve,} \begin{bmatrix} x \\ y \end{bmatrix} \rightarrow \begin{bmatrix} \pi & \pi \\[1mm] \dfrac{1}{3}\cdot\pi & \dfrac{1}{3}\cdot\pi \\[2mm] \dfrac{-1}{3}\cdot\pi & \dfrac{-1}{3}\cdot\pi \end{bmatrix}$$

By analysis, we can determine that the solutions found by MATH-CAD form

$(-\pi/3, -\pi/3)$ a minimum

$(\pi/3, \pi/3)$ a maximum

* *approximate solution:*

This we understand to be an *approximate solution* for the *equations* from the *necessary optimality conditions* using the numerical methods specified in Section 16.4.

We arbitrarily choose 5 as starting points for a minimum:

$x := 5$ $y := 5$

given

$$\frac{d}{dx}(\sin(x) + \sin(y) + \sin(x + y)) \equiv 0$$

$$\frac{d}{dy}(\sin(x) + \sin(y) + \sin(x + y)) \equiv 0$$

$$\mathrm{find}(x, y) = \begin{pmatrix} 5.236 \\ 5.236 \end{pmatrix}$$

The found *solution* approximates the necessary and sufficient optimality conditions for a minimum:

$$f(x, y) := \sin(x) + \sin(y) + \sin(x + y)$$

$x := 5.236$

$y := 5.236$

$$\frac{d}{dx} f(x, y) = 3.181 \cdot 10^{-5}$$

$$\frac{d}{dy} f(x, y) = 3.181 \cdot 10^{-5}$$

$$\frac{d^2}{dx^2} f(x, y) \cdot \frac{d^2}{dy^2} f(x, y) - \left(\frac{d}{dx} \frac{d}{dy} f(x, y) \right)^2 = 2.25$$

and returns the *objective function value*

$f(5.236, 5.236) = -2.598$

c2) Use of the **minimize** or **maximize** *numerical functions*, where we also use 5 as starting values:

$$f(x,y) := \sin(x) + \sin(y) + \sin(x + y)$$

$$x := 5 \qquad y := 5$$

$$\text{minimize}(f, x, y) = \begin{bmatrix} 5.236 \\ 5.236 \end{bmatrix} \qquad \text{maximize}(f, x, y) = \begin{bmatrix} 1.047 \\ 1.047 \end{bmatrix}$$

$$f(5.236, 5.236) = -2.598 \qquad\qquad f(1.047, 1.047) = 2.598$$

We can see that a minimum and a maximum is calculated for the used starting values.

♦

Because the *necessary optimality conditions* normally contain nonlinear equations, which are difficult to solve, MATHCAD quickly reaches its limits for the *exact solution* of *extreme value problems*, i.e., there does not normally exist any finite solution algorithm (see Section 16.3).

Consequently, you are restricted to *numerical methods* for most practical *extreme value problems*; two methods are available:

* *Numerical solution* of the *necessary optimality conditions* (*indirect method*), as shown in Example 25.1c2.

* Use of *direct numerical methods*. The simplest and also best known of these is the *gradient method*, which we discuss in Section 25.4.

  MATHCAD provides with **minimize** and **maximize** *numerical functions* a means of obtaining an *approximate solution* (see Examples 25.1a2, b3 and c2).

25.2 Linear Programming

The classical methods for the exact solution discussed in Section 25.1 cannot be used when *inequalities* occur as *constraints* for optimization problems.

The simplest problems of this form result when the *objective function* f and the functions g_j of the *inequality constraints* are *linear*, i.e., when the *optimization problem* has the form

$$c_1 \cdot x_1 + c_2 \cdot x_2 + \ldots + c_n \cdot x_n \; \rightarrow \; \underset{x_1, x_2, \ldots, x_n}{\text{Minimum / Maximum}}$$

$$a_{11} \cdot x_1 + a_{12} \cdot x_2 + \ldots + a_{1n} \cdot x_n \leq b_1$$
$$\vdots$$
$$a_{m1} \cdot x_1 + a_{m2} \cdot x_2 + \ldots + a_{mn} \cdot x_n \leq b_m$$

$$x_j \geq 0 \, , \; j = 1, \ldots, n$$

that then can be written in the following *matrix notation*:

$$\mathbf{c} \cdot \mathbf{x} \rightarrow \underset{x_1, x_2, \ldots, x_n}{\text{Minimum / Maximum}}$$

$$\mathbf{A} \cdot \mathbf{x} \leq \mathbf{b} \quad , \quad \mathbf{x} \geq 0$$

These are called *linear programming* problems. They are used in many applications, primarily in economics.

Solution methods exist for linear optimization problems that return a solution in a finite number of steps. The *simplex method* is the best known of these .

MATHCAD provides the following solution methods:

- Use of the **minimize** or **maximize** *numerical function* that we have already discussed in Section 25.1.

  The following *solve block* is required when they are used for linear optimization problems:

  * First *define* the *objective function* and *assign starting values* (*initial estimates*) to the *variables*.

  * Then, use the **given** command to start entering the *constraints*. The completion of the solve block initiates the call to **minimize** or **maximize** .

- Use of the **simplx** *numerical function* from the **Numerical Recipes** *electronic book* that does not require any starting values for the variables. We reproduce the *explanatory text* from this book here:

8.6   **Linear Programming and the Simplex Method**

simplx

C: 439
F: 432

uses the simplex algorithm to maximize an objective function subject to
a set of inequalities. Its arguments are:

- the submatrix of the first tableau containing the original
objective function and inequalities written in restricted normal form
- nonnegative integers **m1** and **m2**, giving the number of
constraints of the form $\leq$ and $\geq$; the number of equality constraints
is equal to **r - m1 - m2 - 1** , where **r** is the number of rows in the
matrix

The output is either:

- a vector containing the values of the variables that maximize the
objective function; or
- an error message if no solution satisfies the constraints or the objective
function is unbounded.

The following examples illustrate the use of MATHCAD functions to solve
linear programming problems.

Example 25.2:

Note that we use *variables* with a *literal index* in all examples (see Section
8.2):

a) Consider the simple problem

$$-2 \cdot x_1 + 2 \cdot x_2 + 1 \; \rightarrow \; \underset{x_1, x_2}{\text{Minimum}}$$

$$- 2 \cdot x_1 + x_2 \; \leq \; 0$$

$$x_1 + x_2 \; \leq 2 \; , \; x_2 \geq 0$$

that has the *solution*

$$x_1 = 2 \; , \; x_2 = 0$$

which, because the solution is assumed at a corner of the feasible region, we can easily obtain from the *graphical display* of the straight lines
that define the constraints:

$$x_1 := 0, 0.001 \ .. \ 2.5$$

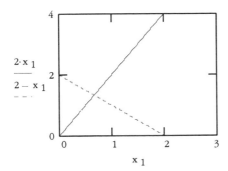

a1) Use of the **simplx** *numerical function* whose three arguments consist of the

* *matrix* of *coefficients* of the *result function* to be maximized (constants have been omitted) and the *constraints*

* *number*

   m_1

   of *inequalities* with $\leq$

* *number*

   m_2

   of inequalities with $\geq$

where the *first column* of the *matrix* for the *right hand sides* of the *inequalities* is reserved (enter zero in the first row here).
We accept the computational part from the provided **Numerical Recipes** *electronic book* and substitute the coefficients for our example:

Objective function: $\mathrm{Ob}\!\left(x_1, x_2\right) := -2 \cdot x_1 + 2 \cdot x_2 + 1$

Constraints: $-2 \cdot x_1 + x_2 \leq 0$

$$x_1 + x_2 \leq 2$$

$$x_2 \geq 0$$

$$v := \mathbf{simplx}\left(\begin{pmatrix} 0 & 2 & -2 \\ 0 & 2 & -1 \\ 2 & -1 & -1 \\ 0 & 0 & -1 \end{pmatrix}, 2, 1\right) \qquad v = \begin{pmatrix} 2 \\ 0 \end{pmatrix}$$

Maximum value of objective function: $\text{Ob}\left(v_1, v_2\right) = -3$

It is easy to see how the *matrix* needs to be *formed* in the *argument* of the **simplx** *function* for a given problem.

* The first row contains the two coefficients of the objective function with reversed sign, because the **simplx** *function* always maximizes the objective function but the minimum is required for our example.

* The first column contains the right-hand sides of the constraints. Enter zero as the first element, because the objective function is contained in the first row.

* The type of the input for the coefficients of the constraints (with reversed sign) can be obtained directly from the comparison of these conditions with the second to fourth row of the matrix.

The two remaining arguments of the **simplx** *function* represent the number of inequalities with $\leq$ (2) and $\geq$ (1).

The obtained *solution*

$x_1 = 2$, $x_2 = 0$

is contained in *vector* **v**; *finally* calculate the *optimal value* (here minimum) -3 of the *objective function*.

a2) Use of the **minimize** *numerical function*. The *solve block* has the following form here (we use zero as *starting values* for the variables):

$$f\left(x_1, x_2\right) := -2 \cdot x_1 + 2 \cdot x_2 + 1$$

$$x_1 := 0 \qquad x_2 := 0$$

given

$$-2 \cdot x_1 + x_2 \leq 0$$

$$x_1 + x_2 \leq 2 \qquad x_2 \geq 0$$

$$\text{minimize}\left(f, x_1, x_2\right) = \begin{bmatrix} 2 \\ 0 \end{bmatrix}$$

b) Use *linear programming* to solve a simple *mixing problem:*
We are provided with three different types of grain G1, G2 and G3 to be used to mix animal feed. Each of these grains contains a different amount of the required nutrients A and B, from which the animal feed must contain at least 42 and 21 quantity units, respectively. The follow-

ing table provides the proportions of the nutrients in each grain type and the price/quantity unit:

	G1	G2	G3
Nutrient A	6	7	1
Nutrient B	1	4	5
Price/unit	6	8	18

The cost for the animal feed is to be minimized. This produces the following *linear programming* problem:

$$6 \cdot x_1 + 8 \cdot x_2 + 18 \cdot x_3 \;\to\; \underset{x_1, x_2, x_3}{\text{Minimum}}$$

$$6 \cdot x_1 + 7 \cdot x_2 + x_3 \geq 42$$

$$x_1 + 4 \cdot x_2 + 5 \cdot x_3 \geq 21$$

$$x_1 \geq 0 \quad x_2 \geq 0 \quad x_3 \geq 0$$

b1)The use of the **simplx** function using the same method as Example a yields:

Objective function: $\text{Ob}\!\left(x_1, x_2, x_3\right) := 6 \cdot x_1 + 8 \cdot x_2 + 18 \cdot x_3$

Constraints: $6 \cdot x_1 + 7 \cdot x_2 + x_3 \geq 42$

$$x_1 + 4 \cdot x_2 + 5 \cdot x_3 \geq 21$$

$$x_1 \geq 0 \quad x_2 \geq 0 \quad x_3 \geq 0$$

$$v := \mathbf{simplx}\left(\begin{pmatrix} 0 & -6 & -8 & -18 \\ 42 & -6 & -7 & -1 \\ 21 & -1 & -4 & -5 \\ 0 & -1 & 0 & 0 \\ 0 & 0 & -1 & 0 \\ 0 & 0 & 0 & -1 \end{pmatrix}, 0, 5\right) \qquad v = \begin{pmatrix} 1.235 \\ 4.941 \\ 0 \end{pmatrix}$$

Maximum value of objective function: $\text{Ob}\!\left(v_1, v_2, v_3\right) = 46.941$

b2)When we set the *starting values* for the variables to zero, the *use* of the **minimize** *function* yields:

$$f\left(x_1, x_2, x_3\right) := 6 \cdot x_1 + 8 \cdot x_2 + 18 \cdot x_3$$

$$x_1 := 0 \qquad x_2 := 0 \qquad x_3 := 0$$

given

$$6 \cdot x_1 + 7 \cdot x_2 + x_3 \geq 42$$

$$x_1 + 4 \cdot x_2 + 5 \cdot x_3 \geq 21 \qquad x_1 \geq 0 \qquad x_2 \geq 0 \qquad x_3 \geq 0$$

$$\text{minimize}\left(f, x_1, x_2, x_3\right) = \begin{bmatrix} 1.235 \\ 4.941 \\ 0 \end{bmatrix}$$

The *result* of cost minimization shows that the animal feed contains 1.235, 4.941 and 0 quantity units for the grains G1, G2 and G3, respectively, and the price of the animal feed is 46.94.

♦

25.3 Nonlinear Programming

If at least one *inequality* occurs as *constraint* for an optimization problem, this is called a problem of the *nonlinear programming*, for which the theory has been intensively developed since the 1950s. These problems have the following form:

An *objective function* f is to be *minimized/maximized*, i.e.,

$$f(x_1, x_2, \ldots, x_n) \underset{x_1, x_2, \ldots, x_n}{\rightarrow} \text{Minimum / Maximum}$$

where, in addition to *equality constraints*, *inequality constraints* are to be considered, i.e.,

$$g_j(x_1, x_2, \ldots, x_n) \leq 0, \quad j = 1, 2, \ldots, m$$

Because we can replace any equation with two inequalities, we normally write constraints only in the inequality form in the *nonlinear programming*.

If both the *objective function* and the *constraint functions* are *linear* in *nonlinear programming* problems, this produces the *special case* of *linear programming* discussed in Section 25.2.

♦

In reality, many *extreme value problems* are in fact *nonlinear programming problems*, because constraints required in practice have been omitted.

Thus, the problems from *Example 25.1a* and *b* are nonlinear programming problems, because the x and y variables cannot be negative. Thus, we must add the inequalities

$$x \geq 0 \quad \text{and} \quad y \geq 0$$

for Example 25.1b and so obtain a *nonlinear programming problem*:

$$F(x,y) = x \cdot y \rightarrow \underset{x,y}{\text{Maximum}}$$

with the *constraints*

$$2 \cdot (x+y) = U \quad , \quad x \geq 0 \ , \quad y \geq 0$$

If we make a small change to this problem by requiring that the perimeter of the rectangle does not exceed a specified value U, we obtain only inequalities as constraints:

$$F(x,y) = x \cdot y \rightarrow \underset{x,y}{\text{Maximum}}$$

with the *inequality constraints*

$$2 \cdot (x+y) \leq U \quad , \quad x \geq 0 \ , \quad y \geq 0$$

♦

Because the *optimality conditions* (Kuhn-Tucker conditions) that exist for nonlinear programming problems are only suitable for the solution of very simple problems, we are forced to use *numerical methods*. MATHCAD provides the **minimize** and **maximize** *numerical functions* that we discussed in the previous sections. We use these in the following example to solve a simple nonlinear programming problem.

♦

Example 25.3:

Solve the *problem*

$$f(x,y) = \frac{x^2}{2} + \frac{(y-2)^2}{2} \rightarrow \underset{x,y}{\text{Minimum}}$$

with the *inequality constraints*

$$y - x - 1 \le 0 \quad \text{and} \quad y + x - 1 \le 0$$

using the **minimize** *numerical function,* where we use zero as the *starting values* for the variables. We must write the following solve block:

$$f(x,y) := \frac{x^2}{2} + \frac{(y-2)^2}{2}$$

$$x := 0 \qquad y := 0$$

given

$$y - x - 1 \le 0$$

$$y + x - 1 \le 0$$

$$\text{minimize}(f,x,y) = \begin{bmatrix} 0 \\ 1 \end{bmatrix}$$

MATHCAD calculates the *solution* x=0 and y=1.

In Example 25.5b we calculate the same problem using *penalty functions* and *gradient methods.*

◆

If the MATHCAD **minimize** and **maximize** *numerical functions* fail, you can write your own programs. In the following Section 25.4, we write two simple programs using the MATHCAD programming facilities.

◆

25.4 Numerical Methods

Nonlinear programming problems normally cannot be solved with a finite algorithm and so we cannot use computer algebra methods. We are restricted to *numerical methods.*

In addition to the **minimize** and **maximize** functions used previously, MATHCAD can also use the **minerr** *function* for the *numerical solution* of *systems* of *nonlinear equations:*

If the *function* f(**x**) to be *minimized* consists of *quadratic expressions,* i.e., has the following form

$$f(\mathbf{x}) = u_1^2(\mathbf{x}) + u_2^2(\mathbf{x}) + \dots$$

you can use the MATHCAD **minerr** function for the *numerical calculation* of a *minimum*, which returns the („generalized") solution of a system of equations of the form (see Section 16.4)

$$u_1(\mathbf{x}) = 0 \ , \ u_2(\mathbf{x}) = 0 \ , \ ...$$

for *minimizing* the *sum* of *squares* , i.e., for

$$f(\mathbf{x}) = u_1^2(\mathbf{x}) + u_2^2(\mathbf{x}) + ... \ \rightarrow \ \underset{x}{\text{minimum}}$$

You can also write your own *programs* in MATHCAD or use the **Numerical Recipes** *electronic book* for the *numerical solution* of *optimization problems.*

The *table of contents* for Chapter 8 of the **Numerical Recipes** *electronic book* is reproduced below:

Chapter 8	Minimization or Maximization of Functions
8.1	*Parabolic Interpolation and Brent's Method in One Dimension* *brent*
8.2	*Downhill Simplex Method in Multidimensions* *amoeba*
8.3	*Direction Set Methods in Multidimensions* *powell*
8.4	*Conjugate Gradient Methods in Multidimensions* *frprmn*
8.5	*Variable Metric Methods in Multidimensions* *dfpmin*
8.6	*Linear Programming and the Simplex Method* *simplx*
8.7	*Simulated Annealing Methods* *anneal*

You can already see from this table of contents, that other than the simplex methods, MATHCAD provides in this electronic book only methods to solve problems without constraints.

In the following examples, we write two *programs* for the classical numerical optimization methods

* *Gradient method*

* *Penalty function method*

to stimulate readers to write their own programs in MATHCAD.♦

From the range of numerical methods available for the solution of optimization problems, we use the *classical gradient method* to *determine* a *local minimum* of a function (objective function) f(**x**) for n variables; no constraints are present.

We write the n variables as *row vector* **x**, i.e.,

$$\mathbf{x} = (x_1, x_2, ..., x_n)$$

The *gradient method* belongs to the *iteration methods*, from which we have already discussed the Newton method (see Example 10.6b). The *gradient method* calculates, beginning with a *starting vector*

$$\mathbf{x}^1$$

further vectors

$$\mathbf{x}^2, \ \mathbf{x}^3, \ \mathbf{x}^4, ...$$ with $f(\mathbf{x}^1) \geq f(\mathbf{x}^2) \geq f(\mathbf{x}^3) \geq f(\mathbf{x}^4) \geq \ ...$

using the *formula*

$$\mathbf{x}^{k+1} = \mathbf{x}^k - \alpha_k \cdot \mathbf{grad} \, f(\mathbf{x}^k) \ , \qquad k = 1, 2, 3, ...$$

where **grad** f represents the *gradient* (see Section 22.2) of the function f(**x**) and

$$\alpha_k$$

represents the *step size*, for which there are several *methods* of *calculation*; we show three of these methods:

I.   Determination using the one-dimensional minimization problem

$$f(\mathbf{x}^k - \alpha_k \cdot \mathbf{grad}(\mathbf{x}^k)) = \underset{\alpha \geq 0}{\text{Minimum}} \ f(\mathbf{x}^k - \alpha \cdot \mathbf{grad}(\mathbf{x}^k))$$

This method is longwinded, because the minimization for $\alpha \geq 0$ normally has to be performed numerically.

II. $0 < \alpha_k \to 0$ and $\displaystyle\sum_{k=1}^{\infty} \alpha_k = \infty$

This can be used to prove the convergence of the gradient method for convex functions.

This calculation method can, for example, be realized using

$$\alpha_k = \frac{1}{k}$$

III. A frequently used *heuristic method* consists of the specification of a fixed step size α. If the associated function value does not become smaller, you reduce it yourself (etc.). This method is justified by the property that the negative gradient points in the direction of the largest decrease of the function.

The *gradient method* converges under certain assumptions to a local or global minimum, we do not discuss this here. We only make use of the fact that the value of the objective function f decreases relative to the starting value.

The *gradient method* ends after a finite number of steps only when the *gradient* is zero in an iteration point, i.e., the necessary optimality condition is satisfied. However, because this is not the general case, we must use a *termination criterion* (*stopping rule*) for the *iteration*. There are several *possibilities* here:

For example, you can *terminate* the gradient method, when

I. The difference between two successive iteration points or objective function values is sufficiently small.

II. The absolute value of the gradient is sufficiently small.

III. A given number of iteration steps has been reached (provided no convergence occurs).

In the following example we write a MATHCAD *program* to realize the provided *gradient method* for functions of two variables, where the step size and the termination criterion are determined using Method II.

Example 25.4:

a) *Search for a local minimum for the function*

$$f(x,y) := (1-x)^2 + x^2 + y^2 - 4 \cdot y$$

We *use* the **until** statement and the **while** operator from *operator palette no. 6* (see Chapter 10) to write a MATHCAD *program* for the *gradient method*. The *created programs* are *generally usable*; you only need to *substitute* the appropriate *function definition* and, if necessary, use *other starting values* and *termination criteria*.

a1) The MATHCAD program for the *gradient method* using the

operators from *operator palette no. 6* (see Chapter 10) is *shown* in the *following worksheet:*

Gradient method for functions of two variables
z = f(x,y)

We seek a local minimum for the following function

$$f(x,y) := (1-x)^2 + x^2 + y^2 - 4 \cdot y$$

We use

$$\alpha_k = 1 / k$$

as step size.

Calculation of the gradient for the function $z = f(x,y)$ *using MATH-CAD*

$$\text{grad\_f}(x,y) := \begin{pmatrix} \dfrac{d}{dx} f(x,y) \\ \dfrac{d}{dy} f(x,y) \end{pmatrix} \qquad \text{grad\_f}(x,y) \rightarrow \begin{pmatrix} -2 + 4 \cdot x \\ 2 \cdot y - 4 \end{pmatrix}$$

The **min** *function subroutine that is written to perform the gradient method has the following parameters*

* x_0 *and* y_0 *the starting values for* x *and* y

*   *the termination bound* eps

$$\text{min}(x_0, y_0, \text{eps}) := \begin{vmatrix} x \leftarrow x_0 \\ y \leftarrow y_0 \\ k \leftarrow 1 \\ \text{while} \quad |\text{grad\_f}(x,y)| > \text{eps} \\ \quad \begin{vmatrix} x \leftarrow x - \dfrac{\text{grad\_f}(x,y)_1}{k} \\ y \leftarrow y - \dfrac{\text{grad\_f}(x,y)_2}{k} \\ k \leftarrow k + 1 \end{vmatrix} \\ \begin{pmatrix} x \\ y \end{pmatrix} \end{vmatrix}$$

Display the result for the starting values

$x_0 = 0$, $y_0 = 0$

and the termination bound

eps = 0.001:

$$\text{min}(0, 0, 10^{-3}) = \begin{pmatrix} 0.5 \\ 1.999999999999999 \end{pmatrix} .$$

i.e., we have obtained the solution x = 0.5 , y = 2.

a2) The *following worksheet shows* the MATHCAD program written using the **until** statement for the *gradient method*:

Gradient method for functions of two variables
z = f(x,y)

We seek a local minimum for the following function

$f(x,y) := (1-x)^2 + x^2 + y^2 - 4 \cdot y$

We use the vector

$$\begin{pmatrix} x_1 \\ y_1 \end{pmatrix} := \begin{pmatrix} 0 \\ 0 \end{pmatrix}$$

as starting value for the method and

$\alpha_k = 1 / k$

as step size.

The maximum number of performed iterations is

$N := 20$

Calculation of the *gradients* of the *function* $z = f(x,y)$ using *MATH-CAD:*

$$\text{grad\_f}(x,y) := \begin{pmatrix} \dfrac{d}{dx} f(x,y) \\ \dfrac{d}{dy} f(x,y) \end{pmatrix} \qquad \text{grad\_f}(x,y) \rightarrow \begin{pmatrix} -2 + 4 \cdot x \\ 2 \cdot y - 4 \end{pmatrix}$$

Specification of the termination criterion for the iteration:

$eps := 10^{-3}$

$\text{break}(x,y) := |\text{grad\_f}(x,y)| - eps$

*Perform the gradient method using the **until** statement and the definition of the index (array index) as range variable* (see Chapter 10):

$k := 1 .. N$

$$\begin{pmatrix} x_{k+1} \\ y_{k+1} \end{pmatrix} := \text{until}\left(\text{break}(x_k, y_k) , \begin{pmatrix} x_k \\ y_k \end{pmatrix} - \frac{\text{grad\_f}(x_k, y_k)}{k} \right)$$

*Display of the **x** and **y** vectors whose components contain the results of the iterations:*

$$x = \begin{pmatrix} 0 \\ 2 \\ -1 \\ 1 \\ 0.5 \end{pmatrix} \qquad y = \begin{pmatrix} 0 \\ 4 \\ 2 \\ 2 \\ 2 \end{pmatrix}$$

Graphical display of the iteration points

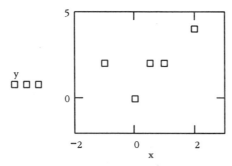

The result obtained after 4 Iterations

$$\begin{pmatrix} x_M \\ y_M \end{pmatrix} = \begin{pmatrix} 0.5 \\ 2 \end{pmatrix}$$

The objective function has the following value for this calculated point:

$$f(x_M, y_M) = -3.5$$

We must use indexed variables for the **until**-statement, because the *following simpler form* without indexed variables does not *operate.*

We use the vector

$$\begin{pmatrix} x \\ y \end{pmatrix} := \begin{pmatrix} 0 \\ 0 \end{pmatrix}$$

as starting value and

$$\alpha_k = 1 / k$$

as step size for the method

The maximum of performed iterations is

$$N := 20$$

Calculation of the termination criterion for the iteration

$$\text{eps}:=10^{-3}$$

$$\text{break}(x,y):=\left|\,\text{grad\_f}(x,y)\,\right|-\text{eps}$$

Perform the gradient method

$$k := 1\;..\;N$$

$$\binom{x}{y} := \text{until}\left[\,\text{break}(x,y),\binom{x}{y}-\frac{\text{grad\_f}(x,y)}{k}\,\right]$$

non-scalar value

If you use the two programs shown in Example a1 and a2 to evaluate the gradient method, you can see that the program written in a1 is better, because you do not require any loop with a fixed number of passes in this case. However, to ensure that the method ends even in the case of non-convergence, an additional termination criterion must be included in the program a1. We leave this as exercise for the reader. You can use a similar method as we used in Section 10.6.2 for the programs for the Newton method (Example 10.6b).

b) We now solve an additional problem with the program from Example a1. This can be used for minimization problems for functions with two variables. You must just change the function f, and possibly the starting values and the *termination bound* eps.

We wish to determine a local minimum of the function $z = \sin xy$. We just enter the program quantities from Example a1 that need to be changed:

$$f(x,y) := \sin(x\cdot y)$$

$$\binom{x}{y} := \min\left(1,1,10^{-3}\right) \qquad \binom{x}{y} = \binom{-1.195234977917725}{1.313744823814854}\;\blacksquare$$

$$f(x,y) = -0.999999841762389 \;\blacksquare$$

A *minimum* of the function at the point x=−1.195 and y=1.314 is determined for the *starting values*

$$x_0 = 1\,,\; y_0 = 1$$

and the *termination bound*

$$\text{eps} = 0.001$$

◆

Numerical methods exist for the *optimization problems* considered in Section 25.3:

$z = f(x_1, x_2, ..., x_n) \rightarrow \underset{x_1, x_2, ..., x_n}{\text{Minimum / Maximum}}$

with *equality constraints*

$g_j(x_1, x_2, ..., x_n) = 0 \quad , \quad j=1, 2, ..., m$

or *inequality constraints*

$g_j(x_1, x_2, ..., x_n) \leq 0 \quad , \quad j=1, 2, ..., m$

to *transform* these to *problems without constraints.*

These include the *penalty function methods* that add the constraints in the form of suitable *penalty terms* to the objective function. The following *problems without constraints* supply a *simple version* of these *penalty function methods:*

* for *equality constraints:*

$$F(x_1, x_2, ..., x_n, \lambda) = f(x_1, x_2, ..., x_n) + \lambda \cdot \sum_{j=1}^{m} g_j^2(x_1, x_2, ..., x_n)$$

$\rightarrow \underset{x_1, x_2, ..., x_n}{\text{Minimum / Maximum}}$

* for *inequality constraints*

$$F(x_1, x_2, ..., x_n, \lambda) = f(x_1, x_2, ..., x_n) + \lambda \cdot \sum_{j=1}^{m} (\max\{g_j(x_1, x_2, ..., x_n), 0\})^2$$

$\rightarrow \underset{x_1, x_2, ..., x_n}{\text{Minimum / Maximum}}$

Although numerical analysis provides theorems for the *convergence* of these *penalty function methods* for monotonously increasing *penalty parameters* λ (>0), we will not discuss these here.

We show two examples to illustrate how the *gradient method* programmed in Example 25.4 can be used to solve the resulting problems without constraints.

Example 25.5:

a) Solve the *problem*

$$f(x,y) = \frac{(x-2)^2}{2} + \frac{(y-2)^2}{2} \rightarrow \underset{x,y}{\text{Minimum}}$$

with the *equality constraint*

$x + y = 1$

by using the *gradient method* from Example 25.4a1 on the *penalty func-tion*

$$F(x, y, \lambda) = \frac{(x-2)^2}{2} + \frac{(y-2)^2}{2} + \lambda \cdot (x+y-1)^2$$

that consists of the *objective function* to which the *penalty term* has been added.

Gradient method for penalty functions $F(x,y,\lambda)$ (equality constraints)

We seek a local minimum for the following function

$$F(x, y, \lambda) := \frac{1}{2} \cdot (x-2)^2 + \frac{1}{2} \cdot (y-2)^2 + \lambda \cdot (x+y-1)^2$$

Calculate the gradient for the function

$z = F(x,y,\lambda)$:

$$\text{grad\_}F(x, y, \lambda) := \begin{pmatrix} \dfrac{d}{dx} F(x, y, \lambda) \\ \dfrac{d}{dy} F(x, y, \lambda) \end{pmatrix}$$

Perform the gradient method where the penalty parameter λ is included in the argument of the **min** *function subroutine:*

$$\min\left(x_0, y_0, \lambda, eps\right) := \begin{vmatrix} x \leftarrow x_0 \\ y \leftarrow y_0 \\ k \leftarrow 1 \\ \text{while} \quad \left| \text{grad\_F}(x, y, \lambda) \right| > eps \\ \quad \begin{vmatrix} x \leftarrow x - \dfrac{\text{grad\_F}(x, y, \lambda)_1}{100} \\ y \leftarrow y - \dfrac{\text{grad\_F}(x, y, \lambda)_2}{100} \\ k \leftarrow k + 1 \end{vmatrix} \\ \begin{pmatrix} x \\ y \end{pmatrix} \end{vmatrix}$$

A *minimum* of the function at the point x=0.51 and y=0.51 is determined

for the *starting values*

$x_0 = 0$, $y_0 = 0$

the *termination bound*

eps = 0.001

and the *penalty parameter*

$\lambda = 50$

that also satisfy approximately the equality constraints:

$$\begin{pmatrix} x \\ y \end{pmatrix} := \min\left(0,0,50,10^{-3}\right) \qquad \begin{pmatrix} x \\ y \end{pmatrix} = \begin{pmatrix} 0.507956615848763 \\ 0.506973746984577 \end{pmatrix}$$

Value of the objective function : $f(x,y) := \dfrac{(x-2)^2}{2} + \dfrac{(y-2)^2}{2}$

$f(x,y) = 2.227660426191374$ ∎

Because the method is not successful for the *step size selection*

$\alpha = 1/k$

we chose the *constant step size*

$\alpha = 1/100$

that worked correctly.

It is desirable to experiment with various values for the *penalty parameter* λ and the *step size* α.

b) Solve the *problem*

$$f(x,y) = \frac{x^2}{2} + \frac{(y-2)^2}{2} \rightarrow \underset{x,y}{\text{Minimum}}$$

with the *inequality constraints*

$y - x - 1 \leq 0$ and $y + x - 1 \leq 0$

that we have already discussed in Example 25.3. We now use the *penalty function*

$$F(x,y,\lambda) = \frac{x^2}{2} + \frac{(y-2)^2}{2} +$$

$$+ \lambda \cdot \left((\max\{y-x-1,0\})^2 + (\max\{y+x-1,0\})^2 \right)$$

that consists of the *objective function* to which the *penalty term* has been added. We use the *gradient method* from Example 25.4a1 to numerically solve this resulting problem without constraints.

The MATHCAD **max** function, which determines the maximum of the elements of a matrix (see Section 15.2), can be used to realize the *max* *function* required for the penalty function that determines the maximum of two numbers. We write the following *program:*

Gradient method for penalty functions F(x,y,λ) (inequality constraints)

We seek a local minimum for the following function

$$F(x,y,\lambda) := \frac{x^2}{2} + \frac{(y-2)^2}{2} + \lambda \cdot \left(\begin{array}{l} \max\left(\binom{y-x-1}{0}\right)^2 \cdots \\ + \max\left(\binom{y+x-1}{0}\right)^2 \end{array} \right)$$

Calculate the gradient of the function z = F(x,y,λ):

$$\text{grad\_}F(x,y,\lambda) := \left(\begin{array}{l} \dfrac{d}{dx}F(x,y,\lambda) \\ \dfrac{d}{dy}F(x,y,\lambda) \end{array} \right)$$

Because the gradient method is performed using the same method as in Example a, we only need to consider the result:

For the *starting values*

$x_0 = 1$, $y_0 = 1$

the *termination bound*

eps = 0.001

and the *penalty parameter*

λ=10

the method determines a *minimum* of the function at the point x=1.37 and y=1.02

$$\binom{x}{y} := \min\left(1, 1, 10, 10^{-3}\right) \qquad \binom{x}{y} = \binom{1.368119736441957 \cdot 10^{-5}}{1.024379492569818} \quad \blacksquare$$

Value of the objective function : $f(x,y) := \dfrac{x^2}{2} + \dfrac{(y-2)^2}{2}$

$f(x,y) = 0.475917687352751$ ∎

However, the calculated minimum only approximates the constraints. This effect can be seen for the penalty function methods. You can then try to increase the penalty parameter.

Because the method is not successful for the *selected step size*

$\alpha=1/k$

we select the *constant step size*

$\alpha=1/100$

that produces the required result.

It is desirable to experiment with various values for the penalty parameter λ and the step size α.

♦

The classical gradient and penalty function methods used in this section do not have the best convergence properties. The programs shown here should stimulate readers to write their own programs using the more effective methods described in the literature.

♦

26 Probability

The *outcome* for *deterministic events* is *determined* uniquely. However, *events* adopt a major role in many *practical applications* that *depend* on *random factors*, i.e., their *outcome* is *indeterminate*. Such events are called *random events*.

Mathematicians understand a *random event* to be the possible *realization* of a *random experiment* (*trial*) that can be *characterized* as follows:

* They are performed under *random conditions* and can be *repeated* as often as required.
* Several *different results* are possible.
* The *occurrence* or *non-occurrence* of an *event* is *random*.

♦

Example 26.1:

Examples of *random experiments* are

* *tossing a coin*
* *throwing dice*
* *drawing lotto numbers*
* *selecting products* for *quality control*.

♦

Probability theory investigates *random events* using mathematical methods by using the notations *probability, random variable* and *distribution function* to provide quantitative *propositions* about *random events*.

♦

In this book, we discuss how MATHCAD can be used to *solve standard problems* for *probability theory* and *statistics*.

Because of the quantity of material invoked, a comprehensive discussion of probability theory and statistics using the MATHCAD electronic books for this task must be left to a separate book.

♦

☞

There are a number of *special systems* such as SAS, UNISTAT, STATGRAPH-ICS, SYSTAT and SPSS, that have been expressly developed for the *solution* of *problems* for *probability theory* and *statistics* and so generally offer significantly more comprehensive capabilities.

However, this does not mean that computer algebra systems, and thus also MATHCAD, are not suitable for this type of problem. We will see in the course of this chapter, and in the next chapter, that MATHCAD can also be used to successfully solve *standard problems* from *probability theory* and *statistics*.

♦

Before we solve problems from probability theory in Section 26.2-26.5, we first consider in Section 26.1 the basic formulae of combinatorics that you require for the calculation of probabilities.

26.1 Combinatorics

We require the *binomial coefficients*

$$\binom{a}{k} = \begin{cases} \dfrac{a \cdot (a-1) \cdot \ldots \cdot (a-k+1)}{k!} & \text{für } k > 0 \\[2ex] 1 & \text{für } k = 0 \end{cases}$$

for the *formulae* of *combinatorics,* where a and k represent a real and a natural number, respectively, and k! designates the *factorial.*

For the case that a=n is also a natural number, the above formula can be written in the following form:

$$\binom{n}{k} = \frac{n!}{k! \cdot (n-k)!}$$

MATHCAD can be used to perform the *calculation*

- for the *factorial* k!

  using one of the following ways, after *entering* k! into the *worksheet* using the keyboard or the button

  from operator palette no.1 and *marking* it with *editing lines*:

  * *exact calculation*

- activation of the *menu sequence*

   Symbolics ⇒ **Evaluate** ⇒ **Symbolically**

- activation of the *menu sequence*

   Symbolics ⇒ **Simplify**

- input of the *symbolic equal sign* → and then pressing the ⏎-key

* *numerical calculation*

   by entering the *numerical equal sign* =

- of the *binomial coefficients* $\binom{n}{k}$

by the evaluation of the given formula, where the product

$a \cdot (a-1) \cdots (a-k+1)$

is calculated using the *product operator* (see Chapter 14), because
MATHCAD, in contrast with other computer algebra systems, does not
provide any integrated function for the binomial coefficient.
Consequently, we define the **Binomial** *function* in the following Ex
ample 26.2 that can be used to calculate the *binomial coefficient.*

Example 26.2:

a) The following **Binomial** *function* defined in MATHCAD calculates the
   binomial coefficients for real a and integer k≥1:

$$\textbf{Binomial}\,(a,k) := \frac{\displaystyle\prod_{i=0}^{k-1}(a-i)}{k!}$$

as the *sample calculations* for the *numerical calculation*

Binomial (6 , 1) = 6   **Binomial** (25 , 3) = $23 \cdot 10^3$

Binomial (5 , 2) = 10   **Binomial** (12 , 11) = 12

or the *exact calculation*

Binomial (6 , 1) → 6   **Binomial** (25 , 3) → 2300

Binomial (5 , 2) → 10   **Binomial** (12 , 11) → 12

show.

b) If we use the **if**-statement described in Section 10.4, we can calculate the
   binomial coefficients for k≥0, i.e., this includes the case k=0:

$$\textbf{Binomial}\,(a,k) := \textbf{if}\,(\,k=0\,,\,1\,,\,\frac{\displaystyle\prod_{i=0}^{k-1}(a-i)}{k!}\,)$$

Binomial $(5 , 0) = 1$ **Binomial** $(0 , 0) = 1$

♦

We require *factorials* and *binomial coefficients* in *combinatorics* to calculate the *formulae* for

- *Permutations* (sequential arrangement of n different elements):

  n!

- *Variations/selections* (*selection* of *k elements* from n given elements *taking account of the order*)

  * *without repetition:* $\dfrac{n!}{(n - k)!}$

  * *with repetition:* n^k

- *Combinations* (*selection* of *k elements* from n given elements *without taking account of the order*)

  * *without repetition:* $\dbinom{n}{k}$

  * *with repetition:* $\dbinom{n + k - 1}{k}$

26.2 Probability and Random Variables

For *random events* we require a *measure* (the so-called *probability*) that describes the *chance* for the *occurrence* of an *event*. For practical reasons, this number is chosen to lie between 0 and 1, where the probability 0 represents the *impossible* and 1 represents the *certain event*.

We have the first contact with the notion *probability* for a *random event* with the

* *classical definition* of the *probability* as *quotient* from the *number* of *favorable cases* and the *number* of *possible cases*. These *probabilities* can be *calculated using* the *formulae* from *combinatorics*.

* *relative frequency* as *quotient* from the m-times *occurrence* of an *event* for *random experiments* and the *number* n of performed *random experiments* ($n \geq m$).

These *pragmatic definitions* are adequate for simple cases. *In general,* we use an *axiomatic definition* of the *probabilities* that can be found in text-books.

♦

The notion *random variable* plays a *fundamental role* in *probability theory* and *statistics.* It has been introduced in order to permit the calculation using random events. An exact definition is an advanced mathematical topic.

For an *application,* it suffices to know that the *random variable* is *defined as a function* that *assigns real numbers* to the *events* of a *random experiment.*

We differentiate between two *types* of *random variables:*

*   a *discrete random variable:*
    can only assume a *finite* (or *countable*) *number* of *values.*

*   a *continuous random variable:*
    can assume an arbitrary number of values.

Together with the *distribution functions* (Section 26.4), *probabilities* and *random variables* belong to the *fundamental notions* of *probability theory.*

♦

We use several examples to explain the *probability* and *random variable* terms.

Example 26.3:

a) Consider the *standard example* of *throwing* an ideal *dice.* The *probability* of *throwing* a specific *number* between 1 and 6 is determined using *classical probability* as the *quotient* of the *favorable cases* (1) and the *possible cases* (6) giving 1/6.

    We use the function that assigns just this number to the random event for throwing a specific number as *discrete random varible* X for this *random experiment,* i.e., X is a function that can assume the values 1, 2, 3, 4, 5, 6.

b) The *number* of the *parts produced* daily in a company within a specific time can be considered as being a *discrete random variable* that can assume all integer values within an interval.

c) The *temperature* of a *workpiece* being *processed* can be considered as being a *continuous random variable* that can assume all values within a certain *temperature interval* required for the processing.

d) The *fuel consumption* of a car can be considered as being a *continuous random variable.*

    ♦

The formulae for combinatorics can be used to consider the *classic probabilities* of the occurrence of events. We consider several examples.

Example 26.4:

a) Make a random selection of 2 parts from a population of 12 parts that contains 4 defective parts. Calculate the *probabilities* for the *following events* for this experiment:

  * *Event A:*
    The two selected parts are defective.

  * *Event B:*
    The two selected parts are not defective.

  * *Event C:*
    At least one of the selected parts is defective.

The *number* of the

  * *possible cases*

    to select 2 parts from 12 parts is calculated using the formula for combinations without repetitions. This yields:

    Binomial $(12,2) = 66$

  * *favorable cases*

    for the *events* A and B are calculated from:

    Binomial $(4,2) = 6$ or   **Binomial** $(8,2) = 28$

for which we use the **Binomial** function defined in Example 26.2.

This yields the *probabilities* for the *events* A and B:

$$P(A) := \frac{6}{66} \quad \textit{simplifies to} \quad P(A) := \frac{1}{11}$$

$$P(B) := \frac{28}{66} \quad \textit{simplifies to} \quad P(B) := \frac{14}{33}$$

This permits the calculation of the *probability* $P(C) = 1 - P(B)$ for the *event* C:

$$P(C) := 1 - \frac{14}{33} \quad \textit{simplifies to} \quad P(C) := \frac{19}{33}$$

b) Find the probability p that n people have their birthday on various days. We make the assumption that the birthday of a person has the same probability for any day.
Assume that a year has 365 days and thus the n people have 365^n possibilities for birthdays. The following number specifies the possibilities that the n people have birthday on different days

$$365 \cdot 364 \cdot 363 \cdots (365-n+1)$$

Thus, the required probability p is calculated from

$$p = \frac{365 \cdot 364 \cdot 363 \cdots (365 - n + 1)}{365^n}$$

$$= \frac{365}{365} \cdot \frac{364}{365} \cdot \frac{363}{365} \cdots \frac{365 - n + 1}{365}$$

This formula shows that the probability p decreases monotonically for increasing n. You can check this in MATHCAD by calculating using various values of n and displaying p(n) as function of n:

$$p(n) := \frac{\displaystyle\prod_{k=1}^{n}(365 - k + 1)}{365^n}$$

n := 1 .. 30

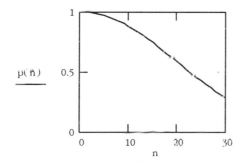

p(10) = 0.883

p(20) = 0.589

p(30) = 0.294

◆

26.3 Expected Value and Variance

The *distribution* of a *random variable* can be determined completely through the knowledge of its *distribution function* (or *density function*). *Moments,* for which we discuss only *expected value* and *variance* as the two most important, provide additional important information for a distribution:

- The *expected value* μ =E(X) of a
  - * *discrete random variable* X

with the values

$$x_i \qquad (\,i = 1\,,\,2\,,\,...\,)$$

and the *probabilities*

$$p_i \;=\; P(X = x_i)$$

is calculated from

$$\mu \;=\; E(X) \;=\; \sum_{i=1}^{n} x_i \cdot p_i \qquad \text{(for a \textit{finite} number of \textit{values} } x_i\,)$$

$$\mu \;=\; E(X) \;=\; \sum_{i=1}^{\infty} x_i \cdot p_i \qquad \text{(for a \textit{countable finite} number of \textit{values} } x_i\,)$$

* *continuous random variable*

  with the *density function* f(x) is calculated from

$$\mu \;=\; E(X) \;=\; \int_{-\infty}^{\infty} x \cdot f(x)\,dx$$

- The *variance* σ^2

  of a *random variable* X can be calculated using the given *expected value*
  E(X):

$$\sigma^2 \;=\; E(X - E(X))^2$$

  where the quantity σ is called the *standard deviation* or *dispersion*.

The formulae for the *expected value* assume the *convergence* of the infinite
series and the indefinite integral.

♦

Although MATHCAD does not provide *any functions* for the calculation of
the *expected value* and the *variance*, you can easily calculate the specified
formulae using series (Section 21.1) or indefinite integrals (Section 20.3).

♦

26.4 Distribution Functions

The *distribution function* F(x) for a *random variable* X is defined by

F(x) = P (X ≤ x)

(sometimes also with P (X < x)), where the associated *probability*

$P(X \leq x)$

specifies that the *random variable* X assumes a value less than or equal to x (real number).

☞

Inverse distribution functions also play an important role in probability theory. The x_s value is called the *s-quantile*, for which

$$F(x_s) = P(X \leq x_s) = s$$

i.e., x_s is determined from

$$x_s = F^{-1}(s)$$

where s is a given number from the interval (0,1).

♦

For *discrete random variables* X with the values

$$x_1, x_2, \ldots, x_n, \ldots$$

yields

$$F(x) = \sum_{x_k \leq x} p_k$$

for the defined *distribution function* where

$$p_k = P(X = x_k)$$

is the associated probability, for which the *random variable* X assumes the value x_k.

The *graph* of a discrete distribution function has the form of a *step function*. The *most important discrete distributions* for practical applications are:

- *Binomial distribution* B (n, p):

  with the *probability*

  $$P(X = k) = \binom{n}{k} \cdot p^k \cdot (1-p)^{n-k}$$

  for n independent experiments *with replacement* that can have only the result A (with probability p) or $\overline{A}$ (with probability 1–p), the result A occurs k times (k = 0, 1 ,..., n).

- *Hypergeometric distribution* H (N, M, n):

  with the *probability*

  $$P(X = k) = \frac{\binom{M}{k} \cdot \binom{N-M}{n-k}}{\binom{N}{n}}$$

for n experiments of the random removal of an element *without replacement* from a population of N elements for which M have a required property E, k elements occur with this property E, where

k = 0 , 1 , ... , min { n , M }

- *Poisson distribution* P (λ):

  with the *probability*

  $$P(X = k) = \frac{\lambda^k}{k!} \cdot e^{-\lambda} \qquad (k = 0, 1, 2, \dots)$$

  This distribution can be used as a good *approximation* for the *binomial distribution*, if n is large and p is small and λ is set to n·p, i.e., n·p remains constant.

MATHCAD provides a number of functions for *discrete distributions*, from which we have mentioned only some of the most important:

* **dbinom** (k , n , p)

  calculates the *probability* P(X=k) for the *binomial distribution* B(n,p).

* **pbinom** (k , n , p)

  calculates the *distribution function* F(k) for the *binomial distribution* B(n,p).

* **dhypergeom** (k , M , N–M , n)

  calculates the *probability* P(X=k) for the *hypergeometric distribution* H(N,M,n).

* **phypergeom** (k , M , N–M , n)

  calculates the *distribution function* F(k) for the *hypergeometric distribution* H(N,M,n).

* **dpois** (k , λ)

  calculates the *probability* P(X=k) for the *Poisson distribution* P(λ).

* **ppois** (k , λ)

  calculates the *distribution function* F(k) for the *Poisson distribution* P(λ).

☞

Although MATHCAD provides all *discrete distribution functions*, for the purposes of an exercise, we use the programming capabilities of MATHCAD and define in Example 10.6 and in the following Example 26.5 some of the functions.

♦

Example 26.5:

As an exercise, we use the MATHCAD programming capabilities to define the *distribution function* for the *binomial distribution* **pbinom**.

a) The *distribution function* for *discrete distributions*

$$F(x) = \sum_{x_k \le x} p_k$$

has the following form for the *binomial distribution* B(n,p)

$$F(x) = \begin{cases} 0 & \text{for } x < 0 \\[2em] \sum_{k=0}^{m} \binom{n}{k} p^k (1-p)^{n-k} & \text{for } x \ge 0 \end{cases}$$

where m is the largest integer less than or equal to x (>0). The sum operators and the binomial coefficients calculated in Example 26.2 can be used to calculate this *distribution function* as follows:

$$\text{Binomial}(a, k) := \text{if}\left(k = 0, 1, \frac{\prod_{i=0}^{k-1}(a-i)}{k!}\right)$$

$$F(x,n,p) := \text{if}\left(x<0, 0, \text{if}\left(\text{floor}(x) \le n, \sum_{k=0}^{\text{floor}(x)} \text{Binomial}(n,k) \cdot p^k \cdot (1-p)^{n-k}, 1\right)\right)$$

We have specified as arguments the parameters n and p required for the definition of the binomial distribution. As the following Example b shows, these must be entered for the specific calculation.

b) We use a classical *urn problem* as an application for which we calculate the values of the associated *distribution function*:

An urn contains 10 white and 6 black balls. A ball is removed (with replacement) twenty-five times in succession. The *binomial distribution* B(25, 3/8) determines the probability that k of the removed balls are black.

The probability p as the quotient is calculated from the number 6 of black balls (favorable cases) and the total number 16 of balls (possible cases).

We can now use the distribution function defined for the binomial distribution in Example a to calculate the probabilities and draw the associated distribution function.

The following *value tables* contain the calculated values for the defined *distribution function* F (x , 25 , 3/8) and the **pbinom** (x , 25 , 3/8) *distribution function* for the *binomial distribution* integrated in MATH-CAD at the values

x = 0, 1, 2, ... , 25

where we have defined x as range variable:

$x := 0 .. 25$

$$F\left(x, 25, \frac{3}{8}\right)$$

$$\text{pbinom}\left(x, 25, \frac{3}{8}\right)$$

$F\left(x, 25, \frac{3}{8}\right)$	$\text{pbinom}\left(x, 25, \frac{3}{8}\right)$
0	$7.889 \cdot 10^{-6}$
$1.262 \cdot 10^{-4}$	$1.262 \cdot 10^{-4}$
$9.782 \cdot 10^{-4}$	$9.782 \cdot 10^{-4}$
0.005	$4.897 \cdot 10^{-3}$
0.018	0.018
0.05	0.05
0.116	0.116
0.222	0.222
0.365	0.365
0.527	0.527
0.683	0.683
0.811	0.811
0.9	0.9
0.954	0.954
0.981	0.981
0.994	0.994
0.998	0.998
1	1
1	1
1	1
1	1
1	1
1	1
1	1
1	1
1	1

The comparison of both results shows the good match.

The *graph* of the defined *distribution function* F for the *binomial distribution* shows that it has the form of a *step function:*

x := 0, 0.01 .. 25

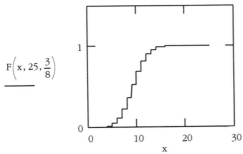

$$F\left(x, 25, \frac{3}{8}\right)$$

The use of the **pbinom** MATHCAD function for the *binomial distribution* naturally returns the same *graph:*

x := 0, 0.01 .. 25

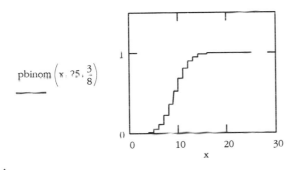

$$\text{pbinom}\left(x, 25, \frac{3}{8}\right)$$

♦

$$F(x) = \int_{-\infty}^{x} f(t)\, dt$$

specifies the *distribution function* of a *continuous random variable* X, where f(t) represents the *probability density.*

The *normal distribution*

N (μ , σ)

with the *density function*

$$f(t) = \frac{1}{\sigma \cdot \sqrt{2\pi}}\, e^{-\frac{1}{2}\left(\frac{t-\mu}{\sigma}\right)^2}$$

and the *distribution function*

$$F(x) = \frac{1}{\sigma \cdot \sqrt{2\pi}} \int_{-\infty}^{x} e^{-\frac{1}{2}\left(\frac{t-\mu}{\sigma}\right)^2} dt$$

frequently used for the *continuous distribution functions* plays a dominating role, where

* μ – expected value
* σ^2 – variance
* σ – standard deviation

If

$\mu = 0$ and $\sigma = 1$

this produces the *standard normal distribution* N (0,1), whose *distribution function*

$$\Phi(x) = \frac{1}{\sqrt{2\pi}} \int_{-\infty}^{x} e^{-\frac{1}{2}t^2} dt$$

we designate as $\Phi(x)$ and for which:

$\Phi(-x) = 1 - \Phi(x)$

The *error integral*

$$Fi(x, y) = \frac{2}{\sqrt{\pi}} \int_{y}^{x} e^{-t^2} dt = 2 \cdot (\Phi(\sqrt{2} \cdot x) - \Phi(\sqrt{2} \cdot y))$$

is also required for a number of calculations.

It is possible to transform any *normal distribution* to the *standard normal distribution*:
The *transformation*

$$Y = \frac{X - \mu}{\sigma}$$

can transform an N(μ , σ) distributed *random variable* X to an N(0 , 1) distributed *random variable* Y.

This permits the $\Phi(x)$ *distribution function* for the *standard normal distribution* to be used to calculate the F(x) *distribution function* of an N(μ , σ) distributed random variable X:

$$F(x) = P(X \le x) = P\left(\frac{X - \mu}{\sigma} \le \frac{x - \mu}{\sigma}\right) = P\left(Y \le \frac{x - \mu}{\sigma}\right) = \Phi\left(\frac{x - \mu}{\sigma}\right)$$

♦

☞

The *Chi-square distribution*, the *Student distribution* and the *F-distribution* are other important distributions (in particular for statistics).

◆

We describe here only a few of the functions that MATHCAD provides for *continuous distributions*:

* **pnorm** (x , μ , σ)

  calculates the *distribution function* for the *normal distribution* with the *expected value* μ and the *standard deviation* σ at point x.

* **cnorm** (x)

  calculates the *distribution function* of the *standard normal distribution* Φ(x) at point x.

* **erf** (x)

  calculates the *error integral* Fi(x, 0).

* **pchisq** (x , d)

  calculates the *distribution function* for the *Chi-square-distribution* with d *degrees of freedom* at point x

* **pt** (x , d)

  calculates the *distribution function* for the *Student distribution* with d *degrees of freedom* at point x.

* **pF** (x , d1 , d2)

  calculates the *distribution function* for the *F-distribution* with d1, d2 *degrees of freedom* at point x.

☞

The *Statistics* topic in the MATHCAD *help* contains the *complete list* of the *discrete* and *continuous distribution functions* integrated in MATHCAD. This also applies to the *inverse distribution functions* that MATHCAD provides for all distributions and which you can recognize with the initial letter q. Example 26.6c shows the *use* of the *inverse distribution function* for the normal distribution.

◆

Example 26.6:

a) *Draw* the *standard normal distribution* and the *error integral* for x from the interval [-3,3] in a coordinate system:

$$x := -3, -2.999 .. 3$$

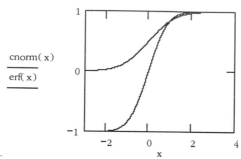

cnorm(x)
———
erf(x)

b) Consider a simple problem for the practical *use* of the *normal distribution*:

The length of some accessory part has the distribution N(50, 0.2). The part cannot be used when its length deviates by more than 0.25 mm from the specified value 50 mm. We are interested in the *probability* that some randomly selected part is defective.

We use the *standard normal distribution* $\Phi(x)$:

$$P(|X - 50| > 0.25) = 1 - P(|X - 50| \le 0.25)$$

$$= 1 - P(50 - 0.25 \le X \le 50 + 0.25) = 1 - P(50.25) + P(49.75)$$

$$= 1 - P\left(\frac{50 - 0.25 - 50}{0.2} \le \frac{X - 50}{0.2} \le \frac{50 + 0.25 - 50}{0.2} \right)$$

$$= 1 - (\Phi(1.25) - \Phi(-1.25))$$

$$= 1 - (2 \cdot \Phi(1.25) - 1) = 2 - 2 \cdot \Phi(1.25)$$

and can use the MATHCAD functions **pnorm** or **cnorm** to complete the calculation:

$$1 - \textbf{pnorm}\,(50.25\,,\,50\,,\,0.2) - \textbf{pnorm}\,(49.75\,,\,50\,,\,0.2) = 0.211$$

$$2 - 2 \cdot \textbf{cnorm}\,(1.25) = 0.211$$

Thus the *required probability* is 0.211.

c) Consider a simple problem for the practical *use* of the *inverse distribution function* for the *normal distribution*:

The weight (in grams) of packets of coffee has the distribution N(1000, 5). We require the maximum weight that a randomly selected packet has with the probability of 0.9 (90%).

Solve the F(x) = 0.9 equation for x, where F represents the *distribution function* for the *normal distribution.*

MATHCAD can use the **qnorm** *inverse distribution function* to solve this problem:

$$\text{qnorm}(\,0.9\,,\,1000\,,\,5\,) = 1.006 \cdot 10^3 \quad \blacksquare$$

This calculation shows that the weight of a randomly selected packet with a probability of 0.9 has the maximum weight of 1006 grams.

♦

26.5 Random Numbers

A number of mathematical methods require *random numbers*, such as for the *Monte Carlo methods* (see Section 27.3).
The computer can create random numbers, which, because they are produced by a program and thus have a certain regularity, are then called *pseudo random numbers.*

Because most practical applications require not one but several random numbers from a given interval [0,a], MATHCAD also has the capability to produce vectors of random numbers.

♦

MATHCAD provides a number of *functions* to calculate *random numbers* that satisfy various probability distributions. We consider some of the most important ones:

* **rnd** (a)

   creates an *equally distributed random number* that lies between 0 and a (a > 0).

* **runif** (n , a , b)

   creates a *vector* with n *equally distributed random numbers* that lie between a and b.

* **rnorm** (n , μ , σ)

   creates a *vector* with n *normally distributed random numbers* with *expected value* μ and *standard deviation* σ .

The *calculation* of *random numbers* using the given functions is initiated once this function with the appropriate arguments has been entered in the worksheet, has been marked with editing lines and then the *numerical equal sign* = entered.

☞

Assign a *starting value* to the *functions* used to *create random numbers*:

* The same random number is created if you *reset* this *starting value* (integer) to the current value for every invocation by clicking the OK button in the

Math Options

dialogue box that appears for

Built-In Variables

after activating the *menu sequence*.
This is called *resetting* MATHCAD's *random number generator*.

* If you do not reset the starting value (random number generator), a different random number is normally created for every invocation.

* Enter a different integer in the *dialogue box* for

Seed value for random numbers

if you want to *change* the *starting value*.

♦

Example 26.7:

a) Create an *equally distributed random number* from the interval [0,2]. An entry of the form:

$$\text{rnd}(2) = 2.536838874220848 \cdot 10^{-3} \quad \blacksquare$$

appears in the worksheet.

The repeated use of **rnd** without resetting the random number generator could, for example, produce the following values:

$$\text{rnd}(2) = 0.386646039783955$$

$$\text{rnd}(2) = 1.170012198388577$$

$$\text{rnd}(2) = 0.700616206973791$$

$$\text{rnd}(2) = 1.645675450563431$$

b) The **runif** function can be used to produce, for example, 15 *equally distributed random numbers* from the interval [0,5].
A vector **x** with the 15 produced random numbers as components is returned as *result*. The MATHCAD worksheet then has the following appearance:

$$\text{runif}(\,15\,,0\,,5\,) = \begin{bmatrix} 6.342 \cdot 10^{-3} \\ 0.967 \\ 2.925 \\ 1.752 \\ 4.114 \\ 0.871 \\ 3.552 \\ 1.52 \\ 0.457 \\ 0.737 \\ 4.943 \\ 0.595 \\ 0.045 \\ 2.658 \\ 3.009 \end{bmatrix} \cdot$$

c) We now use the **rnorm** function to create 15 *normally distributed random numbers* with the *expected value* 0 and the *standard deviation* 1. A vector **x** with the 15 produced random numbers as components is returned as *result*. The MATHCAD worksheet then has the following appearance:

$$\text{rnorm}(15, 0, 1) = \begin{bmatrix} -0.508 \\ 1.241 \\ 0.071 \\ 1.345 \\ -0.656 \\ -0.254 \\ -1.244 \\ 0.547 \\ -0.031 \\ -0.097 \\ 0.228 \\ 0.169 \\ -0.847 \\ 0.834 \\ -0.082 \end{bmatrix} \quad \blacksquare$$

◆

☞

The *Statistics* topic in the MATHCAD *help* contains the *complete list* of the functions integrated in MATHCAD that are used to create random numbers.

◆

☞

From the wide range of applications of random numbers, we present in Section 27.3 a Monte Carlo method used to calculate definite integrals.

◆

27 Statistics

Although *statistical* methods are gaining in importance in all scientific areas, it is difficult to provide a comprehensive definition of *statistics*.

We consider *statistics* as the theory of description and examination of a large number of events and mass processes in nature and society based on available data (in form of numbers). These data (numbers) are obtained by

* *observations* (counting, measurements)

* *questioning* (of people)

* *experiments*.

Because the obtained *data* represents only a (small) part of the considered *population*, it is called a (*random*) *sample*. If m *characteristics* are considered in the population, this is called an *m-dimensional sample*. The *sample* is said to be of *size* n, if n values are taken.

In this book, we consider

one-dimensional samples (n numbers)

$x_1, x_2, \ldots, x_n$

two-dimensional samples (n number pairs)

$(x_1, y_2), (x_2, y_2), \ldots, (x_n, y_n)$

each of size n.

♦

Statistics differentiate between

* *Descriptive statistics*:

  The provided data (e.g., a *sample*) *is prepared* or displayed in a usable form using

  * *point graphs*

  * *diagrams*

  * *histograms*

  and *characterized* using

  * *mean (expected value)* and *variance*.

The descriptive statistics here provide *only conclusions* about the provided *data*.

A *typical example* is the analysis of an election. Tables and graphs are used to display the vote distribution for the individual parties, the distribution of seats in parliament, changes compared with the previous election, etc.

* *Mathematical statistics:*

  Probability theory is used here on the *provided data* that are taken as *samples* to form general *conclusions* concerning the considered *population*. *Typical examples* are

  * *Quality control:*

    A company wishes to use the characteristics of a *sample* taken from the day's production for some manufactured product to make a predication about the characteristics of the day's production, which then represents the considered population.

  * *Election forecasts:*

    A sample taken before the election is used to make predictions about the vote distribution of each party.

Because we are not capable of collecting all information for many mass processes or events, the examples show the *large importance* of *mathematical statistics*. We can only *take samples* and then make conclusions for the *population*, i.e., the *basic principle* behind *mathematical statistics* is the use of a *part* to make a prediction for the *total*. In *mathematical terms*, this means using a sample to obtain *conclusions* about

* *unknown parameters* (*moments*) expected value, variance,...

* *unknown distribution functions.*

The associated *methods* are provided by the

* *theory of estimating* (estimate values for the parameters)

* *theory of testing* (validation of hypotheses for the distribution function and the parameters).

   ♦

Special program systems exist to solve *problems* from *descriptive* and *mathematical statistics*. If you mainly need to solve *statistical problems*, you can make use of *systems* such as SAS, UNISTAT, STATGRAPHICS, SYSTAT, SPSS, etc., that have been specially developed for these problems and are consequently more effective than MATHCAD and the other computer algebra systems.

However, this does not mean that MATHCAD and the other computer algebra systems are not suitable for statistical problems.

Although MATHCAD does not contain any functions to estimate parameters and to test hypotheses, you can use the provided functions to calculate distribution functions, statistical parameters, etc, and so perform a number of statistical calculations.

If you have the MATHCAD *electronic books* provided for *statistics*, you can use MATHCAD to effectively solve the standard statistical problems; we shown an example in Section 27.4.

27.1 Parameters of a Sample

We consider only the following *parameters (characteristics)* for *descriptive statistics*

* *empirical (arithmetic) mean*
* *median*
* *empirical variance*

that MATHCAD can calculate. We restrict ourselves here to *one-dimensional samples* of size n, i.e., to n number values

$$x_1, x_2, \ldots, x_n$$

for which these *parameters* are defined as follows:

* the *arithmetic mean* $\bar{x}$ using

$$\bar{x} = \frac{1}{n} \sum_{i=1}^{n} x_i$$

* the *median* $\tilde{x}$ using

$$\tilde{x} = \begin{cases} x_{k+1} & \text{falls} & n = 2k + 1 \,(\text{ungerade}) \\[2ex] \dfrac{x_k + x_{k+1}}{2} & \text{falls} & n = 2k \,(\text{gerade}) \end{cases}$$

where the number values

$$x_i$$

are ordered according to size, i.e.,

$$x_1 \le x_2 \le \ldots \le x_n$$

* the (unbiased) *variance* from

$$s^2 = \frac{1}{n-1} \sum_{i=1}^{n} (x_i - \bar{x})^2$$

where s is called the *standard deviation.*

MATHCAD provides the following *functions* to calculate these *statistical parameters/characteristics:*

If the *values*

$$x_1, x_2, \ldots, x_n$$

from a *one-dimensional sample* of *size* n are *saved* in a *vector* (column vector) **x**

$$\mathbf{x} := \begin{pmatrix} x_1 \\ \vdots \\ x_n \end{pmatrix}$$

- **mean (x)**

calculates the *arithmetic mean* $\bar{x}$

- **median (x)**

calculates the *median* $\tilde{x}$, where MATHCAD orders the components of the vector **x** according to size before performing the calculation

- **var (x)**

calculates the *variance* in the form that is divided by n rather than n−1, i.e.,

$$\frac{1}{n} \sum_{i=1}^{n} (x_i - \bar{x})^2$$

- **Var (x)**

calculates the *unbiased variance*

$$\frac{1}{n-1} \sum_{i=1}^{n} (x_i - \bar{x})^2$$

i.e., between **Var** and **var** consist the *relationship*

$$\mathbf{Var}\ (\mathbf{x}) = \frac{n}{n-1} \cdot \mathbf{var}\ (\mathbf{x})$$

- **stdev (x)**

calculates the *standard deviation* for **var**

- **Stdev (x)**

calculates the *standard deviation* for **Var**

The calculation is initiated by typing the numerical equal sign = after entering the appropriate function. Thus, the quantities are *calculated numerically.* An exact calculation is not provided in MATHCAD. Because the data are normally determined numerically, an exact calculation is not necessary.

☞

MATHCAD can also use the given functions to calculate the *mean, median, variance* and *standard deviation* for *m-dimensional samples* of *size n* when you enter these as m×n-matrices in the argument of the functions.

◆

Example 27.1:

Calculate the *mean, median, variance* and *standard deviation* for the *number values*

3.6 4.7 5.1 6.3 7.5 7.2 4.9 3.9 5.5 4.1 5.9 3.3

of a given *one-dimensional sample* of *size* 12. This is performed in MATH-CAD as follows:

* *First, assign* the *number values* for the given *sample* to a *column vector.* As described in Section 9.1, this can also be done by reading from data media (diskette, hard disk):

$$\mathbf{x} := \begin{pmatrix} 3.6 \\ 4.7 \\ 5.1 \\ 6.3 \\ 7.5 \\ 7.2 \\ 4.9 \\ 3.9 \\ 5.5 \\ 4.1 \\ 5.9 \\ 3.3 \end{pmatrix}$$

* *Then,* use the appropriate MATHCAD functions to calculate the required parameters:

mean (x) = 5.167   **median (x)** = 5

var (x) = 1.707   **stdev (x)** = 1.307

Var (x) = 1.862   **Stdev (x)** = 1.365

To *test* the *use* of the *built-in* (*integrated*) *functions* (Section 17.1), we write our own **med** *function* that calculates the *median* of the components of a column vector in which the elements are not ordered according to their size:

$$\mathbf{x} := \begin{pmatrix} 3.6 \\ 4.7 \\ 5.1 \\ 6.3 \\ 7.5 \\ 7.2 \\ 4.9 \\ 3.9 \\ 5.5 \\ 4.1 \\ 5.9 \\ 3.3 \end{pmatrix}$$

$$\mathbf{med}(\mathbf{x}) := \mathbf{if}\left(\mathbf{floor}\left(\frac{\mathbf{rows}(\mathbf{x})}{2}\right) \cdot 2 = \mathbf{rows}(\mathbf{x}), \frac{x_{\frac{\mathbf{rows}(\mathbf{x})}{2}} + x_{\frac{\mathbf{rows}(\mathbf{x})}{2}+1}}{2}, x_{\mathbf{floor}\left(\frac{\mathbf{rows}(\mathbf{x})}{2}\right)+1}\right)$$

$\mathbf{med}\,(\mathbf{x}) = 6.05$

The starting value for the indexing of the vector for this function must have been set to 1, i.e., **ORIGIN** := 1.

♦

MATHCAD permits the *graphical display* of the provided *data* for one-dimensional, two-dimensional and three-dimensional samples. We have already discussed the display in point form in Section 18.3 (*Scatter Plots*).

Because MATHCAD provides other *display forms* for *one-dimensional samples*, we explain the procedure again:

- First assign the *data* to be *displayed* to a *vector* (*column vector*) **x**.
- Enter the *index* (e.g., i) in the middle *placeholder* of the x-axis and the *components* x_i of the *vector* **x** in the middle *placeholder* of the y-axis in the *graphic window* for curves (see Section 18.1).
- Then define the *index* as *range variable* with step size 1 above the graphic window.
- Finally, click with the mouse outside the graphic window or press the ⏎-key in automatic mode to initiate drawing the data:
  * In the *standard setting*, MATHCAD connects the drawn points with straight lines.

* After a double-click on the graphic, a *dialogue box* appears in which you can set *another form* of the *graphical display* for

Traces ⇒ Type

For example

- Display in *point form*
- Display in *bar form*
- Display in *step form*

The *procedure* can be *seen* from *the following Example 27.2*, in which we graphically display the sample from Example 27.1.

Example 27.2:

Consider the *one-dimensional sample* from Example 27.1 whose number values in the worksheet have been assigned to the following *column vector* **x**:

$$x := \begin{pmatrix} 3.6 \\ 4.7 \\ 5.1 \\ 6.3 \\ 7.5 \\ 7.2 \\ 4.9 \\ 3.9 \\ 5.5 \\ 4.1 \\ 5.9 \\ 3.3 \end{pmatrix}$$

MATHCAD provides the following *graphical display capabilities:*

a) Display in *point form*

i := 1 .. 12

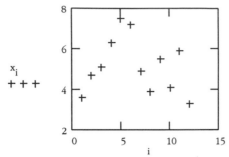

Several forms are *available* to *display* the number values as *points*. We have chosen crosses for the display.

b) Display in *bar form*

i := 1 .. 12

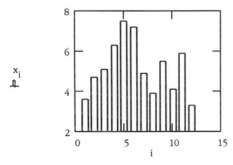

c) Display in *step form*

i := 1 .. 12

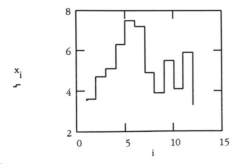

◆

☞

MATHCAD also contains the **hist** *function* to *calculate frequency distributions* for a given one-dimensional sample. These are displayed graphically using a *histogram*:

* A set of *number values* is assigned in *classes* K_i with *class width* b_i.
* The relative frequency h_i of the class K_i is displayed in a so-called *histogram* using the surface area of a rectangle constructed over b_i.

After *entering* the *column vectors*

* **x**

   that contains the number values of the sample as components

* **u**

   from whose components the specified *class widths* b_i can be calculated in the form

   $b_i = u_{i+1} - u_i$ ($u_{i+1} > u_i$ assumed)

MATHCAD returns with the

hist (u,x)

function by entering the numerical equal sign = a column vector whose i-th component contains the number of components (number values) from **x** which lie within the interval $[u_i, u_{i+1}]$.

◆

Example 27.3:

Now calculate the frequency distribution for the number values of the one-dimensional sample from Example 27.1 where the class widths are given in the vector **u**:

$$
\mathbf{x} := \begin{pmatrix} 3.6 \\ 4.7 \\ 5.1 \\ 6.3 \\ 7.5 \\ 7.2 \\ 4.9 \\ 3.9 \\ 5.5 \\ 4.1 \\ 5.9 \\ 3.3 \end{pmatrix}
\qquad
\mathbf{u} := \begin{pmatrix} 3 \\ 4 \\ 5 \\ 6 \\ 7 \\ 8 \end{pmatrix}
$$

$$\mathbf{h} := \mathbf{hist}(u, x) \qquad \mathbf{h} = \begin{pmatrix} 3 \\ 3 \\ 3 \\ 1 \\ 2 \end{pmatrix}$$

The *result vector* **h** returns the result for the number values of the sample

* 3 values in the interval [3,4]
* 3 values in the interval [4,5]
* 3 values in the interval [5,6]
* 1 value in the interval [6,7]
* 2 values in the interval [7,8]

The *display* of the *histogram* in the form of a *bar chart* can, for example, be made in the following form:

Histogramm

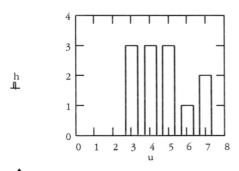

◆

27.2 Correlation and Regression

The *correlation* and *regression analysis* investigate the *type* of the *relationship* between various *characteristics* (*random variables*) of a considered population using *probability theory / statistics* and produces *functional relationships*.

In this book, we consider only the relationships between two *characteristics* (*random variables*) X and Y, i.e., functional relationships realized by functions of one variable y = f(x). The behaviour is similar when several characteristics (random variables) are considered.

27.2.1 Introduction

The *analytical representation* of *functions* plays an important role in those practical applications in which their equation is not known, i.e., the functions are only present in the *form* of *function values* (*tables*):

* This situation always occurs when the investigations between certain characteristics (random variables) show a *functional relationship* for which, however, you do not have any analytical expression.

* You are then dependent on the *measurements* of the associated *characteristics* (random variables) and so obtain the *relationship* in *tabular form*, which, for *two characteristics* (random variables), assumes the *form* of a *table* with *number pairs*

  $(x_1, y_1), (x_2, y_2), ..., (x_n, y_n)$

* This *function table* is not easy to comprehend. Example 18.6 shows a graphical display that provides a first attempt at producing a more easily understood representation. However, we are also interested in an approximate *analytical representation* using a *function equation*, because this is easier to use in calculations.

Numerical analysis provides *various methods* that can be used to *approximate* a *function*

$f(x)$

given by *n number pairs* (*points in the plane*)

$(x_1, y_1), (x_2, y_2), ..., (x_n, y_n)$

by a *function expression* (e.g., a polynomial).

Because we only know the *function values*

$y_i = f(x_i)$

in the *x-values*

$x_i \qquad (i = 1, ..., n)$

we must construct *functions* that *approximate* these *points*

$(x_1, y_1), (x_2, y_2), ..., (x_n, y_n)$

in accordance with a given *criterion*.

♦

The best known methods to approximate given points (scatter diagram) by a function, include

* *interpolation*
* *least squares.*

We discussed *interpolation* in Section 17.2.4. We use the *method of least squares* in the following discussion to *construct regression curves*.

♦

The *correlation* and *regression analysis*

* is based on n *number pairs*

  $(x_1, y_1), (x_2, y_2), \ldots, (x_n, y_n)$

  that have been obtained as a *sample* (e.g., by *measurements*),

  if for *two characteristics* (*random variables*)

  X   and   Y

  is *expected* a *functional relationship*.

* investigate using a taken sample the *form* of the *relationship* between the *characteristics* (random variables) X and Y using methods of probability theory/statistics and *construct* a *functional relationship*. Where

  * the *correlation analysis* supplies conditions about the *relationship* between the *characteristics* (random variables) X and Y

  * the *regression analysis* constructs an *approximate function* (*regression curve*) for the functional relationship using the *method* of *least squares*.

The use of *interpolation* in Section 17.2.4 assumes that a functional relationship exists between the considered characteristics, which, for example, is obtained from knowledge of certain laws. Consequently, no investigation of the relationship is made here and just an approximate function using the *interpolation principle* is constructed for the obtained sample.

♦

In contrast to *interpolation*, the given points for the *method* of *least squares* used in the *regression analysis* do not need to satisfy the constructed function. The principle here is to construct the approximate function that *minimizes* the *sum* of the *squares of the differences* between the function and the given points.

♦

27.2.2 Linear Regression Curves

If a *linear functional relationship* (one says *correlation*) between two *characteristics* (*random variables*)

X and Y

is assumed, we have the following to perform:

- First perform the *correlation analysis*. It uses methods from probability theory / statistics to provide conditions concerning the *strength* of the assumed *linear relationship* (one says *correlation*) between the two *characteristics* that are regarded as being *random variables*:

  * The *correlation coefficient*

    ρ_{XY}

    is used as a measure for the linear relationship between the two random variables X and Y. For

    $$\left|\rho_{XY}\right| = 1$$

    this *linear relationship* has the *probability* 1.

  * Because we only have the number pairs (points in the plane)

    $$(x_1, y_1), (x_2, y_2), \dots, (x_n, y_n)$$

    of the sample, we can use the *empirical correlation coefficient*

    $$r_{XY} = \frac{\sum_{i=1}^{n}(x_i - \bar{x})(y_i - \bar{y})}{\sqrt{\sum_{i=1}^{n}(x_i - \bar{x})^2}\sqrt{\sum_{i=1}^{n}(y_i - \bar{y})^2}}$$

    and statistical tests to obtain propositions about the linear relationship and, if necessary, construct the empirical regression line.

- The *linear regression* is then used to establish a *linear relationship* of the form

  $$Y = a\,X + b$$

  between the *random variables* X and Y, provided that the calculated *correlation coefficient* permits this. For the given number pairs of the sample, this produces a problem that we can approximate with a *straight line*

  $$y = a\,x + b \qquad\qquad (\textit{empirical regression line})$$

  The *Gaussian method* of *least squares* is used here:

  $$F(a,b) = \sum_{i=1}^{n}(y_i - a \cdot x_i - b)^2 \rightarrow \underset{(a,b)}{\text{Minimum}}$$

The unknown *parameters* a and b are determined to minimize the sum of the squares of differences for the given points (number pairs) to the regression line.

In order to determine whether the degree of a linear relationship is sufficient, it is important that we perform a *correlation analysis before* making a *linear regression*. If we do not want to perform this correlation analysis, we can still use the *empirical regression line* as approximation, provided that the calculated absolute value of the *empirical correlation coefficient* lies near to 1.

♦

After entering the *number pairs*

$$(x_1, y_1), (x_2, y_2), ..., (x_n, y_n)$$

of the given *sample* as *column vectors*

$$\mathbf{x} := \begin{pmatrix} x_1 \\ \vdots \\ x_n \end{pmatrix} \qquad \mathbf{y} := \begin{pmatrix} y_1 \\ \vdots \\ y_n \end{pmatrix}$$

the following *functions* can be used in MATHCAD for the *linear correlation* and *regression analysis*:

- **corr (x, y)**

  calculates the *empirical correlation coefficient*

- **slope (x, y)**

  calculates the *slope* a of the *empirical regression line*

- **intercept (x, y)**

  calculates the *intercept* b of the *empirical regression line* with the y-*axis*

The calculation is initiated when you input the numerical equal sign = after you enter the corresponding function.

Example 27.4:

We now investigate the five *number pairs*

(1,2) , (2,4) , (3,3) , (4,6) , (5,5)

of a *sample* using *correlation* and *regression analysis* for which we determined in Example 17.6 an approximation function using interpolation.

We proceed as follows in MATHCAD:

* First assign the first coordinates of the number pairs to a column vector **vx** and the second coordinates to a column vector **vy**:

$$\mathbf{vx} := \begin{pmatrix} 1 \\ 2 \\ 3 \\ 4 \\ 5 \end{pmatrix} \qquad \mathbf{vy} := \begin{pmatrix} 2 \\ 4 \\ 3 \\ 6 \\ 5 \end{pmatrix}$$

* Then calculate the *empirical correlation coefficient*:

  corr (vx , vy) = 0.8

Because the calculated value lies near to 1, we accept the approximation by an *empirical regression line*.

* Now calculate the *slope* a and *intercept* b for the *empirical regression line*

  y = a·x + b

* Finally, draw the calculated regression line and the given number pairs in a coordinate system:

  a := slope(vx , vy)   b := intercept (vx , vy)

  x := 0 .. 6

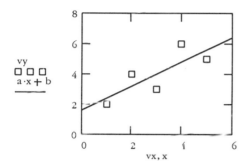

27.2.3 Nonlinear Regression Curves

In *nonlinear regression*, the given *number pairs* (*points*)

$(x_1, y_1), (x_2, y_2), ..., (x_n, y_n)$

for a one-dimensional *sample* are approximated by *nonlinear regression curves* of the form

$y = f(x; a_1, a_2, ..., a_m) = a_1 \cdot f_1(x) + a_2 \cdot f_2(x) + ... + a_m \cdot f_m(x)$

in which the *functions*

$f_1(x)$, $f_2(x)$, ... , $f_m(x)$

are given and the *parameters*

a_1 , a_2 , ... , a_m

can be chosen arbitrarily. Similar to linear regression, the *parameters*

a_i

are determined using the *method* of *least squares*. It remains to substitute the equation of the given regression curve for the equation of the regression line.

Apparently the *linear regression* is a *special case* of *nonlinear regression*. We can obtain it by setting

$$m = 2 \quad , \quad f_1(x) = 1 \quad \text{and} \quad f_2(x) = x$$

i.e., by using the *regression curve*

$$y = f(x; a_1, a_2) = a_1 \cdot f_1(x) + a_2 \cdot f_2(x)$$

(see Example 27.5a).

♦

MATHCAD performs the *nonlinear regression* using the *regression function*

linfit (vx, vy, F)

and uses the *method* of *least squares* to return as the result a vector with the values for the *calculated parameters*

$$a_1 , a_2 , \ldots , a_m$$

In the argument of the *regression function*, **F** designates a column vector in which the given functions

$$f_i(x)$$

from the used regression curve must have been assigned in the following form

$$\mathbf{F}(x) := \begin{pmatrix} f_1(x) \\ f_2(x) \\ \vdots \\ f_m(x) \end{pmatrix}$$

The **vx** and **vy** vectors in the argument have the same meaning as for linear regression, i.e., they contain the x and y-coordinates of the given number pairs.

The *regression function* **linfit** functions only when the number n of number pairs is larger than the number m of the functions

$$f_i(x)$$

(see Example 27.5d).

♦

MATHCAD also has the *regression function*

regress (vx, vy, n)

that can be used to calculate *regression polynomials of degree* n (*polynomial regression*), whose **vx** and **vy** vector arguments have the same meaning as for linear regression, i.e., they contain the x and y-coordinates of the given number pairs.

This function returns a *vector*, which we designate as **vs**, that requires the *interpolation function*

interp (**vs** , **vx** , **vy** , x)

in order to calculate the *regression polynomial* of *degree n* at point x.

Thus, the **regress** function, together with the **interp** function, returns the *regression curve* when we use *polynomials of degree n*. This also permits the *calculation* of the *regression lines* when *n = 1* is set (see Example 27.5a).

Example 27.5:

a) The *regression line*

   y(x) = a·x + b

   from *Example 27.4* for the five *number pairs* (points in the plane)

   (1, 2), (2, 4), (3, 3), (4, 6), (5,5)

   can also be constructed in MATHCAD using the *regression functions* **lin-fit** or **regress**:

   * Using **linfit :**

$$vx := \begin{pmatrix} 1 \\ 2 \\ 3 \\ 4 \\ 5 \end{pmatrix} \qquad vy := \begin{pmatrix} 2 \\ 4 \\ 3 \\ 6 \\ 5 \end{pmatrix} \qquad F(x) := \begin{pmatrix} 1 \\ x \end{pmatrix}$$

   a := linfit (vx , vy , F)   y(x) := F(x)· a

   x := 0 .. 6

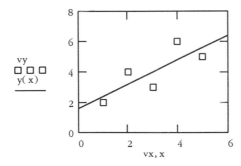

   * Using **regress:**

$$vx := \begin{bmatrix} 1 \\ 2 \\ 3 \\ 4 \\ 5 \end{bmatrix} \qquad vy := \begin{bmatrix} 2 \\ 4 \\ 3 \\ 6 \\ 5 \end{bmatrix} \qquad vs := regress(vx, vy, 1)$$

$$x := 0 .. 6$$

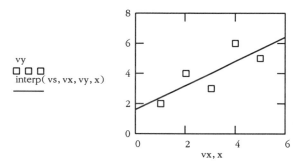

b) Approximate the number pairs from Example a using a *regression parabola* in which we use the **linfit** and **regress** *functions* of MATHCAD:

* Using **linfit**:

$$vx := \begin{pmatrix} 1 \\ 2 \\ 3 \\ 4 \\ 5 \end{pmatrix} \qquad vy := \begin{pmatrix} 2 \\ 4 \\ 3 \\ 6 \\ 5 \end{pmatrix} \qquad F(x) := \begin{pmatrix} 1 \\ x \\ x^2 \end{pmatrix}$$

$$a := linfit(vx, vy, F) \qquad y(x) := F(x) \cdot a$$

$$x := 0 .. 6$$

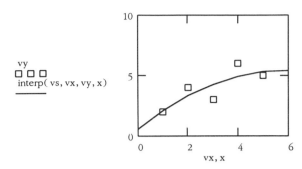

vy
□ □ □
interp(vs, vx, vy, x)

*   Using **regress**:

$$vx := \begin{bmatrix} 1 \\ 2 \\ 3 \\ 4 \\ 5 \end{bmatrix} \qquad vy := \begin{bmatrix} 2 \\ 4 \\ 3 \\ 6 \\ 5 \end{bmatrix} \qquad vs := \text{regress}(\, vx, vy, 2)$$

$$x := 0 .. 6$$

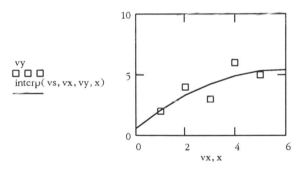

vy
□ □ □
interp(vs, vx, vy, x)

c) We now assume that the number pairs from Example a and b are provided as a structured ASCII file *points.prn* on the *diskette* in *drive* A. After *reading* these into a *matrix* **A**, we use a *regression curve* of the form

$$y = a_1 + a_2 \cdot \cos x + a_3 \cdot \sin x + a_4 \cdot e^x$$

and the *regression function* **linfit** to approximate these number pairs:

A := **READPRN** ("A:\points.prn")

$$A := \begin{pmatrix} 1 & 2 \\ 2 & 4 \\ 3 & 3 \\ 4 & 6 \\ 5 & 5 \end{pmatrix} \qquad\qquad F(x) := \begin{pmatrix} 1 \\ \cos(x) \\ \sin(x) \\ e^x \end{pmatrix}$$

$$a := \text{linfit}(A^{<1>}, A^{<2>}, F) \qquad y(x) := F(x) \cdot a$$

$$x := 0, 0.001 .. 6$$

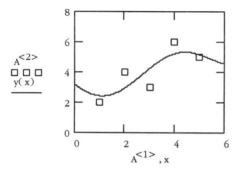

d) Approximate the five number pairs for the previous examples using a *regression polynomial* of degree 4. In accordance with the theory, this must return the same result as for *interpolation* using *polynomials* (see Section 17.2.4). We use the *regression functions* **linfit** and **regress**:

*   Using **linfit**:

$$\mathbf{vx} := \begin{pmatrix} 1 \\ 2 \\ 3 \\ 4 \\ 5 \end{pmatrix} \qquad \mathbf{vy} := \begin{pmatrix} 2 \\ 4 \\ 3 \\ 6 \\ 5 \end{pmatrix} \qquad F(x) := \begin{pmatrix} 1 \\ x \\ x^2 \\ x^3 \\ x^4 \end{pmatrix}$$

$$a := \text{linfit}(\mathbf{vx}, \mathbf{vy}, F)$$

need more data points than parameters

As previously indicated, this shows that MATHCAD rejects the calculation using **linfit** when the same number of number pairs (points) and functions $f_i(x)$ are used. In this case, we can use the **polint** *interpolation function* (see Example 17.6) or the **regress** *function* for *polynomial regression* to return the same result if we take *polynomials.*

* Using **regress**:

$$\mathbf{vx} := \begin{pmatrix} 1 \\ 2 \\ 3 \\ 4 \\ 5 \end{pmatrix} \qquad \mathbf{vy} := \begin{pmatrix} 2 \\ 4 \\ 3 \\ 6 \\ 5 \end{pmatrix}$$

$$vs := regress(\,vx\,,vy\,,4) \qquad x := 0\,,0.001\,..\,6$$

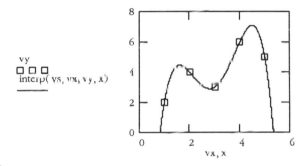

◆

🖙

We have previously discussed only the functional relationship between two random variables using regression analysis. This is called *simple regression.* It is assumed here that the random variable Y depends only on one random variable X. If Y depends on several random variables

$$X_1\,,\,X_2\,,\,\ldots\,,\,X_n$$

this is called *multiple regression.* The procedure for multiple regression is similar to simple regression. The **Statistics** (Volume II) *electronic book* contains detailed calculations and explanations for both types of regression (see Section 27.4).

◆

27.3 Simulations

This book obviously cannot provide a detailed discussion of the complex topic of *simulation methods*. We restrict this discussion to an example of a *Monte Carlo method* to show how you can use MATHCAD to solve such problems.

In general, you can regard *simulation* as being the representation of a real practical system/process using a formal (mathematical) model in which a computer is normally used to solve this model.

♦

Monte Carlo methods designate a *class* of *simulation methods* that are based on *methods* of *probability theory* and *statistics*. They have the following *characteristics*:

* *Approximation* of a given practical (material) deterministic or stochastic model using a formal *stochastic model*.

* *Perform random experiments* using *random numbers* based on this stochastic model.

* The *analysis* of the *results* of *random experiments* provides *approximate values* for the *given problem*.

You can also use *Monte Carlo methods* to *solve many mathematical problems*, such as

* to *solve* algebraic *equations* and *differential equations*

* to *calculate integrals*

* to *solve optimization problems*.

However, their use can only be *recommended* for *higher-dimensional problems*, which, for example, is the case for multiple integrals. The Monte Carlo methods in some cases are better as deterministic numerical methods.

♦

Because MATHCAD can easily generate random numbers, it is suitable to perform *Monte Carlo methods*. The use of MATHCAD's programming capabilities enable algorithms for the Monte Carlo methods to be realized without difficulty. We now illustrate this for the *calculation* of *definite integrals*.

♦

We use a *Monte Carlo method* to *calculate* the *definite integrals* of the form

$$I = \int_a^b f(x)\, dx$$

- For a simple application, the *integral* must be *transformed* into a form in which the interval [0,1] specifies the integration interval and the function values of the integrand f(x) lie between 0 and 1.

- Thus, we require the *integral* in the *form*

$$\int_0^1 h(x)\, dx \qquad \text{with } 0 \le h(x) \le 1$$

- Assuming that the *integrand* f(x) is *continuous* over the interval [a,b], we can use the calculation of

$$m = \underset{x \in [a,b]}{\text{Minimum}} f(x) \quad \text{and} \quad M = \underset{x \in [a,b]}{\text{Maximum}} f(x)$$

to bring the given integral I into the *following form*:

$$I = (M - m) \cdot (b - a) \cdot \int_0^1 h(x)\, dx + (b - a) \cdot m$$

where the function

$$h(x) = \frac{f(a + (b - a) \cdot x) - m}{M - m}$$

satisfies the condition $0 \le h(x) \le 1$.

- The resulting *integral*

$$\int_0^1 h(x)\, dx$$

determines *geometrically* the *surface area* under the function curve of h(x) in the *unit square* $x \in [0,1]$, $y \in [0,1]$.

- The *simulation* with *equally distributed random numbers* permits the use of this *geometric behaviour* in order to calculate the integral:
Create n *number pairs*

$$(zx_i, zy_i)$$

of *equally distributed random numbers* in the interval [0,1] and count how many z(n) of the number pairs lie within the area determined by h(x), i.e., for which

$$zy_i \le h(zx_i)$$

applies.

- Thus, the quotient z(n)/n (statistical empirical probability) can be used to provide an *approximation* for the *integral* to be calculated, i.e.,

$$\int_0^1 h(x)\, dx \approx \frac{z(n)}{n}$$

☞

When a Monte Carlo method is used, the repeated use of the same number of random numbers normally returns a different result unless the MATHCAD random number generator is reset (see Section 26.5). The following example demonstrates this.

♦

Example 27.6:

MATHCAD *cannot exactly solve* the definite *integral*

$$I = \int_1^3 x^x \, dx$$

$$\int_1^3 x^x \, dx \rightarrow \int_1^3 x^x \, dx$$

The MATHCAD *numerical calculation* returns:

$$\int_1^3 x^x \, dx = 13.725 \quad \blacksquare \qquad \int_1^3 x^x \, dx = 13.725 \quad \blacksquare$$

We now *calculate* this *integral* using the given *Monte Carlo method*.

a) To determine information about the minimum m and maximum M of the integrand in the interval [1,3], we first draw the integrand:

   $$x := 1, 1.001 \,.. \, 3$$

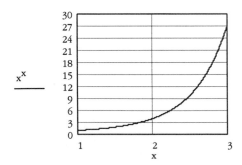

The graph provides us with the values m = 1 and M = 27. We can solve the given integral (with the integration limits a=1 and b=3) using the derived *Monte Carlo method*, for which we write the following MATHCAD program. The number n indicates the specified number of random numbers to be generated (100,000 in the program). The created program can be used to calculate any definite integral; you only need to change the first two lines of the program appropriately:

$$a := 1 \quad b := 3 \qquad M := 27 \qquad m := 1$$

$$n := 100000 \qquad f(x) := x^x$$

$$h(x) := \frac{f(a + (b - a) \cdot x) - m}{M - m} \qquad i := 1 \,..\, n$$

$$zx_i := \mathrm{rnd}(1) \qquad\qquad zy_i := rnd(1)$$

$$z(n) := \sum_{i=1}^{n} if\left(zy_i \le h\left(zx_i\right), 1, 0\right)$$

$$I := (M - m) \cdot (b - a) \cdot \frac{z(n)}{n} + (b - a) \cdot m$$

$$I = 13.703$$

The following table contains the results of the calculation for additional values of n:

n	I
10	12.4
100	12.92
1000	13.856
10,000	13.731
100,000	13.703

For $n = 10$, take the randomly created number pairs from the following graph:

The graphical display shows that from the 10 generated number pairs, 2 lie in the area to be calculated, so that integral I is calculated from

$$I = 26 \cdot 2 \cdot 2 / 10 + 2 = 12.4$$

$x := 0 , 0.01 .. 1$

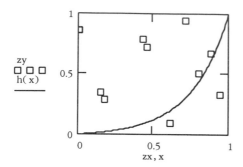

b) *Calculate* the *integral* from Example a for n = 1000 through the repeated generation of 1000 random numbers without resetting the random number generator:

Attempt	Value of the integral
1.	13.128
2.	13.44
3.	13.804
4.	13.076
5.	14.376
6.	13.856

♦

The last example shows that the *Monte Carlo methods* does not offer any advantage over MATHCAD's numerical integration for the calculation of simple definite integrals. These methods become useful only for higher-dimensioned problems, e.g., for multiple integrals.

♦

27.4 Electronic Books for Statistics

You can find the MATHCAD *electronic books* for *statistics* under the designation:

*   **Practical Statistics**
*   **The Mathcad Treasury of Statistics**, *Volume I , II*

* **Statistics**

We now show the title pages of these books that appear when they are opened and their tables of contents:

- The following title page (the first page of the book) appears when **Practical Statistics** is *invoked*:

This electronic book is provided free with the MATHCAD program CD-ROM. Whereas in Version 8 it must be specifically installed from the program CD-ROM (see Chapter 2), in Version 7 it is contained in the **Resource Center** after MATHCAD has been installed.

A mouse click displays the following *table of contents:*

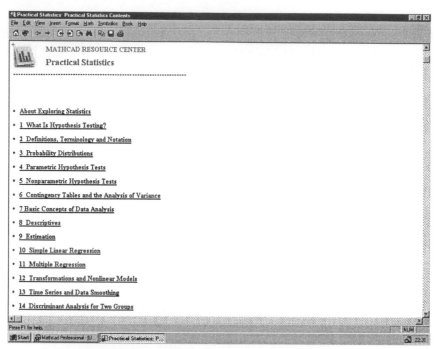

As can be seen from the table of contents, *tests, regression* and *time series* are included in the discussed topics.

*   The *invocation* of **The Mathcad Treasury of Statistics** (*Volume I*) displays the following title page (first page of the book). This shows that *hypothesis testing* is performed:

Click on the TOC button in the EB bar to display the following *table of contents:*

About the Mathcad Treasury of Statistics

Chapter 1 What is Hypothesis Testing?

Chapter 2 Definitions, Terminology and Notation

Chapter 3 Probability Distributions

Chapter 4 Parametric Hypothesis Tests

Chapter 5 Non-Parametric Hypothesis Tests

Chapter 6 Contingency Tables and the Analysis of Variance

Appendix: Summary of Statistical Built-ins

Bibliography

Index

The table of contents indicates that *probability distributions* and *tests* are included in the discussed topics.

- The *invocation* of **The Mathcad Treasury of Statistics** (*Volume II*) displays the following title page (first page of the book). This shows that data analysis is performed:

Click on the TOC button to display the following *table of contents* for Volume II:

The table of contents indicates that *estimation, regression analysis* and *time series* are included in the discussed topics.

• The *invocation* of **Statistics** displays the following title page (first page of the book):

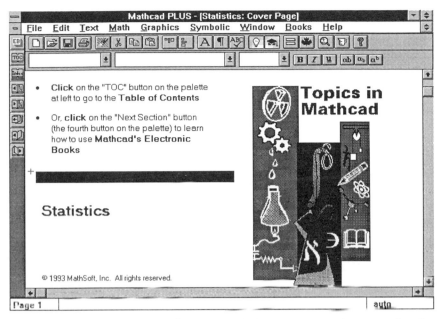

Click on the TOC button to display the following *table of contents* for the
Statistics book:

About Mathcad Electronic Books

About this Electronic Book

Mathcad Statistical Functions

Testing

Section   1:   Choosing the Sample Size for a t Test on Means

Section   2:   Kendall's Rank Correlation Coefficient

Section   3:   Spearman's Rank Correlation Coefficient

Section   4:   The Kolmogorov-Smirnov Test

Section   5:   Counting Runs

Estimation

Section   6:   Estimating The Mean of a Normal Population

Section   7:   Jackknife Estimates

Section   8:   Probability Estimation Through Monte Carlo Methods

Modeling

Section   9:   Operating Characteristic Curves

Section 10:   Principal Component Analysis

The table of contents indicates that *testing, estimation* and *simulation* are included in the discussed topics.

☞

The methods provided in the *electronic books* for *statistics* can be used to perform the most important calculations that arise in practical problems. Because these books provide detailed descriptions and contain many explanations, the user can work with them without difficulty.

However, we cannot provide a detailed discussion of these books here. The description of the electronic books for the comprehensive subject of statistics must be left to a separate book.

♦

The following *Example 27.7* shows part of Section 6 of the **Statistics** *electronic book.*

Example 27.7:

Section 6 from the **Statistics** *electronic book* discusses the *estimation* of the *mean* for a *normal distributed population*. The *explanatory text* is italicized. The non-italicized *formulae* are used for the calculation and can be passed directly to the worksheet. The **READPRN** function (see Section 9.1) is used to read the *edata.prn* file that contains *sample* to be *analysed*:

Section 6   **Estimating the Mean of a Normal Population**

This application calculates confidence limits for an estimate of the mean of a normal population when the population variance is unknown and must be estimated from the sample. The approximation used for the inverse of the t *distribution gives accurate results for samples of size* **12** *or larger.*

■   *This application is set up to use data stored in a file.* **Mathcad** *reads the data and calculates the sample mean* **m** *and sample standard deviation* **s**.

■   *You can read in your own data by assigning the name of your data file to the file variable* **edata**. *To do so, choose* **Associate Filename** *from the* **File** *menu. Then assign the filename of your data to the* **Mathcad** *variable* **edata**. *Be sure to point* **Mathcad** *to the proper directory for your file.*

Note: *This procedure for calculating confidence limits assumes that the distribution of the sampled population is normal.*

Background

For samples of size **n** *from a normal population with mean* **μ**, *the statistic*

$$\frac{m - \mu}{\left(\dfrac{s}{\sqrt{n}}\right)}$$

has a t *distribution with* **n-1** *degrees of freedom. Here* **m** *is the sample mean and* **s** *is the sample standard deviation.*

When you use the sample mean as an estimate for the population mean, you can use the distribution to measure the probability that the mean will be close to your estimate. One way to express this probability is to construct a **confidence interval**. *For example, to construct a symmetric* **99%** *confidence interval, you do the following:*

1.   Find the point **x** on the **t** distribution with **n-1** degrees of
        freedom such that the probability of a larger **t** is **.005**. Since **t**
        is symmetric, the probability of a value less than   **-x** is also
        .005.

2.   The interval from **m - x·s/√n** to **m + x·s/√n** is a **99%**
        confidence interval for the population mean. The endpoints
        of the interval are **confidence limits** .

Over a long series of sampling experiments,   **99%** of the confidence
intervals constructed in this way using the sample mean **m** and the
sample standard deviation **s** will contain the actual population mean
μ.

For more background on the construction of confidence intervals see
General Statistics, by Haber and Runyon (Addison-Wesley, 1977).

Mathcad Implementation

This model uses some **Mathcad** statistical functions to find a
confidence interval for the mean of a normal population, based on
the mean and variance of a sample of size at least **12**.

Read the data from the file **edata.prn** :

$$X := \text{READPRN}(\text{edata})$$

The equations below calculate some sample statistics using   **Mathcad**'s
built-in **mean** and **var** functions. Note that **var** returns the mean
squared deviation of the sample values from their mean. For samples
of size **n** this value must be multiplied by **n/(n-1)** to yield an
unbiased estimate for the population variance. The sample standard
deviation is the square root of this estimate.

Sample size: $n := \text{length}(X)$ $n =$

Sample mean $m := \text{mean}(X)$ $m =$

Sample standard $s := \sqrt{\dfrac{n}{n-1} \cdot \text{var}(X)}$ $s =$
deviation

Choose a value for **a**. The confidence level for your mean estimate will
be **1** minus the value you choose for **a**. For example, to find **95%**
confidence limits, set **a** equal to **.05**.

$$\alpha := .01$$

The confidence level for the estimate of the mean will be

$$1 - \alpha = 99 \cdot \%$$

Below are the equations for calculating percentage points of the **t** *distribution.*

- **w(a)** *is the value such that the tail of the standard normal distribution to the right of* **w(a)** *has area* **a** . *It is computed using the* **root** *function.*

- *The function* **t(a,d)** *returns the* **a** *point on the* **t** *distribution with* **d** *degrees of freedom. The probability of drawing a value larger than this point is* **a** .

Percentage points for **t** *are computed with the following approximation:*

$$\text{TOL} := .00001$$

$$p := 2 \qquad\qquad w(a) := \text{root}(1 - \text{cnorm}(p) - a, p)$$

$$t(a,d) := w(a) + \frac{1}{4 \cdot d} \cdot (w(a)^3 + w(a)) \ \ldots$$
$$+ \frac{1}{96 \cdot d^2} \cdot (5 \cdot w(a)^5 + 16 \cdot w(a)^3 + 3 \cdot w(a)) \ \ldots$$
$$+ \frac{1}{384 \cdot d^3} \cdot (3 \cdot w(a)^7 + 19 \cdot w(a)^5 + 17 \cdot w(a)^3 - 15 \cdot w(a))$$

Note: *For* **d > 10** *and* **a** *between* **.001** *and* **.1** *this approximation,* **t(a,d)**, *is good to within* **.01** . *The number of degrees of freedom,* **d**, *is one less than the sample size, so the size of the confidence interval will be correct to within* **1** *percent for samples of size* **12** *or larger.*

The definitions for the confidence limits **L** *and* **U** *use the function* **t** *to calculate the percentage point*

$$t_{\alpha/2,\ n-1}$$

Below are the calculations for the confidence limits for estimating the population mean.

$$L := m - t\left(\frac{\alpha}{2}, n - 1\right) \cdot \frac{s}{\sqrt{n}} \qquad\qquad U := m + t\left(\frac{\alpha}{2}, n - 1\right) \cdot \frac{s}{\sqrt{n}}$$

The confidence limits are:

L =   U =

*The probability that a confidence interval constructed in this way
contains the true population mean is*

$$1 - \alpha = 0.99$$

The empty right-hand side after the numerical equal sign results because no
edata.prn file has been read.

♦

The considered section of the electronic book well illustrates the structure
of such books. Because comprehensive explanations accompany all the calculations, its self-explanatory form is apparent.

♦

28 Summary

In this book, we have used MATHCAD as an example of how a computer can be used to effectively solve mathematical problems.

Compared with other popular computer algebra systems, such as MATHEMATICA and MAPLE, MATHCAD has shown itself to be an equal system, but has advantages in the display of the performed calculations and for numerical mehods (see [3,4]).

The following provides a *summarized evaluation* of working with MATHCAD:

- Although wonders cannot be expected for the *exact solution* of mathematical problems, MATHCAD can calculate many standard problems for which mathematics provides a finite solution algorithm. Consequently, you should always begin with the exact solution for a given problem.

- If MATHCAD cannot calculate an exact solution for a problem, it provides a number of *methods* for the *numerical (approximate) solution*, for which we have gained an impression of their effectiveness in the course of this book.

- We have often mentioned in this book that even a powerful system like MATHCAD cannot replace mathematics. An effective use of MATHCAD is possible only when you have a solid mathematical knowledge. However, MATHCAD frees the user from longwinded calculations and so provides users with the capability to improve their mathematical models and to experiment with them.

- Even such a comprehensive book cannot describe all the capabilities of MATHCAD. In this book, I have tried to provide a clear explanation of the *important background* for the mathematical solution of standard problems and special problems that frequently occur. If you do not happen to find any references in this book for your problem, you should,

  * use the MATHCAD *help functions*; the help also includes the MATHCAD *user's guide*

  * *experiment* with the capabilities of MATHCAD provided in the book; this permits you to *gain experience* and to *investigate* further *capabilities*

  * if you have an *Internet connection*, use the **Resource Center** to access the *www page* for MathSoft to obtain *explanations* and *help*, or

use the free MathSoft *Internet forum* to make worldwide contact with other MATHCAD users (see Section 5.1).

- Because the user interface for Version 8 described in the book has only minor changes compared with Version 7, most of the problems discussed in the book can also be calculated without difficulty using Version 7. The same applies for versions 5 and 6. You can also find additional details in the author's previous books [2,3,4]. However, do not forget that the earlier versions are less powerful and provide fewer functions than in Version 8.

- To keep this book compact, the mathematical background has been discussed only to the extent required to use MATHCAD. If users have mathematical questions, they should refer to the comprehensive literature available for mathematics; the Bibliography mentions some of these books.

- Obviously, *errors* can occur when you work with a complex system like MATHCAD. These errors can be caused by both MATHCAD and the user. We describe some of these *error sources* below and also provide some *correction measures:*

  * You cannot blindly trust the exact or numerical results that MATHCAD calculates. Although MATHCAD works reliably in most cases, *errors* can also occur because these cannot be avoided in such a comprehensive system. We have seen some examples in the course of the book. Consequently, we recommend that the user always checks the returned results. This can be done in many ways, for example, with a graphical display or with tests.

  * If unexplained effects (errors) occur after extensive work has been performed in a MATHCAD worksheet, it is often better to close this worksheet and to reattempt the calculations in a new worksheet. One cause for these effects is that during a long session you can lose the overview of the variables and the functions you have used, with the result being double usage of variables and functions.

  * The *user* rather than MATHCAD is often the cause of the *errors* that occur. We now discuss some of the most important user errors and which we have already described in the course of the book:

    - Formulae are used in which a *division* by *zero* can occur if you do not explicitly assign values to the variables.

    - The *numerical equal sign* for numerical calculations is used instead of the *equal operator* to display equations.

    - The *symbolic* and *numerical equal signs* are *used incorrectly* when you perform exact or numerical calculations.

- The *initial value* for the *indexing* is set *incorrectly*, because MATHCAD uses 0 as standard value, whereas applications normally require 1.

- The *same designations* are used for *functions* and *variables*, with the consequence that the previous ones are no longer available.

- You use *variables* for *exact (symbolic) calculations* for which you have previously *assigned numerical values*. This calculation then can no longer be performed. You must use either different variable designations or redefine these variables, such as x := x.

To summarize, MATHCAD normally represents an effective system to solve common mathematical problems with a computer; its unexcelled display capabilities represent a significant advantage over other systems.

Bibliography

MATHCAD

[1]   Benker: Mathematik mit dem PC, Vieweg Verlag Braunschweig, Wiesbaden 1994,

[2]   Benker: Mathematik mit MATHCAD, Springer Verlag Berlin, Heidelberg, New York 1996,

[3]   Benker: Wirtschaftsmathematik mit dem Computer, Vieweg Verlag Braunschweig, Wiesbaden 1997,

[4]   Benker: Ingenieurmathematik mit Computeralgebra-Systemen, Vieweg Verlag Braunschweig, Wiesbaden 1998,

[5]   Benker: Mathematik mit MATHCAD, 2.neubearbeitete Auflage, Springer Verlag Berlin, Heidelberg, New York 1999,

[6]   Born, Lorenz: MathCad – Probleme, Beispiele, Lösungen –, Int. Thomson Publ. Bonn 1995,

[7]   Born: Mathcad Version 3.1 und 4, Int.Thomson Publ. Bonn 1994,

[8]   Desrues: Explorations in MATHCAD, Addison-Wesley 1997,

[9]   Donnelly: MathCad for Introductory Physics, Addison-Wesley 1992,

[10] Hill, Porter: Interactive Linear Algebra, Springer Verlag Berlin, Heidelberg, New York 1996,

[11] Hörhager, Partoll: Mathcad 5.0/PLUS 5.0, Addison-Wesley Bonn 1994,

[12] Hörhager, Partoll: Problemlösungen mit Mathcad für Windows, Addison-Wesley Bonn 1995,

[13] Hörhager, Partoll: Mathcad 6.0/PLUS 6.0, Addison-Wesley Bonn 1996,

[14] Hörhager, Partoll: Mathcad, Version 7, Addison-Wesley Bonn 1998,

[15] Miech: Calculus with Mathcad, Wadsworth Publishing 1991,

[16] Weskamp: Mathcad 3.1 für Windows, Addison-Wesley Bonn 1993,

[17] Wieder: Introduction to MathCad for Scientists and Engineers McGraw-Hill New York 1992,

Mathematics

[18] Barrett, Wylie: Advanced Engineering Mathematics, McGraw-Hill New York 1995,

[19] Beltzer: Engineering Analysis with Maple/Mathematica, Academic Press 1996,

[20] Bence, Hobson, Riley: Mathematical Methods for Physics and Engineering, Cambridge University Press 1998,

[21] Betounes: Partial Differential Equations for Computational Science, Springer Verlag New York 1998,

[22] Bronshtein, Semendyayev: Handbook of Mathematics, Van Nostrand Reinhold Company New York 1985,

[23] Croft, Davison: Mathematics for Engineers, Addison-Wesley Longman 1999,

[24] Davies, Hicks: Mathematics for Scientific and Technical Students, Addison-Wesley Longman 1998,

[25] Evans: Engineering Mathematics, Chapman & Hall London 1997,

[26] Gerald, Wheatley: Applied Numerical Analysis, Addison-Wesley Longman 1999,

[27] Greenberg: Advanced Engineering Mathematics, Prentice Hall 1998,

[28] Harris, Stocker: Handbook of Mathematics and Computational Science, Springer-Verlag New York 1998,

[29] James: Modern Engineering Mathematics, Addison-Wesley 1996

[30] James: Advanced Modern Engineering Mathematics, Addison-Wesley 1993,

[31] Kreyszig: Advanced Engineering Mathematics, John Wiley & Sons New York 1993,

[32] Malek-Madani: Advanced Engineering Mathematics with Mathematica and Matlab, Addison-Wesley New York 1997,

[33] Mustoe: Engineering Mathematics, Addison-Wesley 1997,

[34] Nagle, Saff: Fundamentals of Differential Equations and Boundary Value Problems, Addison-Wesley 1996,

[35] Spiegel: Theory and Problems of Advanced Mathematics for Engineers and Scientists, McGraw-Hill New York 1973,

[36] Stroud: Engineering Mathematics, Macmillan Press 1995,

[37] Stroud: Further Engineering Mathematics, Macmillan Press 1996.

Computer Algebra Systems

[38] Adams, Tocci: Applied MAPLE for Engineers and Scientists, Artech House Boston, London 1996,

[39] Abell, Braselton, Rafter: Statistics with Mathematica, Academic Press 1999,

[40] Bahder: Mathematica for Scientists and Engineers, Addison-Wesley Publishing Company 1995,

[41] Campbell, Redfern: The MATLAB 5 Handbook, Springer Verlag New York 1998,

[42] Heck: Introduction to Maple, Springer Verlag New York 1997,

[43] Nikolaides, Walkington: MAPLE, A Comprehensive Introduction, Cambridge University Press 1996,

[44] Redfern: The Maple Handbook, Springer Verlag New York 1996,

[45] Redfern: The Practical Approach Utilities for Maple, Springer Verlag New York 1995,

[46] Wagon: Mathematica in Action, Springer Verlag New York 1999,

[47] Wolfram: The MATHEMATICA Book, Cambridge University Press 1996,

[48] Zachary: Introduction to Scientific Programming, Springer Verlag (TELOS) 1998.

Index